T0145364

Lecture Notes in Electrical Engineering

Volume 449

About this Series

"Lecture Notes in Electrical Engineering (LNEE)" is a book series which reports the latest research and developments in Electrical Engineering, namely:

- Communication, Networks, and Information Theory
- Computer Engineering
- Signal, Image, Speech and Information Processing
- Circuits and Systems
- Bioengineering

LNEE publishes authored monographs and contributed volumes which present cutting edge research information as well as new perspectives on classical fields, while maintaining Springer's high standards of academic excellence. Also considered for publication are lecture materials, proceedings, and other related materials of exceptionally high quality and interest. The subject matter should be original and timely, reporting the latest research and developments in all areas of electrical engineering.

The audience for the books in LNEE consists of advanced level students, researchers, and industry professionals working at the forefront of their fields. Much like Springer's other Lecture Notes series, LNEE will be distributed through Springer's print and electronic publishing channels.

More information about this series at http://www.springer.com/series/7818

Kuinam J. Kim · Hyuncheol Kim
Nakhoon Baek

Editors

IT Convergence and Security 2017

Volume 1

 Springer

Editors
Kuinam J. Kim
iCatse, B-3001, Intellige 2
Kyonggi University
Seongnam-si, Kyonggi-do
Korea (Republic of)

Nakhoon Baek
School of Computer Science
    and Engineering
Kyungpook National University
Daegu
Korea (Republic of)

Hyuncheol Kim
Computer Science
Namseoul University
Cheonan, Chungcheongnam-do
Korea (Republic of)

ISSN 1876-1100   ISSN 1876-1119   (electronic)
Lecture Notes in Electrical Engineering
ISBN 978-981-13-4881-5   ISBN 978-981-10-6451-7   (eBook)
DOI 10.1007/978-981-10-6451-7

Printed on acid-free paper

This Springer imprint is published by Springer Nature
The registered company is Springer Nature Singapore Pte Ltd.
The registered company address is: 152 Beach Road, #21-01/04 Gateway East, Singapore 189721, Singapore

Preface

This LNEE volume contains the papers presented at the iCatse International Conference on IT Convergence and Security (ICITCS 2017) which was held in Seoul, South Korea, during September 25 to 28, 2017.

The conferences received over 200 paper submissions from various countries. After a rigorous peer-reviewed process, 69 full-length articles were accepted for presentation at the conference. This corresponds to an acceptance rate that was very low and is intended for maintaining the high standards of the conference proceedings.

ICITCS2017 will provide an excellent international conference for sharing knowledge and results in IT Convergence and Security. The aim of the conference is to provide a platform to the researchers and practitioners from both academia and industry to meet the share cutting-edge development in the field.

The primary goal of the conference is to exchange, share and distribute the latest research and theories from our international community. The conference will be held every year to make it an ideal platform for people to share views and experiences in IT Convergence and Security-related fields.

On behalf of the Organizing Committee, we would like to thank Springer for publishing the proceedings of ICITCS2017. We also would like to express our gratitude to the 'Program Committee and Reviewers' for providing extra help in the review process. The quality of a refereed volume depends mainly on the expertise and dedication of the reviewers. We are indebted to the Program Committee members for their guidance and coordination in organizing the review process and to the authors for contributing their research results to the conference.

Our sincere thanks go to the Institute of Creative Advanced Technology, Engineering and Science for designing the conference Web page and also spending countless days in preparing the final program in time for printing. We would also like to thank our organization committee for their hard work in sorting our manuscripts from our authors.

We look forward to seeing all of you next year's conference.

Kuinam J. Kim
Nakhoon Baek
Hyuncheol Kim
Editors of ICITCS2017

Organizing Committee

General Chairs

Hyung Woo Park	KISTI, Republic of Korea
Nikolai Joukov	New York University and modelizeIT Inc, USA
Nakhoon Baek	Kyungpook National University, Republic of Korea
HyeunCheol Kim	NamSeoul University, Republic of Korea

Steering Committee

Nikolai Joukov	New York University and modelizeIT Inc, USA
Borko Furht	Florida Atlantic University, USA
Bezalel Gavish	Southern Methodist University, USA
Kin Fun Li	University of Victoria, Canada
Kuinam J. Kim	Kyonggi University, Republic of Korea
Naruemon Wattanapongsakorn	King Mongkut's University of Technology Thonburi, Thailand
Xiaoxia Huang	University of Science and Technology Beijing, China
Dato' Ahmad Mujahid Ahmad Zaidi	National Defence University of Malaysia, Malaysia

Program Chair

Kuinam J. Kim	Kyonggi University, Republic of Korea

Publicity Chairs

Miroslav Bureš   Czech Technical University, Czech Republic
Dan (Dong-Seong) Kim   University of Canterbury, New Zealand
Sanggyoon Oh   BPU Holdings Corp, Republic of Korea
Xiaoxia Huang   University of Science and Technology Beijing,
                                     China

Financial Chair

Donghwi Lee   Dongshin University, Republic of Korea

Publication Chairs

Minki Noh   KISTI, Republic of Korea
Hongseok Jeon   ETRI, Republic of Korea

Organizers and Supporters

Institute of Creative Advanced Technologies, Science and Engineering
Korea Industrial Security Forum
Korean Convergence Security Association
University of Utah, Department of Biomedical Informatics, USA
River Publishers, Netherlands
Czech Technical University, Czech Republic
Chonnam National University, Republic of Korea
University of Science and Technology Beijing, China
King Mongkut's University of Technology Thonburi, Thailand
ETRI, Republic of Korea
KISTI, Republic of Korea
Kyungpook National University, Republic of Korea
Seoul Metropolitan Government

Program Committee

Bhagyashree S R   ATME College of Engineering, Mysore,
                                     Karnataka, India
Richard Chbeir   Université Pau & Pays Adour (UPPA), France
Nandan Mishra   Cognizant Technology Solutions, USA

Reza Malekian	University of Pretoria, South Africa
Sharmistha Chatterjee	Florida Atlantic University, USA
Shimpei Matsumoto	Hiroshima Institute of Technology, Japan
Sharifah Md Yasin	University Putra Malaysia, Malaysia
C. Christober Asir Rajan	Pondicherry Engineering College, India
Chin-Chen Chang	Feng Chia University, Taiwan
Danilo Pelusi	University of Teramo, Italy
Necmi Taspinar	Erciyes University, Kayseri, Turkey
Alvaro Suarez	University of Las Palmas de G.C., Spain
Wail Mardini	Jordan University, Jordan
Josep Domingo-Ferrer	Universitat Rovira i Virgili, Spain
Yaxin Bi	Ulster University at Jordanstown, UK
Jie Zhang	Newcastle University, UK
Miroslav N. Velev	Aries Design Automation, USA
Johann M. Marquez-Barja	CONNECT Research Centre, Trinity College Dublin, Ireland
Nicholas Race	Lancaster University, UK
Gaurav Sharma	Université libre de Bruxelles, Belgium
Yanling Wei	Technical University of Berlin, Germany
Mohd Fairuz Iskandar Othman	Universiti Teknikal Malaysia Melaka (UTeM), Malaysia
Harikumar Rajaguru	Bannari Amman Institute of Technology, Sathyamangalam, India
Chittaranjan Pradhan	KIIT University, India
Frank Werner	Otto-von-Guericke University Magdeburg, Germany
Suranga Hettiarachchi	Indiana University Southeast, USA
Sa'adah Hassan	Universiti Putra, Malaysia
Frantisek Capkovic	Institute of Informatics, Slovak Academy of Sciences, Slovakia
Oscar Mortagua Pereira	University of Aveiro, Portugal
Filippo Gaudenzi	Università degli Studi di Milano, Italy
Virgilio Cruz Machado	Universidade Nova de Lisboa-UNIDEMI, Portugal
Pao-Ann Hsiung	National Chung Cheng University, Taiwan
M. Iqbal Saripan	Universiti Putra Malaysia, Malaysia
Lorenz Pascal	University of Haute Alsace, France
Helmi Zulhaidi Mohd Shafri	Universiti Putra Malaysia, Malaysia
Harekrishna Misra	Institute of Rural Management Anand, India
Nuno Miguel Castanheira Almeida	Polytechnic of Leiria, Portugal
Bandit Suksawat	King Mongkut's University, Thailand
Jitender Grover	IIIT Hyderabad, India
Kwangjin Park	Wonkwang University, Korea
Ahmad Kamran Malik	COMSATS Institute of IT, Pakistan

Shitala Prasad	NTU Singapore, Singapore
Hao Han	The University of Tokyo, Japan
Anooj P.K.	Al Musanna College of Technology, Oman
Hyo Jong Lee	Chonbuk National University,Korea
D'Arco Paolo	University of Salerno, Italy
Suresh Subramoniam	CET School of Management, India
Abdolhossein Sarrafzadeh	Unitec Institute of Technology, New Zealand
Stelvio Cimato	University of Milan, Italy
Ivan Mezei	University of Novi Sad, Serbia
Terje Jensen	Telenor, Norway
Selma Regina Martins Oliveira	Federal Fluminense University, Brazil
Firdous Kausar	Imam Ibm Saud University, Saudi Arabia
M. Shamim Kaiser	Jahangirnagar University, Bangladesh
Maria Leonilde Rocha Varela	University of Minho, Portugal
Nadeem Javaid	COMSATS Institute of Information Technology, Pakistan
Urmila Shrawankar	RTM Nagpur University, India
Yongjin Yeom	Kookmin University, Korea
Olivier Blazy	Université de Limoges, France
Bikram Das	NIT Agartala, India
Edelberto Franco Silva	Universidade Federal de Juiz de Fora, Brazil
Wing Kwong	Hofstra University, USA
Dae-Kyoo Kim	Oakland University, USA
Nickolas S. Sapidis	University of Western Macedonia, Greece
Eric J. Addeo	DeVry University, USA
T. Ramayah	Universiti Sains Malaysia, Malaysia
Yiliu Liu	Norwegian University, Norway
Shang-Ming Zhou	Swansea University, UK
Anastasios Doulamis	National Technical University, Greece
Baojun Ma	Beijing University, China
Fatemeh Almasi	Ecole Centrale de Nantes, France
Mohamad Afendee Mohamed	Universiti Sultan Zainal Abidin, Malaysia
Jun Peng	University of Texas, USA
Nestor Michael C. Tiglao	University of the Philippines Diliman, Philippines
Mohd Faizal Abdollah	University Technical Malaysia Melaka, Malaysia
Alessandro Bianchi	University of Bari, Italy
Reza Barkhi	Virginia Tech, USA
Mohammad Osman Tokhi	London South Bank University, UK
Prabhat K. Mahanti	University of New Brunswick, Canada
Chia-Chu Chiang	University of Arkansas at Little Rock, USA
Tan Syh Yuan	Multimedia University, Malaysia
Qiang (Shawn) Cheng	Southern Illinois University, USA
Michal Choras	University of Science and Technology, Korea

El-Sayed M. El-Alfy	King Fahd University, Saudi Arabia
Abdelmajid Khelil	Landshut University, Germany
James Braman	The Community College of Baltimore County, USA
Rajesh Bodade	Defence College of Telecommunication Engineering, India
Nasser-Eddine Rikli	King Saud University, Saudi Arabia
Zeyar Aung	Khalifa University, United Arab Emirates
Schahram Dustdar	TU Wien, Austria
Ya Bin Dang	IBM Research, China
Marco Aiello	University of Groningen, Netherlands
Chau Yuen	Singapore University, Singapore
Yoshinobu Tamura	Tokyo City University, Japan
Nor Asilah Wati Abdul Hamid	Universiti Putra Malaysia, Malaysia
Pavel Loskot	Swansea University, UK
Rika Ampuh Hadiguna	Andalas University, Indonesia
Hui-Ching Hsieh	Hsing Wu University, Taiwan
Javid Taheri	Karlstad University, Sweden
Fu-Chien Kao	Da-Yeh University, Taiwan
Siana Halim	Petra Christian University, Indonesia
Goi Bok Min	Universiti Tunku Abdul Rahman, Malaysia
Shamim H Ripon	East West University, USA
Munir Majdalawieh	George Mason University, USA
Hyunsung Kim	Kyungil University, Korea
Ahmed A. Abdelwahab	Qassim University, Saudi Arabia
Vana Kalogeraki	Athens University, Greece
Joan Ballantine	Ulster University, UK
Jianbin Qiu	Harbin Institute of Technology, China
Mohammed Awadh Ahmed Ben Mubarak	Infrastructure University Kuala Lumpur, Malaysia
Mehmet Celenk	Ohio University, USA
Shakeel Ahmed	King Faisal University, Saudi Arabia
Sherali Zeadally	University of Kentucky, USA
Seung Yeob Nam	Yeungnam University, Korea
Tarig Mohamed Hassan	University of Khartoum, Sudan
Vishwas Ruamurthy	Visvesvaraya Technological University, India
Ankit Chaudhary	Northwest Missouri State University, USA
Mohammad Faiz Liew Abdullah	University Tun Hussein Onn, Malaysia
Francesco Lo Presti	University of Rome Tor Vergata, Italy
Muhammad Usman	National University of Sciences and Technology (NUST), Pakistan
Kurt Kurt Tutschku	Blekinge Institute of Technology, Sweden

Ivan Ganchev	University of Limerick, Ireland/University of Plovdiv "Paisii Hilendarski"
Mohammad M. Banat	Jordan University, Jordan
David Naccache	Ecole normale supérieure, France
Kittisak Jermsittiparsert	Rangsit University, Thailand
Pierluigi Siano	University of Salerno, Italy
Hiroaki Kikuchi	Meiji University, Japan
Ireneusz Czarnowski	Gdynia Maritime University, Poland
Lingfeng Wang	University of Wisconsin-Milwaukee, USA
Somlak Wannarumon Kielarova	Naresuan University, Thailand
Chang Wu Yu	Chung Hua University, Taiwan
Kennedy Njenga	University of Johannesburg, Republic of South Africa
Kok-Seng Wong	Soongsil University, Korea
Ray C.C. Cheung	City University of Hong Kong, China
Stephanie Teufel	University of Fribourg, Switzerland
Nader F. Mir	San Jose State University, California
Zongyang Zhang	Beihang University, China
Alexandar Djordjevich	City University of Hong Kong, China
Chew Sue Ping	National Defense University of Malaysia, Malaysia
Saeed Iqbal Khattak	University of Central Punjab, Pakistan
Chuangyin Dang	City University of Hong Kong, China
Riccardo Martoglia	FIM, University of Modena and Reggio Emilia, Italy
Qin Xin	University of the Faroe Islands, Faroe Islands, Denmark
Andreas Dewald	ERNW Research GmbH, Germany
Rubing Huang	Jiangsu University, China
Sangseo Parko	Korea
Mainguenaud Michel	Insitut National des sciences Appliquées Rouen, France
Selma Regina Martins Oliveira	Universidade Federal Fluminense, Brazil
Enrique Romero-Cadaval	University of Extremadura, Spain
Noraini Che Pa	Universiti Putra Malaysia (UPM), Malaysia
Minghai Jiao	Northeastern University, USA
Ruay-Shiung Chang	National Taipei University of Business, Taiwan
Afizan Azman	Multimedia University, Malaysia
Yusmadi Yah Jusoh	Universiti Putra Malaysia, Malaysia
Daniel B.-W. Chen	Monash University, Australia
Wuxu Peng	Texas State University, USA
Noridayu Manshor	Universiti Putra Malaysia, Malaysia
Alberto Núñez Covarrubias	Universidad Complutense de Madrid, Spain

Contents

Machine Learning and Deep Learning

Image-Based Content Retrieval
via Class-Based Histogram Comparisons

John Kundert-Gibbs$^{(\boxtimes)}$

The Institute for Artificial Intelligence,
The University of Georgia, Athens, GA 30604, USA
`jkundert@uga.edu`

Abstract. Content-based image retrieval has proved to be a fundamental research challenge for disciplines like search and computer vision. Though many approaches have been proposed in the past, most of them suffer from poor image representation and comparison methods, returning images that match the query image rather poorly when judged by a human. The recent rebirth of deep learning neural networks has been a boon to CBIR, producing much higher quality results, yet there are still issues with many recent uses of deep learning. Our method, which makes use of a pre-trained deep net, compares class-based histograms between the known image database and query images. This method produces results that are significantly better than baseline methods we test against. In addition, we modify the base network in two ways and then use a weighted voting system to decide on images to display. These modifications further improve image recall quality.

Keywords: Deep learning · Image retrieval · Content-Based image retrieval · Image based recall · CBIR · IBR · Information retrieval · Computer vision

1 Introduction

In the last few years, users' desire to find more images like the one they are currently viewing has increased dramatically. From personal photo libraries to personal and business searches, vast numbers of image consumers are interested in finding images that "look like" the one they are viewing at the moment. As the quantity of stored images has expanded to a number far beyond what any team of humans could examine, classify, and catalogue, we have turned to machines running Artificial Intelligence searches to do the work for us.

The industry terms for recovering images that are visually and semantically similar to the search image are Content-Based Image Recall (CBIR) or Image-Based Recall (IBR). The major IBR breakthrough in the past few years has been the use of deep convolutional neural networks. Even with major advances in IBR, however, the area is an ongoing topic of research as results are not consistently appropriate. We propose a new system that can outperform publically available IBR packages on a reasonable size database of images. Our system utilizes a class histogram approach (described in Sect. 3) to compare a query image to scores from an image database, producing quality results rapidly. Though evaluating IBR can prove challenging as the results are

© Springer Nature Singapore Pte Ltd. 2018
K.J. Kim et al. (eds.), *IT Convergence and Security 2017*,
Lecture Notes in Electrical Engineering 449,
DOI 10.1007/978-981-10-6451-7_1

generally qualitative, we can make some quantitative assessment as well. By comparing two off-the-shelf IBR solutions, as well as an un-retrained and a retrained network using our method, we show that our system works better than the available systems, and that further training increases the accuracy of our method.

2 Research in Image-Based Recall

Substantial work has been done on the topic of IBR for more than two decades. Most of the traditional methods [1–5] require a large number of training instances. Until recently most training sets were relatively small, thus IBR engines did not have much to work with. Even with the advent of large image databases like Imagenet, and new techniques like Support Vector Machines [6, 7] and active learning [8], results have been only marginal. More recent approaches have made use of ensemble learning. These ensemble schemes have been successful at improving classification accuracy through bias or variance reduction, but they do not help reduce the number of samples and the time required to learn a query concept. An approach based on Support Vector Machines (SVMs) is proposed in [6], but this approach requires seeds to start, which is not practically feasible, especially for large database queries.

Conventional IBR approaches usually choose rigid distance functions on some extracted low-level features for their similarity search mode, such as Euclidean distance. However, a fixed rigid similarity/distance function may not be optimal for the complex visual image retrieval tasks. As a result recently there has been a surge of research into designing various distance/similarity measures on low-level features by exploring machine learning techniques [9–12]. Distance metric learning for image retrieval has been extensively studied [13–21]. In some instances like [16], class labels are used to train DML.

Over the past half decade, a rich family of deep learning techniques has been applied to the field of computer vision and machine learning. Just a few examples are Deep Belief Networks [22], Boltzmann Machines [23], Restricted Boltzmann Machines [24], Deep Boltzmann Machines [25], and Deep Neural Networks [26, 33]. The deep convolutional neural networks (CNNs) proposed in [27] got first place in the 2012 image classification task, ILSVRC-2012, proving the worth of this rejuvenated network architecture. For our method, we make use of a pre-trained VGG- 16 model.

3 IBR Packages

While a number of IBR packages exist, we found two packages based on MATLAB that are good experimental candidates because they utilize MATLAB as a basis and are consistent in their underpinnings, using scripts that are open to examination. These two IBR implementations serve to provide baseline results for comparison with our IBR method, which is also implemented in MATLAB.

The first package examined is cbires, developed primarily by Joani Mitro. cbires uses either k-nearest-neighbors (knn) or Support Vector Machines (SVM) plus feature extraction to perform IBR [28]. The second package, CBIR, was developed by Amine

Ben Khalifa and Faezeh Tafazzoli [29]. CBIR utilizes feature extraction which can either be done locally or globally. Color and texture features can be extracted globally or locally, and different distance measures can be invoked to compare images.

The method we have developed operates differently than the two baseline IBR packages described above. Termed Class-Based Histogram, or CBH-IBR, this system uses a pre-trained deep learning convolutional neural network—in this case trained on the Imagenet database [30]—as the basis for image recall. In our case we use a network trained via matconvnet [31]—a script package for MATLAB that is specifically designed to create and train convolutional neural networks—that is set up to classify the 1,000 categories of images that Imagenet contains. While this network, imagenet-vgg-f.mat, which comes included with the matconvnet download, is intended for use classifying a single output class, we note that the final layer (a fully connected softmax probability layer) produces a 1,000 element vector that contains a probability between 0.0 and 1.0 for each of the classes. We exploit this fact by running a MATLAB script that records the full 1,000 element vector for each image in a resource database (from which images are pulled to match the query image). These vectors create a histogram of each of the 1,000 possible classes. When a query image is submitted via another script, its class vector is calculated and then compared via RMSE to each of the other images, as shown in the following formula.

$$S_{best} = \min\left(\sum_{i=1}^{images} \sum_{j=1}^{classes} (q_j - b_{ij})^2 \right), \quad q = \text{queryimage}, \quad b = \text{baseimages}$$

The n closest matches (smallest differential RMSE) in the resource database are chosen and displayed, as is the error between the query and resource images.

Even for images that do not contain one of the 1,000 Imagenet classes, each histogram turns out to be distinct and therefore can be used to retrieve similar images. The fact that images not featuring an object from the Imagenet classes can be queried and matched is exceptionally useful as it means this method can recall images it was not trained to recognize at all. We have used our CBH method to test and recall many query images that are not classified within the 1,000 image-net classes, and while these images would not produce viable classifications via the network, they produce good results for IBR.

4   Experimental Setup and Methodology

We utilize the F-measure to determine the quality of results in our experiment: we count images that are "very close" to the query image, images that are "pretty close," and images that are "not at all close." From these relatively straightforward metrics we calculate the precision of our results, either using only the correct (very close) images, or both correct and partially correct results. In Table 1, we provide the F-measure for both the correct results and the correct + partially correct results.

We selected two image sets, the Caltech 256 data set [32] and the one included with the CBIR package [29], and combined them into an image database of 29,970 images that fall within 271 classes (many of which are not Imagenet classes). These images contain between 80 and 200 of each image class/descriptor (e.g., sailboat, horses, bear, car).[1] We then selected 50 images from google.com and duckduckgo.com as test query images. The images are chosen to be reasonable images given the source image database; in other words, images that are similar to a large number of images (at least one class of 80+) within the source images. These images are isolated from the query database and any training work, so that they remain completely outside the world that the IBR packages had access to for training or querying.[2] For each engine, after adjusting to find optimum settings, we run a query for each of our 50 test images and request 20 similar images be output. For each of the 50 search results (with 20 images each) we count up the number of correct images, the number of partially correct, and the number of incorrect results, and record them in a spreadsheet. F-measures are computed for each image query as well as a single F-measure result for the entire 50 image query set for each query technique, shown in Table 1.

In our tests, our pretrained Imagenet network works very well, but still has room for improvement. We thus tried numerous methods to retrain/refine the network, including retraining via softmax log loss, top k error, mshinge, and our own modified version of softmax log loss. While our hope was to find one method that outperformed the original network in all cases, this did not occur. We thus created a voting method that utilizes the best three retraining methods—the original network, the network retrained with softmax log, and the network retrained via minimizing sum of squared errors on CBH-IBR—creating a results vector of all three methods combined. We sort this new, combined vector (3 times the length of each return vector, or 60 values in this case) and take the top 20 results. While a few results are actually worse, most are the same or improved, so this method produces the best overall F-measure, as presented in Table 1.

5 Results

While our results are somewhat qualitative, as we have to use human judgment to determine how close IBR results are, we have come up with distinctly differentiated results that are borne out by direct observation. Our IBR engine performs substantially better than the baseline packages using the same source image database and the same query images.

The cbires recall engine performs the worst of the group, as evidenced both by its total F-measure and by observing results. We attempted to improve the results, but there are very few parameters that can be adjusted via its GUI. We tried both knn and SVM methods, and found them about the same. Our results are for the knn method. As Table 1 indicates, the results of cbires are less than adequate. CBIR performed marginally better after some tweaking. Even at the best settings we could find, however, the image query results are also less than adequate, as indicated in Table 1.

---

[1] As the Caltech data set contains many classes with more than 200 images, while others have as few as 80, we removed any images beyond 200 for a class to reduce class imbalance.

[2] cbires requires the query image be in the image database, so we had to place the query image in the database before performing cbires searches.

As opposed to the baseline methods, CBH-IBR produces high quality results, both visually and via F-measure. Without retraining, the only tuning adjustment for this system is whether to ignore small values in the histogram vector, and what the threshold should be for ignoring small values. Empirically we determined that a value of 0.01 works the best. This setting ignores the noise of any very small probabilities, improving results dramatically. Interestingly, a large value for the threshold percentage reduces the quality of the results, indicating that categories of classification with smaller values significantly improve the engine's ability to find similar images. Figure 1 shows two excellent results, while the left-hand side of Fig. 2 shows one that is not wholly adequate.

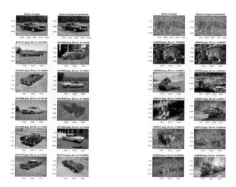

Fig. 1. Using CBH-IBR to query images of a classic car (left) and a tiger (right).

Retraining the CBH-IBR method involves assigning classes to each image in the database, and using loss algorithms to determine the quality of the result. Retraining improves results in many cases, but also creates instances where the results are worse than the original. Thus we have stacked the three best methods—the original network, one retrained via softmax log loss, and one via a custom histogram/class method[3]—and use lowest error scores from amongst the three methods to generate our 20 results. The right-hand side of Fig. 2 shows the results of the same query after retraining. Note that the two images in the second to last row are now images of mandolins, not incorrect images. Though this stacked network method works the best of any we tested, it still produces results that are less than perfect for some query images.

From visual examination, we produce F-measures for each search method, and for both correct and correct + partially correct results. Table 1 shows the results of these more quantitative measures and the numbers parallel our visual observations.

---

[3] Our method minimizes the derivative of the sum of squared errors between class histograms (term 1) added to class error, or softmax log loss (term 2).

$$S_{derivative} = dzdy * \left[-1.5 \left[\sum_{i=1}^{images} \sum_{j=1}^{classes} 2(q_j - b_{ij}) \right] - \sum_{i=1}^{images} \left(\left[\frac{e^{o_i}}{\sum_{k=1}^{classes} e^{o_k}} \right] - c_i \right) \right]$$

.

Table 1. F-measure for each of the four IBR methods tested.

IBR method	F-measure (correct)	F-measure (correct + partial)
cbires	0.242	0.477
CBIR	0.222	0.498
CBH	0.875	0.945
CBH—retrained	**0.928**	**0.979**

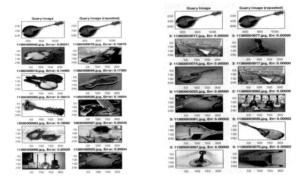

Fig. 2. Using CBH-IBR to query an image of a mandolin (left) and 3 CBH methods to query the same image of a mandolin (right).

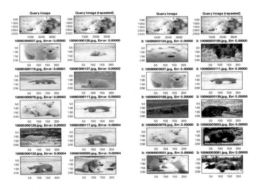

Fig. 3. Using CBH-IBR via original (unmodified) network (left), vs. retrained network (right).

6   Conclusions and Future Work

Our CBH-IBR method produces substantially better results than the baseline CBIR methods we tested. These results indicate that comparing class score histograms produces high quality results without needing much data preprocessing. In addition, retraining the network improves results, though sometimes at the cost of exact matches, as, shown in Fig. 3. While images on the right of Fig. 3 are correct, as they match the class bear, they are visually less correlated than the original network's results, as they output more brown bears. Our retraining method thus needs more granular classes on which to train: with sub classes for various bears, the degraded results would be eliminated. We also note that retraining based on individual classes is likely not the best way to improve performance. Our method allows the network to discover a web of probabilities associating images. Retraining the network to recognize a single class as correct, while improving class-based results, does not necessarily improve the visual quality of the results. Our custom prediction/loss method, which involves minimizing the sum of squared errors between the class histograms of the query image and the training images, works well only if we add to this the standard softmax loss method for classes. While results of this custom loss method are encouraging, research into improving this method is ongoing.

Overall we find our results to be excellent. Using existing networks in a novel way leverages all of the work that has gone into training DLCNNs, and with more robust networks and more research into training techniques, we believe the results will improve further.

References

1. Bishop, C.M.: Neural Networks for Pattern Recognition. Oxford University Press, Oxford (1995)
2. Kearns, M.J., Vazirani, U.V.: An introduction to computational learning theory. MIT press, Cambridge (1994)
3. Mitchell, T.M.: Machine Learning, 1st edn. McGraw-Hill, London (1997)
4. Zhou, X.S., Huang, T.S.: Comparing discriminating transformations and SVM for learning during multimedia retrieval. In: Proceedings of the ninth ACM international conference on multimedia, pp. 137–146. ACM, October 2001
5. Zhou, X.S., Huang, T.S.: Relevance feedback in image retrieval: a comprehensive review. Multimedia Syst **8**(6), 536–544 (2003)
6. Tong, S., Edward, C.: Support vector machine active learning for image retrieval. In: Proceedings of the ninth ACM international conference on Multimedia (MM 2001), pp. 107–118. ACM (2001)
7. Hearst, M.A., Dumais, S.T., Osman, E., Platt, J., Scholkopf, B.: Support vector machines. IEEE Intell Syst Appl **13**(4), 18–28 (1998)
8. Cohn, D.A., Ghahramani, Z., Jordan, M.I.: Active learning with statistical models. J. Artif. Intell. Res. **4**, 129–145 (1996)
9. Wu, J., Rehg, J.M.: Centrist a visual descriptor for scene categorization. IEEE Trans Pattern Anal Mach Intell **33**(8), 1489–1501 (2011)
10. Norouzi, M., Fleet, D.J., Salakhutdinov, R.: Hamming distance metric learning. In: NIPS, pp. 1070–1078 (2012)
11. Chechik, G., Sharma, V., Shalit, U., Bengio, S.: Large scale online learning of image similarity through ranking. J. Mach. Learn. Res. **11**, 1109–1135 (2010)
12. Salakhutdinov, R., Hinton, G.E.: Semantic hashing. Int. J. Approx. Reason. **50**(7), 969–978 (2009)
13. Chang, H., Yeung, D.Y.: Kernel-based distance metric learning for content-based image retrieval. Image Vis. Comput. **25**(5), 695–703 (2007)
14. Domeniconi, C., Peng, J., Gunopulos, D.: Locally adaptive metric nearest-neighbor classification. IEEE Trans. Pattern Anal. Mach. Intell. **24**(9), 1281–1285 (2002)

15. Bar-Hillel, A., Hertz, T., Shental, N., Weinshall. D.: Learning distance functions using equivalence relations. In: ICML, pp. 11–18 (2003)
16. Weinberger, K.Q., Blitzer, J., Saul, L.K.: Distance metric learning for large margin nearest neighbor classification. In: NIPS (2005)
17. Lee, J.E., Jin, R., Jain. A.K.: Rank-based distance metric learning: an application to image retrieval. In: CVPR (2008)
18. Guillaumin, M., Verbeek, J.J., Schmid, C.: Is that you? Metric learning approaches for face identification. In: ICCV, pp. 498–505 (2009)
19. Wang, Z., Hu, Y., Chia, L.-T.: Learning image-to-class distance metric for image classification. ACM TIST **4**(2), 34 (2013)
20. Mian, A.S., Hu, Y., Hartley, R., Owens, R.A.: Image set based face recognition using self-regularized non-negative coding and adaptive distance metric learning. IEEE Trans. Image Process. **22**(12), 5252–5262 (2013)
21. Wang, D., Hoi, S.C.H., Wu, P., Zhu, J., He, Y., Miao, C.: Learning to name faces: a multimodal learning scheme for search-based face annotation. In: SIGIR, pp. 443–452 (2013)
22. Hinton, G.E., Osindero, S., Teh, Y.W.: A fast learning algorithm for deep belief nets. Neural Comput. **18**(7), 1527–1554 (2006)
23. Ackley, D.H., Hinton, G.E., Sejnowski, T.J.: A learning algorithm for boltzmann machines. Cogn. Sci. **9**(1), 147–169 (1985)
24. Salakhutdinov, R., Mnih, A., Hinton, G.E.: Restricted boltzmann machines for collaborative filtering. In: ICML, pp. 791–798 (2007)
25. Salakhutdinov, R., Hinton, G.E.: Deep boltzmann machines. In: AISTATS, pp. 448–455 (2009)
26. Hinton, G., Deng, L., Yu, D., Dahl, G.E., Mohamed, A.R., Jaitly, N., Senior, A., Vanhoucke, V., Nguyen, P., Sainath, T.N., et al.: Deep neural networks for acoustic modeling in speech recognition: the shared views of four research groups. Sig. Process. Mag. **29**(6), 82–97 (2012). IEEE
27. Krizhevsky, A., Sutskever, I., Hinton, G.E.: Imagenet classification with deep convolutional neural networks. In: NIPS, pp. 1106–1114 (2012)
28. Mitro, J.: Content-based image retrieval tutorial, arXiv preprint arXiv:1608.03811 (2016). https://github.com/kirk86/ImageRetrieval
29. Khalifa, A.B., Tafazzoli., F: Content based image retrieval system (2013). https://github.com/aminert/CBIR/blob/master/Report/FeazhAmineCBIR.pdf.   https://github.com/aminert/CBIR
30. http://image-net.org
31. Vedaldi, A.: MatConvNet. Convolutional neural networks for MAT-LAB. In: Proceedings of the ACM international conference on multimedia (MM 2013). http://www.vlfeat.org/matconvnet
32. http://www.vision.caltech.edu/Image_Datasets/Caltech256
33. Krafka, K.J.: Building real-time unconstrained eye tracking with deep learning. Dissertation. Suchendra, B., Don Potter, W., advisors. The University of Georgia (2015)

Smart Content Recognition from Images Using a Mixture of Convolutional Neural Networks

Tee Connie⬛, Mundher Al-Shabi$^{(\boxtimes)}$⬛, and Michael Goh⬛

Faculty of Information Science and Technology,
Multimedia University, Melaka, Malaysia
mundher.ahmed@hotmail.com

Abstract. With rapid development of the Internet, web contents become huge. Most of the websites are publicly available, and anyone can access the contents from anywhere such as workplace, home and even schools. Nevertheless, not all the web contents are appropriate for all users, especially children. An example of these contents is pornography images which should be restricted to certain age group. Besides, these images are not safe for work (NSFW) in which employees should not be seen accessing such contents during work. Recently, convolutional neural networks have been successfully applied to many computer vision problems. Inspired by these successes, we propose a mixture of convolutional neural networks for adult content recognition. Unlike other works, our method is formulated on a weighted sum of multiple deep neural network models. The weights of each CNN models are expressed as a linear regression problem learned using Ordinary Least Squares (OLS). Experimental results demonstrate that the proposed model outperforms both single CNN model and the average sum of CNN models in adult content recognition.

Keywords: NSFW · CNN · Deep learning · Ordinary Least Squares

1 Introduction

The number of Internet users increases rapidly since the introduction of World Wide Web (WWW) in 1991. With the growth of Internet users, the content of the Internet becomes huge. However, some contents such as adult content are not appropriate for all users. Filtering websites and restricting access to adult images are significant problems which researchers have been trying to solve for decades. Different methods have been introduced to block or restrict access to adult websites such as IP address blocking, text filtering, and image filtering. The Internet Protocol (IP) address blocking bans the adult content from being accessed by certain users. This technique works by maintaining a list of IPs or Domain Name Servers (DNS) addresses of such non-appropriate websites. For each request, an application agent compares the requested website IP address or DNS with the restricted list. The request is denied if the two addresses match, and approved otherwise. This method requires manual keeping and maintenance of the

---

T. Connie and M. Al-Shabi—These authors contributed equally to this work.

K.J. Kim et al. (eds.), *IT Convergence and Security 2017*,
Lecture Notes in Electrical Engineering 449,
DOI 10.1007/978-981-10-6451-7_2

restricted list IPs, which is difficult as the number of the adult content websites grows or some websites change their addresses regularly.

Filtering by text is the most popular method to block access to adult content websites. The text filtering method blocks the access to a website if it contains at least one of the restricted words. Another approach is to use a machine learning algorithm to find the restricted words. Sometimes, instead of using the machine learning technique to extract keywords, a classification model is used directly to decide whether the requested webpage is safe [7]. Nonetheless, the text blocking method only understands texts, and it cannot work with images. This problem arises when the webpage does not contain the restricted keywords or does not contain text at all. Worse still, it may block safe webpages such as a medical webpage as it contains some restricted keywords.

Another blocking method uses image filtering [1, 9, 11]. This method works directly on the images, trying to detect if the image contains adult content. Detection directly from images is favorable as it does not require a list of IPs and is scalable to new websites, and is not sensitive to certain keywords. However, detecting adult content from images requires a complex model as the images have different illuminations, positions, backgrounds, resolutions or poses. In addition, the image may contain part of the human body, or the person in the image may be partially dressed.

In this paper, we seek to automatically recognize adult content from images using a mixture of convolutional neural networks (CNNs). Figure 1 shows the architecture of the proposed model in which eight CNNs models, followed by Fully Connected (FC) layers, are used to vote for the possible class of the image. Each model conforms to the same architecture with different weights computed using Ordinary Least Square (OLS). Usually, the training time of the deep CNN is very long. We present a solution to create eight models from a single architecture during training. A checkpoint is set to identify and pick the eight most-performing models during the training session. The solution selects the most optimal model to improve accuracy and helps reduce the training time drastically.

The contributions of this paper are as follows: (1) constructing a mixture of multiple deep CNNs at no extra cost; (2) assigning different weights to every model by applying OLS on all the model's output predictions

2   Related Works

The methods of recognizing adult content images can be divided into four categories: color-based, shape information-based, local features-based, and deep-learning-based.

The first approach analyzes the images based on skin color. This method classifies a region of pixels as either skin or non-skin. The skin color can be detected manually using a color range [1], computed color histograms [4], or parametric color distribution functions [3]. Once a skin color model of the image has been defined, the adult image can be detected by a simple skin color histogram threshold, or by passing the statistics of the skin information to a classifier [11].

Often, the skin areas contain some shape information such as ellipses or color compactness in some parts of the human body. The structure of a group of skin color regions is analyzed to see how they are connected. Several methods have been proposed to detect the shape features such as contour-based features [1] where the outlines of the

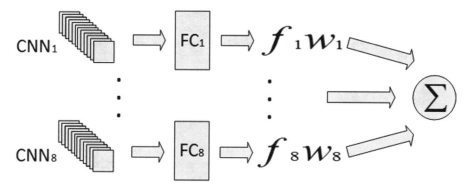

Fig. 1. The mixture of CNNs in which the eight CNNs models are combined linearly.

skin region are extracted and used as a feature, Hu and Zernike moments of the skin distribution [13], and Geometric constraints which model the human body geometry [2].

The third approach based on local features is inspired by the success of local features in other image recognition problems. Scale Invariant Feature Transform (SIFT) was used in conjunction with the bag of words to recognize adult images in [9]. The resulting features were trained using linear Support Vector Machine (SVM). Another local features called Probabilistic Latent Semantic Analysis was proposed in [8] to convert the image into a certain number of topics for adult content recognition.

The fourth and the most recent type of image content recognition technique is the use of deep learning approach. Moustafa [10] adopted AlexNet-based classifier [6] and the GoogLeNet [12] model architecture. Both models were treated as consultants in an ensemble classifier. Simple weighted average with equal weights was used to combine the predictions from the two models. Zhou, Kailong, et al. [14] proposed Another model based on deep learning. A pre-trained caffenet model was developed and the last two layers were fine-tuned with adult images dataset.

3   The Proposed Mixture of CNN Model

The proposed network contains six convolutional layers followed by two fully-connected layers as shown in Fig. 2. The number of filters in each convolutional layer is monotonically increasing from 16 to 128. A 2 × 2 Max-Pooling is inserted after each of the first two layers and after the fourth and the sixth convolutional layer. The size of each filter is 3 × 3 with two-pixel stride. To prevent the network from shrinking after each convolution, one pixel is added to each row and column before passing the image or the feature to the next convolution. After the six convolutional layers, the features bank is flattened and is passed to a fully-connected layer with 128 neurons. The output layer which only contains one neuron is placed after the first fully-connected layer, and before the sigmoid activation function. Except for the last layer, a rectifier linear unit (Relu) is used as the activation function which is less prone to vanishing gradient as the network grows. Another reason to adapt Relu is that it operates very fast as only a simple *max* function is used.

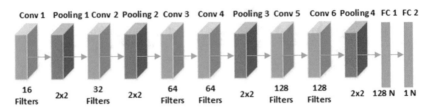

Fig. 2. The architecture of deep convolutional neural network model

To prevent the network from over-fitting the data, two regularization techniques have been applied. The first technique applies *L2* weight decay with 0.01 on the first fully-connected layer. Dropout is also used to prevent over-fitting. Dropout works by randomly zeroing some of the neurons output at training to make the network more robust to small changes. Dropout is placed directly after each Max-Pooling, and also after the first fully-connected layer. The probabilities of these four dropouts are set to 0.1, 0.2, 0.3, 0.4, and 0.4, respectively.

The network is trained for 300 epochs, with each epoch consisting of multiple batches optimized with Adam [5]. The batch size is 128 and is trained with the cross-entropy loss function. As adult image recognition is a binary problem in which the output can be either positive or negative, the binary cross-entropy is used

$$L(f, y) = -\sum f \, logy + (1 - y) \log(1 - f) \tag{1}$$

where *f* is the predicted value and *y* is the true value.

Generally, the training time of deep CNN is very long. To alleviate this problem, we introduce a way to extract eight sub-models with different weights in a single training session.

Algorithm 1 Generating Best Eight Performing Models

Input: Training data, *X*, validation data, *V*,
set of epochs {1,2,…,300}, *Q*
Output: The top 8 models, *top8*
Procedure:
  checkpoints ← empty list
  model ← *Deep CNN model with random weights*
  a ← −∞
  for each epoch ∈ *Q*
    model ← train *model* on *X* using Adam
    accuracy ← validate *model* on *V*
    if *accuracy* > *a*
      a ← *accuracy*
      add *model* to the *checkpoints* list
  end
  top8 ← top 8 models in *checkpoints*
  return *top8*

All the eight models are validated on the validation set, and a $N \times 8$ matrix is constructed from the outputs

$$\mathbf{F} = \begin{bmatrix} f_{11} & \cdots & f_{81} \\ \vdots & \ddots & \vdots \\ f_{1N} & \cdots & f_{8N} \end{bmatrix} \tag{2}$$

where N is the number samples in the validation set, and f_{in} is the predicted output of the i-th model of the N samples. The eight models combined linearly as,

$$z(w) = \begin{cases} 1, & f_1 w_1 + f_2 w_2 + \ldots + f_8 w_8 > 0 \\ 0, & \text{Otherwise} \end{cases} \tag{3}$$

Finding the unknown weights vector $\mathbf{W} = (w_1, \ldots, w_8)$ that minimizes $\min_{\mathbf{W}}(\mathbf{Y} - \mathbf{Z}(\mathbf{W}))^2$ is possible as long as $N \gg 8$. This problem is solved using Ordinary Least Squares (OLS) as,

$$\mathbf{Z} = \begin{bmatrix} f_{11} & \cdots & f_{81} \\ \vdots & \ddots & \vdots \\ f_{1N} & \cdots & f_{8N} \end{bmatrix} \begin{bmatrix} w_1 \\ \vdots \\ w_8 \end{bmatrix} \tag{4}$$

By taking the derivative of $(\mathbf{Y} - \mathbf{Z}(\mathbf{W}))^2$ setting it equal to zero with respect to $\mathbf{W}$

$$\frac{d}{dW}(\mathbf{Y} - \mathbf{Z}(\mathbf{W}))^2 = 0 \tag{5}$$

Finally, we rearrange the equation and solve it for $\mathbf{W}$

$$\mathbf{W} = \left(\mathbf{F}^{\mathsf{T}}\mathbf{F}\right)^{-1}\mathbf{F}^{\mathsf{T}}\mathbf{Y} \tag{6}$$

Usually, a model with a higher accuracy will get a higher weight than a model with lower accuracy.

4   Dataset

Due to the nature of the problem, there is no recognizable public dataset on adult content recognition. In this research, we collected the data manually from the Internet yielding 41,154 adult images. For negative images, images from the ILSVRC-2013 [15] test set were used. These data are separated into training, validation, and testing sets as shown in Table 1.

Each image is resized, centered, and cropped from the middle region to 128×128 pixels in RGB format. After that, all the images are normalized and mean subtracted. RGB images are fed as input to the network as they allow the first convolutional layer

Table 1. Dataset distribution

	Positive (Porn)	Negative (Neutral)	Total
Training	28930	27984	56914
Validation	6092	6104	12196
Testing	6132	6064	12196
Total	41154	40152	81306

to extract the skin color information while the rest of the networks extract high-level features based on body shape or textures.

5   Experimental Results

The CNN model is trained on 56,914 images and the validation set is used to pick the best eight models. In order to increase the generality of the model, the training data is augmented by flipping each image horizontally. This increases the total training images to 113,828.

We observe from Fig. 3 that the accuracy increases in a steady manner during training while the validation accuracy fluctuates around 96%. The best result at 96.43% is achieved at epoch 50.

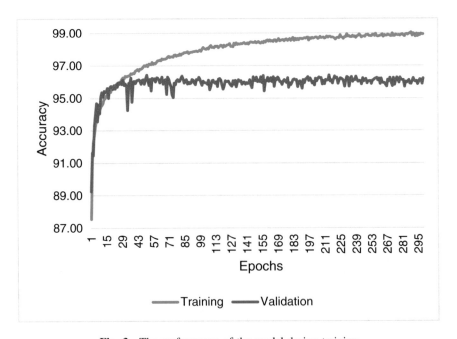

Fig. 3. The performance of the model during training

Table 2. Comparison of the models based on accuracy

	CNN-Mixture	Average Sum	Single CNN
Validation	**96.88%**	96.44%	96.43%
Testing	**96.90%**	96.50%	96.34%

We observe that the CNN-Mixture model gains small but constant accuracy improvement in both validation and testing sets, where it achieves 96.88% in validation set and 96.90% testing set, respectively as shown in Table 2. From the experimental results, we find that CNN-Mixture can effectively distinguish an adult image from a neutral image. Moreover, CNN-Mixture outperforms the single CNN model and even the average sum of multiple CNNs. Compared with the average sum, the CNN-Mixture uses the OLS to find the proper contribution for each model which help to distinguish better the role of each model in extracting useful features for recognizing the image content.

6   Conclusion

In this paper, we propose an adult image recognition system using a mixture of CNN. The proposed model is end-to-end learnable as compared to traditional solutions such as skin color detection. The proposed method is formed as a linear regression problem where the weights are calculated using ordinary least square. The proposed model is tested on a manually collected large dataset consisting of over hundred thousand of training images. Experiment results show that CNN-Mixture is effective against recognition of adult images, yielding an accuracy of over 96% on the testing set.

References

1. Arentz, W.A., Olstad, B.: Classifying offensive sites based on image content. Comput. Vis. Image Underst. **94**(1–3), 295–310 (2004)
2. Fleck, M.M., et al.: Finding naked people. In: Buxton, B., Cipolla, R. (eds.) Computer Vision — ECCV 1996, pp. 593–602. Springer, Heidelberg (1996)
3. Hu, W., et al.: Recognition of pornographic web pages by classifying texts and images. IEEE Trans. Pattern Anal. Mach. Intell. **29**(6), 1019–1034 (2007)
4. Jones, M.J., Rehg, J.M.: Statistical color models with application to skin detection. Int. J. Comput. Vis. **46**(1), 81–96 (2002)
5. Kingma, D., Ba, J.: Adam: a method for stochastic optimization. ArXiv14126980 Cs (2014)
6. Krizhevsky, A., et al.: ImageNet classification with deep convolutional neural networks. In: Pereira, F., et al. (eds.) Advances in Neural Information Processing Systems 25, pp. 1097–1105. Curran Associates, Inc., New York (2012)
7. Lee, P.Y., et al.: Neural networks for web content filtering. IEEE Intell. Syst. **17**(5), 48–57 (2002)
8. Lienhart, R., Hauke, R.: Filtering adult image content with topic models. In: 2009 IEEE International Conference on Multimedia and Expo., pp. 1472–1475 (2009)

9. Lopes, A.P.B., et al.: A bag-of-features approach based on Hue-SIFT descriptor for nude detection. In: 2009 17th European Signal Processing Conference, pp. 1552–1556 (2009)
10. Moustafa, M.: Applying deep learning to classify pornographic images and videos. ArXiv151108899 Cs (2015)
11. Rowley, H.A., et al.: large scale image-based adult-content filtering
12. Szegedy, C., et al.: Going deeper with convolutions. In: 2015 IEEE Conference on Computer Vision and Pattern Recognition (CVPR), pp. 1–9 (2015)
13. Zheng, Q.-F., et al.: Shape-based adult image detection. Int. J. Image Graph. **6**(1), 115–124 (2006)
14. Zhou, K., et al.: Convolutional Neural Networks Based Pornographic Image Classification. In: 2016 IEEE Second International Conference on Multimedia Big Data (BigMM), pp. 206–209 (2016)
15. ImageNet Large Scale Visual Recognition Competition 2013 (ILSVRC 2013). http://image-net.org/challenges/LSVRC/2013/

Failure Part Mining Using an Association Rules Mining by FP-Growth and Apriori Algorithms: Case of ATM Maintenance in Thailand

Nachirat Rachburee$^{(\boxtimes)}$ ⓘ, Jedsada Arunrerk ⓘ,
and Wattana Punlumjeak ⓘ

Department of Computer Engineering, Faculty of Engineering,
Rajamangala University of Technology, Thanyaburi, Pathum Thani, Thailand
{nachirat.r,jedsada.a,wattana.p}@en.rmutt.ac.th

Abstract. This research uses apriori algorithm and FP-growth to discover association rules mining from maintenance transaction log of ATM maintenance. We use ATM maintenance log data file from year 2013 to 2016. In pre-process step, we clean and transform data to symptom failure part attribute. Then, we focus on comparison of association rules between FP-growth and apriori algorithm. The result represents that FP-growth has better execution time than apriori algorithm. Additionally, the result from this paper helps maintenance team to predict symptom of failure or failure parts in future. As the advantage of predict failure parts, maintenance team will prepare a spare parts in store and prevent break down time of machine. The team can add failure parts from rules to preventive maintenance to prevent fail machine.

Keywords: Association · Apriori · FP-growth · ATM · Maintenance

1 Introduction

Many manufacturers used machines to run and handle the business core process. The operation could not be stopped when the machine was running in sequently process. The machine might be malfunction or break apart. Thus, the manufacturers should have plan for machine maintenance periodically but sometime machine might be malfunction before maintenance time. Generally, prediction maintenance used a time series data to predict amount of failure or part of failure by some variable or statistic data. They could predict symptom of failure or part that had been consequently malfunction with other part. Service department could do maintenance or replace part of machine that could be failure before time to malfunction. The analysis could prevent core business process from risk of interruption [1].

Data mining technique had 3 categories consisting of classification, clustering and association rules. This paper used association rules mining technique consist of FP-growth and apriori algorithm in our proposed method.

Currently, Service department had incident log file that had transaction around 3000 transaction per day. Then, they got incident logs data around 100 thousand

© Springer Nature Singapore Pte Ltd. 2018
K.J. Kim et al. (eds.), *IT Convergence and Security 2017*,
Lecture Notes in Electrical Engineering 449,
DOI 10.1007/978-981-10-6451-7_3

transaction from year 2014–2016. Incident log data was a time series data that record maintenance data several times in 1 day. In some cases, the part of failure had been often fail at the same time with one or another part.

This research focused on association rules technique to discover a part of machine that concurrently malfunction. We used the real world time series data set.

This paper organized as follows. In Sect. 2, the related work of association rule technique, In Sect. 3, framework of proposed method is represented. In Sect. 4, the result and discussion is presented. Finally in Sect. 5, the conclusion of this paper is discussed.

2   Relate Work

Knowledge discovery in database (KDD) was a process to explore and analysis a massive data set. Data cleaning and data preprocessing were the significant step. Machine learning was used in KDD to discover the meaningful result [2]. Data mining techniques were divided into two basic groups: unsupervised algorithms and supervised algorithms.

2.1   Association Rules

Association rules were the core process of data mining technique in supervised algorithm group. Association rules used to find frequency and correlation between items set in massive data.

Association rules discovered the significant association rules based on support value and confident value.

The support of rule $X \Rightarrow Y$ was the fraction of transactions in D containing both X and Y. The equation of support and confidence notion were given as below.

$$Support\ (X \Rightarrow Y)\ =\ X \cup Y\ /\ |D| \tag{1}$$

Where | D | was the total number of transaction in data set.

The confidence of rule $X \Rightarrow Y$ was the fraction of transactions in D containing X that also contain Y.

$$Confidence\ (X \Rightarrow Y)\ =\ supp\ (X \cup Y)\ /\ supp(X) \tag{2}$$

Association rules were generated from item set with satisfy both minimum support and minimum confidence.

2.2   Apriori

Apriori algorithm was an association rule technique in classification categories that used breadth first search algorithm or level wise search to count candidate item set in search space [3].

Apriori algorithm had two step to process. First, we had to find frequent item set that had minimum support value by frequent item set in all 1 item and all 2 item. Then, iteratively amount of item should be processed. We created association rules from frequent item set.

Generate association rule step was create rule form frequent item set. If item set was infrequent, all supersets of infrequent should be considered as infrequent that prune the search space (Fig. 1).

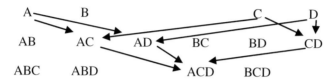

Fig. 1. Generate association rules

Candidate item set generation had 2 steps. First, created all possible candidate frequent item set. Second, from the result of the first step, removed all infrequent supersets.

This research deployed apriori algorithms to find association rules from high dimensional data using QR decomposition. QR decomposition reduced dimensions of data. Research team explored high important association rules from feature selection in high dimension data. QR decomposition selected independent feature and discard dependent feature. The result showed that proposed algorithm outperformed apriori algorithms [4].

This paper introduced new algorithm for association rule technique based on hadoop platform and map reduce. They proposed the interesting of threshold, confident and support values. They improved apriori algorithm by parallelization and interest threshold. The result from experiment showed linear increase of mining time that was suitable for big data mining [5].

Apriori was used in this paper that proposed to use association rules in Retail Company. The retail company was XMART that had 10 million historical transaction data. They discovered frequent item set from sale records and generated association rules with apriori algorithm. The rules were applied to sale department to raise more potential for each store. Additionally, association rules were used to find product layout in the store [6].

Apriori algorithm was used in this research that applied in power plant. They determined association rules from equipment maintenance data. They focused on fault prediction of equipment and optimized process of pruning and database scanning. The result of paper showed that they reduced some of maintenance and cost [7].

This research used apriori algorithm for transformer defect correlation analysis. They determined frequent pattern and dependency between decision attribute and classification for data of defect. They introduced analogous frequent item sets. The large volume of rules were discovered in this experiment. The result satisfied with the improved apriori algorithm by partition data with geographical location and save in array [8].

2.3   FP-Growth

FP-growth was the conventional algorithm that operated to find frequent item set without generating the candidate item set. FP-growth had two steps approach.

First, FP-tree was created from item set. All transactions were read and mapped the item set to construct the FP-tree. The frequent item had a support count. Infrequent items were discard.

Second, FP-growth algorithms extracted frequent item set from FP-tree by bottom-up strategy. This strategy should find frequent item set from ending with particular item (Fig. 2).

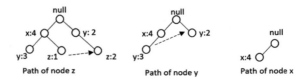

Fig. 2. Decomposition of frequent item set generation

The research approach was to reach the association between characteristic of maintenance and aviation maintenance fault from apriori algorithm. They used minimum confidence 70% and minimum support 10% to find association rules. They found that human characteristic attributes were important and required further analysis [9].

This paper analyzed association rules algorithm technique that were apriori, Eclat, Dclat, FP-growth, FIN, AprioriTID, Relim and H-Mine. The comparative algorithms used different thresholds, number of rules generation and size of data. They found that execute time decreased when minimum support increased. Their result represented the DCLAT algorithm was the best algorithms from the experiment [10].

This paper interested in mining association rules in dynamic huge data set. The important issue was updating of frequent item sets by FP-growth and heap tree. They compared FP-growth with other algorithm. The result showed significant reduce execution time of incremental updating frequent item sets. Additionally, this proposed algorithm had a steady efficiency in continuous data [11].

This research introduced association rules mining by FP-growth and Eclat algorithm in forest fire and land. They found association and pattern of hotspot occurrences. The parameters were set, minimum support was 30% and minimum confidence was 80%. The result showed hotspot occur relation with characteristic of location. The strong rule was precipitation greater than 3 mm/day with confidence 100% and 2.26 of lift value [12].

This research compared apriori and FP-growth algorithm in web usage. They collected data from server log data. They focused on discover pattern of website usage from server log files by fetching, processing efficiency, and memory size etc. From the result, apriori and FP-growth algorithm were appropriate to use in web usage mining, efficient and scalable for frequent pattern [13].

3   The Proposed Model

Pre-process data is the first step before using data in the framework. Hundred thousand of incidents from ATM machine are recorded in incident log data file and linked with ATM asset data by serial number. The combining two table generates 11 attributes (Fig. 3).

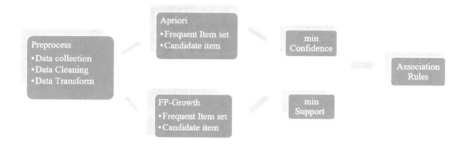

Fig. 3. Framework of proposed model

Data cleaning and transformation: from transaction log file, the data consist of corrective maintenance and preventive maintenance. We select only corrective maintenance and discard preventive maintenance from incident log data. Then, we extract failure part from problem detail attribute. Problem detail attribute saves as a text field that we transform symptom of failure by separating significant keyword of problem detail. Finally, we extract distinct failure words to 33 types of part from failure issue.

Association rule algorithms are used to find frequent item sets using apriori algorithm and FP-growth. We use the same example set to generate association rules by two algorithms. The process of generated rule from data set is iterative generate frequent item set and candidate item set. Then, we get support value and confidence from association rules. We set minimum support and minimum confidence to accept that rules.

4   Result and Discussion

The experiment uses Dell Inspiron 13, 8 GB Ram, Intel Core i7 Processor with RapidMiner version 7.1 and Windows 10. Goal of this experiment is to find association rules that represent symptom of failure or failure part that occur in the same time. Then, we compare performance of association rules between apriori algorithm and FP-growth in maintenance prediction.

From incident log data, we clean and generate new attributes from failure part based on name of failure part. As result, we discover new 33 failure parts from problem detail attribute that show in Table 1.

After cleaning and transforming data, we generate new data set from incident log data. Then, we transform failure part occurrence to binary as true or false. We send new

Table 1. Failure part from transaction log data

Part name		
Reader module	EPP-KB	Filter
Slip	Windows	Vacuum
Printer	Boot	Supply
UPS	URJB	mini-misc-pcb
Log	UTR	Exit sensor
Slot	Deposit	Scanner
Monitor	Modem-commu	Mouse
Dispenser	MoneyBox	SafeC
CCTV	Supervisor	Top-Up
Hard disk	Reject	Stacker
Alarm	EMV	Remote

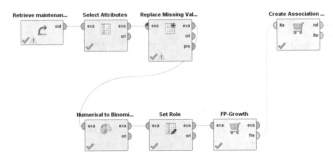

Fig. 4. Process in Rapidminer

data set to FP-growth algorithm to find frequent item set and send to apriori algorithm to create association rules that depend on minimum confidence (Fig. 4).

The results of FP-growth and apriori algorithm produce association rules that show in Tables 2 and 3, respectively. This experiment uses minimum confidence at 0.9 and minimum support at 0.05. The result represents association rules such as symptom of failure parts modem, boot device and power supply that can be concurrently occurred with ups failure.

Table 2. Association rules from FP-growth

Premises	Conclusion	Confidence
Modem-commu, boot, supply	ups	0.96660482
Modem-commu, supply	ups	0.96297989
Reader Module, vacuum	Dispenser Module	0.95369458
Reader Module, filter	Dispenser Module	0.94788274
Reader Module, MoneyBox	Dispenser Module	0.94001579

Table 3. Association rules from apriori

Rule	Confidence
[Commu, boot, supply] → [ups]	0.967
[Commu, supply] → [ups]	0.963
[ReaderModule, vacuum] → [Dispenser]	0.954
[ReaderModule, filter] → [Dispenser]	0.948
[ReaderModule, MoneyBox] → [Dispenser]	0.940
[ups, supply] → [Commu]	0.927
[exit sensor] → [Dispenser]	0.923
[ups, boot, supply] → [Commu]	0.922
[vacuum] → [Dispenser]	0.886
[reject] → [Dispenser]	0.884
[Commu, boot, supply] → [ups]	0.967

5 Conclusion

This research used apriori and FP-growth algorithm to discover association rules from maintenance log data in maintenance department. From the experiment, we found that association rules from both apriori and FP-growth had a same rules. However, FP-growth had better execution time than apriori.

The maintenance department could be alerted for the potential failure part that not show in transaction log data. Therefore, maintenance engineer could plan for a preventive maintenance properly before the problem. Engineer could also prepare the predicted failure part. Additionally, spare part store could be managed more efficiently and effectively.

Acknowledgements. The authors cordially thank Data One Asia (Thailand) Co., Ltd. for their data support.

References

1. Rachburee, N., Jantarat, S., Punlumjeak, W.: Time series analysis for fail spare part prediction: case of ATM maintenance. In: 2016 International Conference on Computer Sciences and Information Technology (2016)
2. Oded, M., Lior, R.: Data mining and knowledge discovery handbook, 2nd edn. Springer Science Business Media, LLC, New York (2010)
3. Tan, P.N., Steinbach, M., Kumar, V.: Introduction to data mining. Person Education. Inc., New Delhi (2006)
4. Harikumar, S., Dilipkumar, D.U.: Apriori algorithm for association rule mining in high dimensional data. In: 2016 International Conference on Data Science and Engineering (ICDSE), pp. 1–6 (2016)
5. Cui, X., Yang, S., Wang, D.: An algorithm of apriori based on medical big data and cloud computing. In: 2016 4th International Conference on Cloud Computing and Intelligence Systems (CCIS), pp. 361–365 (2016)

6. Zulfikar, W.B., Wahana, A., Uriawan, W., Lukman, N.: Implementation of association rules with apriori algorithm for increasing the quality of promotion. In: International Conference on Cyber and IT Service Management, pp. 1–5 (2016)
7. Wei, Y.H., Xue, D.S.: The research of equipment maintenance management in power plant based on data mining. In: 2015 IEEE International Conference on Computational Intelligence & Communication Technology (CICT), pp. 543–547 (2015)
8. Chen, Y., Du, X., Zhou, L.: Transformer defect correlation analysis based on apriori algorithm. In: 2016 IEEE International Conference on High Voltage Engineering and Application (ICHVE), pp. 1–4 (2016)
9. Zhang, R.Q., Yang, J.L.: Association rules based research on man-made mistakes in aviation maintenance: a case study. In: Sixth International Conference on Intelligent Systems Design and Applications, 2006, vol. 1, pp. 545–550 (2006)
10. Vijayarani, S., Sharmila, S.: Comparative analysis of association rule mining algorithms. In: International Conference on Inventive Computation Technologies (ICICT), vol. 3, pp. 1–6 (2016)
11. Chang, H.Y., Lin, J.C., Cheng, M.L., Huang, S.C.: A novel incremental data mining algorithm based on FP-growth for Big Data. In: 2016 International Conference on Networking and Network Applications (NaNA), pp. 375–378 (2016)
12. Arincy, N., Sitanggang, I.S.: Association rules mining on forest fires data using FP-Growth and ECLAT algorithm. In: 2015 3rd International Conference on Adaptive and Intelligent Agroindustry (ICAIA), pp. 274–277 (2015)
13. Singh, A.K., Kumar, A., Maurya, A.K.: An empirical analysis and comparison of apriori and FP-growth algorithm for frequent pattern mining. In: 2014 International Conference on Advanced Communication Control and Computing Technologies (ICACCCT), pp. 1599–1602 (2014)

Improving Classification of Imbalanced Student Dataset Using Ensemble Method of Voting, Bagging, and Adaboost with Under-Sampling Technique

Wattana Punlumjeak$^{(\boxtimes)}$ (ID), Sitti Rugtanom (ID), Samatachai Jantarat (ID), and Nachirat Rachburee (ID)

Department of Computer Engineering, Faculty of Engineering,
Rajamangala University of Technology Thanyaburi, Pathum Thani, Thailand
{wattana.p,sitti.r,samatachai.j,nachirat.r}@en.rmutt.ac.th

Abstract. Student imbalanced data is one of the problems in data mining community. To state the student dropout problem, an ensemble method with under-sampling technique is applied for improved the performance of classification of imbalanced student dataset. Mutual information for feature selection methods is used to find a significant feature. Voting, bagging, and adaboost technique in the ensemble method are used with decision tree (C4.5) and artificial neural network (ANN) classifiers to classify student in point of research objective. The result of this experiment evaluated by overall accuracy, precision, and recall. Bagging technique by random forest gave the best result in terms of overall accuracy is 74.57% and the recall of the prediction in the class (low) which we interested is 95.61%. This experiment extremely useful not only finding a useful knowledge for student and academic planning and management but also improving classification for imbalanced data which is the most effective way to state the classify student performance.

Keywords: Imbalanced data · Ensemble method · Classification

1 Introduction

In Knowledge discovery and Data Mining (KDD) especially in classification approach, imbalanced data problem is one of the problems which emerged many researchers in data mining community. Imbalanced data set is the representative of the relative instance that a large amount of data generated with skewed distribution. The class had known as minority class or rare class when the number of the data are significantly less than 10% of data in these distributions, whereas the remained data are the majority classes.

In classification method, the most classifiers are biased against the minority class, because of most classification learning algorithms are trained and focus on the accuracy of the majority class and ignore the essential patterns from the minority class. In many research and application, class imbalance are most found and observed especially classification approach like anomaly detection, predicting equipment failures, fraud detection, medical screening for cancer, and classify student dropout.

© Springer Nature Singapore Pte Ltd. 2018
K.J. Kim et al. (eds.), *IT Convergence and Security 2017*,
Lecture Notes in Electrical Engineering 449,
DOI 10.1007/978-981-10-6451-7_4

Student dropout problem is one of challenging task in educational data mining. Student failure is a major problem and important for academic advisor and stakeholder. To predicting with high accuracy of student performance in final grade is very useful, not only inform valuable information for students to improve their studies but also for academic planning and management. Conventional classification algorithms which applied to student data are often biased toward majority class and simply attempt to maximize the overall accuracy and minority class had been treated as outliers and then ignored. Although, the accuracy of the classification learning is very high, but it useless. In student data set, the majority of students had been got a good and fair grade. Then, the classifier was trained by the majority data and a few of student data with a low grade or minority data are not trained. In fact, student dropout problem, the academic advisor needs to interested and pay attention to students who had got a low grade or a minority class. Thus, the studies of imbalance data are growing at an explosive rate to understand the classification of learning and state of the art solutions to address the problem.

In this research, an efficient technique in data mining is proposed. The under-sampling technique is used to rebalance data, mutual information in feature selection is used to select the best set of feature and then use an ensemble technique of voting, bagging and adaboost of the decision tree and artificial neural network to find the best result in term of overall accuracy and confusion matrix.

The paper is organized as follow: after introduction, methodology and related work are presented in Sect. 2, follow by the proposed model in Sect. 3. Experiment and the result are shown in Sect. 4. Finally, we provides the conclusive remarks on the research.

2   Methodology and Related Work

2.1   Resampling

Resampling or balancing of class distribution is a task in data preprocessing step. Resampling techniques are used for resolved the problem of class imbalance by modified a dataset to improve its balance. Resampling can be categorized into three groups: under-sampling, over-sampling, and hybrid method which combine both sampling methods. Under-sampling technique creates a subset of the original dataset by randomly choosing instances from majority class which no change the number of sample in minority class. Over-sampling technique creates a superset of the original dataset by replicating some instances from minority class which aims to balance between majority and minority class. Synthetic minority oversampling technique (SMOTE) is one of the popular technique in oversampling which generate synthetic sample data from minority class. The studied from imbalanced data had been subjected for a long time both in academia and industry.

The review of the nature of imbalance learning scenario such as understanding, finding the suitable algorithms, tools, and assessment metric used have been educated and developed to address this problem [1]. In predicting student's final outcome in a particular course, the dataset collected from North South University, Bangladesh which has been imbalanced. The technique of random oversampling (ROS), random under-sampling (RUS), and SMOTE are used. The experiment showed that resampling technique can be solved the problem of imbalanced data by enhanced the performance of the classification model [2], after that they extended their work in preprocess step with optimal equal width binning and SMOTE. The result had clearly shown that the accuracy significantly increased [3].

2.2 Feature Selection

The goal of feature selection is to find out the most intrinsic feature that provides a classifier to reach optimal performance without loss any important information. Feature selection can be categorized into filter, wrapper, and embedded methods.

Mutual information (MI) or information gain is the filter method which measure only intrinsic characteristics of the feature in feature space. The concept of MI is a statistic measurement of the mutual dependence between the two variables in information theory. The MI between two random variables is defined as follow:

$$I(X:Y) = \sum_{x,y} Pxy(x,y) log \frac{Pxy(x,y)}{Px(x)Py(y)} \qquad (1)$$

where, $Pxy(x,y)$ is the combination of the probability distribution of the variable x and y. Respectively, $Px(x)$ and $Py(y)$ are the probability distribution of x and y.

MI is used with ensemble method named mutual information based feature selection for EasyEnsemble (MIEE) [4] which combined adaboost in the final decision. UCI datasets is used and the result shown that MIEE obtained better performance compare with the asymmetric bagging [5]. The imbalanced data research in fraud detection from electronic web payment system in Latin America, the under-sampling method is applied. Gain ratio, correlation-based feature selection, and relief are used to reduce the number of features. The result of the fraud detection compared with the actual scenario presented the financial gain of up to 61% [6]. Many feature selection algorithms have used in many researches in the class imbalanced dataset to solve the problems of imbalanced and also improved the accuracy of the minority class [7, 8].

2.3 Ensemble

The basic idea of ensemble method is to combine the more than one weak classifiers and then constructed the new strong classifier which aims to improve the performance of the classification. The popular methods of ensemble classifier are voting, bagging, and boosting. The vote ensemble method, at least two learners called base learners are used and the prediction of an unknown example uses a majority vote. Bagging or bootstrap aggregate, the basic idea is reduced the variance of the prediction by manipulated data using bootstrapping technique. The base classifier is trained on each bootstrap samples and return the class, and then the maximum is voted. An example of bagging is random forests or random decision forests classification. Boosting, which from adaptive resampling and combining the multiple weak learners and turned to a strong learner. Adaboost is the very well-known method in the boosting family which used on more than two classes of the classification problem.

The problem of student dropout in Thai university, cluster ensemble technique is used as a data transformation framework which the combination of weighted connected triple (WCT-T) with random-k cluster and ANN gave an effective accuracy of 91% [9]. Blogged dataset from UCI machine repository in sentiment classification used bagging to combine with GA, SVM, and NB. All of the bagged ensembles definitely outperformed compared to each single classifier [10]. To increase the accuracy in multispectral satellite images with ensemble technique: bagging, boosting, and adaboost with radial basis function (RBF) and back propagation neural network (BPNN) are used to state the problem. All ensemble technique with BPNN classifier achieved very high performance than RBF ensemble and single classifier [11]. The voting based weighted online sequential extreme learning machine (VWOS-ELM) technique is proposed for multi-class imbalanced classification compared with WOS-ELM. The eight multiclass datasets from UCI are used, the experiment result has shown that the VWOS-ELM more robust than WOS-ELM [12]. 10 iterations used with 15 imbalanced dataset in bioinformatics has been conducted to bootstrapping a sample set in bagging ensemble that have been achieved high quality in the classification performance [13]. The five different boosting based techniques: SMOTEBoost, DATABoost–IM, EUSBoost, DATABoost–IM with SVM, and boosting support vector machines are proposed and the survey conducted that all boosting ensemble have able to improve the minority class classification [14]. BNRAC is a hybrid ensemble model based on adaboost.M2 and SMOTE technique in preprocessed for imbalanced data from Portuguese banking institution. The base classifiers composed of decision tree, tree-J48, reduced-error pruning tree, and bayesian network. The result of BNRAC model gave the high accuracy of 96.3% [15]. Distribution based balanced sampling with the MultiBoosting algorithm (DBMB) is proposed with the gaussian and poisson resampling techniques used with imbalanced data from KEEL repository. The DBMB with both of resampling showed the outperformed comparing with other boosting methods [16, 17].

2.4 Classifier

In Classification method, decision tree is one of the popular method. A decision tree is supervised learning algorithm which partitions data recursively to form groups or classes by the value of information gain and can be used in discrete or continuous data for classification or regression. C4.5 is the popular algorithm in the decision trees family which improved from ID3 or iterative dichotomiser 3.

Artificial neural network (ANN) is a machine learning methodology inspired from biological central nervous systems in a human brain. The connection between one neural and each other will activate when a synapses which located on the dendrites of the neuron received signals, and if the weight of signals are strong enough or meet a threshold. ANN consists of three layer: input layer (the number of attribute/feature), output (the number of class), and hidden layer which can be calculate as follow:

$$no.\ of\ neural = \frac{no.\ of\ attribute + no.\ of\ class}{2} + 1 \qquad (2)$$

2.5   Evaluation

A confusion matrix is a basically used to evaluate the performance of the classification which reported in the tabular form of the actual and predicted classification as shown in Fig. 1 below.

Class	Predicted Positives	Predicted Negative
Actual Positives	True Positives(TP)	False Negative(FN)
Actual Negative	False Positives(FP)	True Negative (TN)

Fig. 1.   The confusion matrix

3   The Proposed Method

Our previous research [18], we have studies about the performance of the classification in the student data and we realized about imbalanced data. Thus, the main ideas of our work as in a proposed model as Fig. 2 as followed.

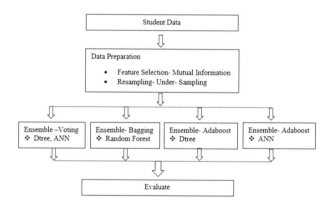

Fig. 2.   The proposed model

3.1   Student Data

In this research, the data is collected from years 2004 to 2010 A.D. from Faculty of Engineering, Rajamangala University of Technology Thanyaburi, Pathumthani, Thailand. The amount of 463,956 records is grouped into 6,882 records by studentID. We interested only the student's admission data and the subjects which registered in first year. Thus, in this experiment, the dataset consists of 15 features.

3.2   Data Preprocessing

In this step, many tasks are done. First, we cleaned the data by correct the inconsistent data, removed noisy, and handle missing data. Secondly, we transformed the data by discretized the student's GPA data into a categorical classes which consisting of high, medium, and low as shown in Table 1. Then, feature selection are applied, we used mutual information (MI) to find the most intrinsic feature from the dataset. After that, under-sampling technique are applied to state the imbalanced data.

Table 1. A categorical classes

Class	Low	Medium	High
Possible value (GPA range)	1.00–1.99	2.00–2.99	3.00–4.00

4 Experiment and Result

In our research, the software we used to implement the proposed model is RapidMiner Studio version 7.10, with window 10 operating system having Intel core i7 and 8 GB of RAM.

We started the experiment with data preprocessing step, composed of data cleaning and transformation. The amount of student records in each classes shown in Table 2. Then, applied MI in feature selection task with a full dataset to find a set of significant features. We interested in 10 significant features from MI method which equal to our previous research [18] features. The amount of student in low classes are 114 records. So, the resampling process with under-sampling technique was applied. After that, the 70% of sample set are selected into a training set and 30% is set to a testing set. Tenfold cross-validation are used for accurate the experimental model. We set four experiment of methods which each method ensemble the variety of classification technique. The ANN parameters we used in this experiment were: learning rate = 0.3, training cycle = 500, and momentum = 0.2. The hidden layer in ANN is set to one layer and the number of neurons in the hidden layer computed from Eq. (2).

Table 2. The amount of student records in each class

Total	Low	Medium	High
6882	114	5908	860

After all the experiment done, the visualization of the performance in each method have shown in Table 3 and Fig. 3, as follow:

Table 3. The experimental result

No.	Method	Overall Accuracy (%)	Precision (%)	Recall (%)
1	Voting with D-Tree(C4.5) and ANN	70.87	75.76	71.43
2	Bagging with random forest	74.57	66.87	95.61
3	AdaBoost with D-Tree(C4.5)	68.99	64.29	86.84
4	AdaBoost with ANN	75.12	74.77	72.81

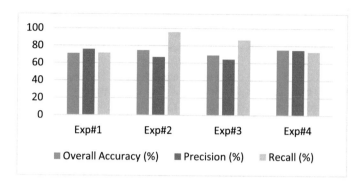

Fig. 3. The comparative results of the experiment

5   Conclusion

To state the student dropout problem, the amount of student with a low grade are imbalanced data. We applied an ensemble technique for improving a performance of classification of imbalanced student dataset. We used mutual information for feature selection methods to find a significant feature. Voting, bagging, and adaboost technique in the ensemble method are used with decision tree (C4.5) and artificial neural network classifiers. The result of this experiment evaluated by overall accuracy, precision, and recall. In terms of prediction in a class (low) we interested, bagging technique with random forest classifier gave the best result of recall at 95.61% and overall accuracy is 74.57%.

Acknowledgements.   We would like to thanks to Rajamangala University of Technology Thanyaburi, Pathumthani, Thailand for providing the student data for conduct this research.

References

1. He, H., Garcia, E.A.: Learning from imbalanced data. IEEE Trans. Knowl. Data Eng. **21**(9), 1263–1284 (2009)
2. Rashu, R.I., Haq, N., Rahman, R.M.: Data mining approaches to predict final grade by overcoming class imbalance problem. In: IEEE, 17th International Conference on Computer and Information Technology (ICCIT), pp. 14–19 (2014)
3. Jishan, S.T., Rashu, R.I., Haque, N., Rahman, R.M.: Improving accuracy of students' final grade prediction model using optimal equal width binning and synthetic minority over-sampling technique. Decis. Anal. **2**(1), 1 (2015)
4. Liu, X.Y., Wu, J., Zhou, Z.H.: Exploratory undersampling for class-imbalance learning. IEEE Trans. Syst. Man Cybern. Part B Cybern. **39**(2), 539–550 (2009)
5. Liu, T.Y.: Easyensemble and feature selection for imbalance data sets. In Bioinformatics. In: IEEE International Joint Conference on Systems Biology and Intelligent Computing IJCBS 2009, pp. 517–520 (2009)

6. Lima, R.F., Pereira, A.C.M.: A fraud detection model based on feature selection and undersampling applied to Web payment systems. In: 2015 IEEE/WIC/ACM International Conference on Web Intelligence and Intelligent Agent Technology (WI-IAT), vol. 3, pp. 219–222 (2015)
7. Yin, H., Gai, K., Wang, Z.: A classification algorithm based on ensemble feature selections for imbalanced-class dataset. In: 2016 IEEE 2nd International Conference on Big Data Security on Cloud (BigDataSecurity), IEEE International Conference on High Performance and Smart Computing (HPSC), and IEEE International Conference on Intelligent Data and Security (IDS), pp. 245–249 (2016)
8. Longadge, R., Dongre, S.: Class imbalance problem in data mining review. arXiv preprint arXiv:1305.1707 (2013)
9. Lam-On, N., Boongoen, T.: Using cluster ensemble to improve classification of student dropout in Thai university. In: IEEE 15th International Symposium on Soft Computing and Intelligent Systems (SCIS), 2014 Joint 7th International Conference on and Advanced Intelligent Systems (ISIS), pp. 452–457 (2014)
10. Govindarajan, M.: Analysis of bagged ensemble classifiers for blogger data. In: IEEE, International Conference in Computing Technologies and Intelligent Data Engineering (ICCTIDE), pp. 1–5 (2016)
11. Kulkarni, S., Kelkar, V.: Classification of multispectral satellite images using ensemble techniques of bagging, boosting and adaboost. In: IEEE 2014 International Conference on Circuits, Systems, Communication and Information Technology Applications (CSCITA), pp. 253–258 (2014)
12. Mirza, B., Lin, Z., Cao, J., Lai, X.: Voting based weighted online sequential extreme learning machine for imbalance multi-class classification. In: IEEE 2015 IEEE International Symposium on Circuits and Systems (ISCAS), pp. 565–568 (2015)
13. Fazelpour, A., Khoshgoftaar, T. M., Dittman, D. J., Naplitano, A.: Investigating the variation of ensemble size on bagging-based classifier performance in imbalanced bioinformatics datasets. In: 2016 IEEE 17th International Conference on Information Reuse and Integration (IRI), pp. 377–383 (2016)
14. Kaur, P., Negi, V.: Techniques based upon boosting to counter class imbalance problem—a survey. In: IEEE, 2016 3rd International Conference on Computing for Sustainable Global Development (INDIACom), pp. 2620–2623 (2016)
15. Ruangthong, P., Jaiyen, S.: Hybrid ensembles of decision trees and Bayesian network for class imbalance problem. In: IEEE 2016 8th International Conference on Knowledge and Smart Technology (KST), pp. 39–42 (2016)
16. Webb, G.I.: Multiboosting: a technique for combining boosting and wagging. Mach. Learn. **40**(2), 159–196 (2000)
17. Mustafa, G., Niu, Z., Yousif, A., Tarus, J.: Distribution based ensemble for class imbalance learning. In: IEEE 2015 Fifth International Conference on Innovative Computing Technology (INTECH), pp. 5–10 (2015)
18. Punlumjeak, W., Rachburee, N., Arunrerk, J.: Big data analytics: student performance prediction using feature selection and machine learning on microsoft azure platform. J. Telecommun. Electron. Comput. Eng. JTEC **9**(1–4), 113–117 (2017)

Reduction of Overfitting in Diabetes Prediction Using Deep Learning Neural Network

Akm Ashiquzzaman[1], Abdul Kawsar Tushar[1], Md. Rashedul Islam[1(✉)], Dongkoo Shon[4], Kichang Im[4], Jeong-Ho Park[3], Dong-Sun Lim[3], and Jongmyon Kim[2(✉)]

[1] Department of CSE, University of Asia Pacific, Dhaka, Bangladesh
zamanashiq3@gmail.com, tushar.kawsar@gmail.com,
rashed.cse@gmail.com
[2] Department of Electrical, Electronics, and Computer Engineering, University of Ulsan, Ulsan,
Republic of Korea
jmkim07@ulsan.ac.kr
[3] Industry IT Convergence Research Group, Intelligent Robotics Research Division, SW
Contents Research Laboratory, Electronics and Telecommunications Research Institute (ETRI),
Daejeon, Republic of Korea
{parkjh,dslim}@etri.re.kr
[4] Safety Center, University of Ulsan, Ulsan, Republic of Korea
{mycom,kichang}@ulsan.ac.kr

Abstract. Accurate prediction of diabetes is an important issue in health prognostics. However, data overfitting degrades the prediction accuracy in diabetes prognosis. In this paper, a reliable prediction system for the disease of diabetes is presented using a dropout method to address the overfitting issue. In the proposed method, deep learning neural network is employed where fully connected layers are followed by dropout layers. The proposed neural network outperforms other state-of-art methods in better prediction scores for the Pima Indians Diabetes Data Set.

Keywords: Dropout · Healthcare · Data overfitting · Diabetes prediction · Neural network · Deep learning

1 Introduction

Diabetes is a common physiological health problem among humans across gender, race and age. The term diabetic is applied when an individual is unable to break down glucose, for lack of insulin. The human organ called pancreas is responsible for generating the hormone called insulin, which is a very important enzyme that regulates the sugar level in human blood stream. It sanctions the human body to utilize sugar to generate energy; sans enough insulin, body cells cannot get the energy they need, consequently the sugar level in the blood gets too high, and many problems can emerge. Diabetes is not a curable disease; although, fortunately, it is treatable. Diabetes and related complications are responsible for the passing away of almost

© Springer Nature Singapore Pte Ltd. 2018
K.J. Kim et al. (eds.), *IT Convergence and Security 2017*,
Lecture Notes in Electrical Engineering 449,
DOI 10.1007/978-981-10-6451-7_5

200,000 Americans every year [1, 2]. In modern healthcare, predicting and properly treating diseases have become of foremost importance in medical prognostics fields. The whole process of determination of diabetes is completely manual, often suggested by the physician.

For diabetic patients, blood sampling by needle is a standard method for measurement of glucose concentration in blood plasma. This method is painful as the diabetic is measured glucose frequently (i.e., more than four times a day) [3]. To address this issue, non-invasive glucose monitoring techniques have been implemented on the wearable product such as GlucoWatchG2 [4], Pendra [5], and GlucoTrack [6]. Recently, Bandodkar et al. [7] developed a needle-free temporary tattoo for glucose monitoring with a reverse iontophoresis operation. In 2016, Lee et al. [8] introduced a wearable patch for sweat-based diabetes monitoring with drug delivery through bioresorbable temperature-responsive microneedles. Non-invasive glucose monitoring systems are generally known to be less accurate than blood-sampling systems. However, several researches have improved the accuracy [9, 10]. If a wearable device for blood glucose measurement provide reasonable accuracy, it can be widely used not only in the healthcare field [11], but also in the industrial safety and health field.

Smith et al. used the perceptron based algorithm called adaptive learning routine (ADAP), which is an early neural network model, to establish a diabetes prediction model for forecasting the arrival of diabetes mellitus. The system's performance measures were done using standard clinical benchmarks as specificity and sensitivity. The results obtained were then compared with those procured from applying linear perceptron models and logistic regression [12]. This method suffers from employing an early and complex structure of neural network which is responsible for performance degradation. On the other hand, Kayaer and Yıldırım proposed three separate neural network structures, which are multilayer perceptron (MLP), general regression neural network (GRNN), and radial basis function (RBF) and afterwards utilized the same data set to evaluate these three models. The performance gained by employing MLP was better than that of RBF method for all spread values tried. Among the three models evaluated, GRNN was able to provide the finest result on the test data [13]. Although not as complex as structures previously used, GRNNs are still convoluted structures; furthermore, this method, too, does not resort to any method for solving data overfitting.

The concept of deep learning is a fast-growing one which is teeming with ideas in recent years. Deep learning techniques are used in a range of diversified fields, including medical prognosis and optical character recognition [14]. In this paper, we will utilize the techniques of deep learning, namely - deep learning neural network, to propose a model for diabetes prognosis with high accuracy. This result is achieved with the help of a regularization layer called Dropout, which addresses the problem of overfitting arising from the use of deep fully connected layers.

The rest of this paper is structured as follows. Section 2 describes the proposed method, where the parts of the complete model are examined closely. Afterward, Section 3 delineates the dataset used in the experiment as well as the experiment procedures, and analyzes the results obtained via the proposed method. Then, Section 4 concludes the paper.

2   Proposed Method

Block diagram of the proposed method is outlined in Fig. 1. Here, the process is started by entering data into the input layer. Then there are two fully connected layers in place, each followed by a dropout layer. Finally, the decision is obtained from the output layer with a single node. Together these layers construct a multilayer perceptron, which is described in detail below.

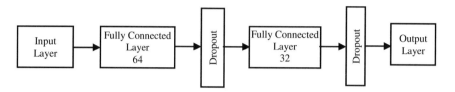

Fig. 1.   Description of proposed method.

2.1   Multilayer Perceptron

Multilayer Perceptron is, in simple terms, a logistic regression classifier with hidden depth. Here, the input data is transformed with non-linearity, or activation functions to output one or more linearly separable classes. These intermediate layers are alluded to as hidden layers. A single hidden layer is sufficient to turn an MLP into a universal approximator. However, as it is learned, more hidden layers make the MLP more adaptive to the data [15].

Figure 2 represents the common architecture of an MLP. In formal notation, an MLP with a sole hidden layer is a function $f: R^A \rightarrow R^B$, where A and B are respectively the

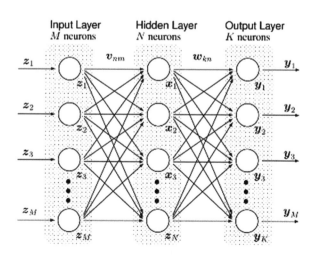

Fig. 2.   A feed forward neural network/MLP.

sizes of input vector x and output vector $f(x)$. Relation between input and output vectors can be coined as:

$$f(x) = \Phi\big(b^{(2)} + W^{(2)}\big(\phi\big(b^{(1)} + W^{(1)}x\big)\big)\big) \qquad (1)$$

with bias vectors $b^{(1)}$, $b^{(2)}$, weight matrices $W^{(1)}$, $W^{(2)}$, and activation functions Φ and ϕ. Here the activation function can be various mathematical threshold functions, i.e. $tanh(x)$, sigmoid, exponential linear unit (ELU), or rectified linear unit (ReLU). To train an MLP, we learn all parameters of the model, and to do that we use Stochastic Gradient Descent or any other relevant algorithm divided into mini-batches. The set of parameters to learn is the set $\theta = \{W^{(1)}, W^{(2)}, b^{(1)}, b^{(2)}\}$. Obtaining the gradients $\dfrac{\delta l}{\delta \theta}$ can be achieved through the backpropagation algorithm which is a special case of the chain-rule derivation [16].

2.2 Dropout

Dropout is nothing but a form of regularization. Srivastava et al. first implemented the dropout in their network to prevent overfitting problems of neural networks, which is depicted in Fig. 3 [17].

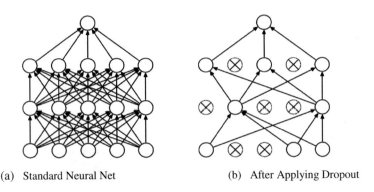

(a) Standard Neural Net   (b) After Applying Dropout

Fig. 3. A demonstration of dropout, adapted from the method of Srivastava et al. [17].

Dropout disables neurons in a neural network in such a random way that, during the learning phase, the network is forced to learn multiple representations of data. These representations are independent of each other and are derived from the same data. In this way, in various layers, neurons are hindered from co-adapting too well and this in turn reduces the possibility of overfitting. The DNN architecture uses a probability distribution in order to randomly exclude a number of neurons in each layer from updating weight. This results the neural network to learn from different representations. Krizhevsky et al. had implemented this regularization in their neural net in 2012, winning the prestigious Imagenet challenge of 2012 [18].

The significant ability of the feed-forward DNN lies in its hidden layer. In our proposed method, we finalize on three layers altogether: the first input layer consists of

64 neurons, which is a multiple of the 8 input attributes. This layer utilizes ELU as its activation function. ELU is a special adaptive form of rectified linear unit which is chosen because of its sustainability toward vanishing gradient problem [19]. The next layer consists of 32 neurons, configured the same as the previous fully connected layer. The final layer is the one neuron output layer for the prediction which has Softplus function [18] as activation.

The numbers of nodes being accommodated in the hidden layers are decided by training several network configurations and choosing the optimal network topology from them. The criteria for optimality is fixed as yielding minimal seam square error (MSE) value and demonstrating the most increased predictability. Both the numbers of attributes as well the of output layer decision are in the powers of 2. Therefore, the numbers of nodes in both hidden layers are preferred to be powers of 2 as well. The point of note here is that, incorporating too many hidden layer neurons or hidden layers in the MLP architecture may sometimes lead to overfitting [21]. On the other hand, if the number of hidden layer neurons is relatively insufficient to capture the complexities of the problem, the issue of under fitting may arise [22].

Addition of dropout results in a generalized neural net rather than an overfitted one, in that it prevents the neurons from learning too much about the input data. This is done when dropout method forces some neurons in random order to be inactive during different phases of forward propagation as well as back-propagation. When neurons are randomly pitched out of the network during training, other neurons are forced to substitute for the missing neurons by handling the representation required to make predictions. This in turn results in multiple independent internal representations being learned by the network. Furthermore, this improves prediction results and validation scores [23, 24]. Each layer of the DNN is configured to have a dropout as a function in learning process. The first two layers of our proposed neural network has a low 25% probability in dropout, but the final layer has a 50% dropout rate to reduce overfitting.

Weights and biases of the entire network are first filled up according to a uniform function for maximum yield in learning. The Mean Squared Error (MSE) method is used as the loss function and Adadelta function is used for optimization. Back-propagation method is used for training the model via forward and backward passes. In forward passes, weights, biases, and inputs are combined to calculate a predicted value for each neuron. In backward passes, loss is calculated from difference of predicted and actual values, and that loss is used to update weights and biases in the model [18].

3   Experimental Results and Analysis

The Pima Indians Diabetes (PID) Data Set [25] is used in experiment. This data set is obtained from the UCI machine learning repository and is a subset of a bigger data set held by the National Institute of Diabetes and Digestive and Kidney Diseases [26]. The patients whose data are present in this database are women of Pima Indian inheritance who were older than 20 years of age and were residents of USA at the time of surveying. The binary output variable takes either 0 or 1, where 1 means testing positive and 0 is a testing negative for diabetes. 268 (34.9%) cases are present in class 1 for positive test and 500 (65.1%) cases in class 0 for negative test. There are eight clinical attributes and a brief overview is given in Table 1, along with specifications of the attributes in Table 2. As we can observe from the data in Table 2, the data could have been normalized. However, in deep learning, the DNN eventually learns the biases and filters the data accordingly. In our experiment, we use the whole dataset as it is, making no change to the data or any attributes.

Table 1. Attributes of PID data set [25]

Class	Attribute number
Pregnancy count	1
Glucose concentration in plasma	2
Blood pressure (diastolic, mmHg)	3
Thickness of triceps skin fold (mm)	4
2-Hour serum insulin (μ U/ml)	5
Body mass index	6
Pedigree function of diabetes	7
Years of age	8

Table 2. A brief description of PID data set [13]

Attribute number	1	2	3	4	5	6	7	8
Mean	3.8	120.9	69.1	20.5	79.8	32.0	0.5	33.2
Standard deviation	3.4	32.0	19.4	16.0	115.2	7.9	0.3	11.8
Minimum	0	0	0	0	0	0	0.078	2.42
Maximum	17	199	122	99	846	67.1	2.42	81

The whole experiment is done in an Intel Core i5-6200U CPU @ 2.30 GHz 4 cores with 4 Gigabytes of DDR4 RAM. The MLP model is implemented with help of python Theano [27], with Keras [28] wrapper at the top.

The results of the previous methods discussed in the paper as well as result of the proposed method are depicted in Table 3. The proposed method has outperformed all the other methods previously described. Smith et al., with their proposed method, yields some good results, but the overall performance is not up to the mark. The specificity and sensitivity of their algorithm is 76% on the test set of 192 instances due to using methods that are now outdated. On the other hand, the three different methods proposed by Kayaer et al. have three superior yields during training phase; however, the accuracy drops significantly in the main cross-validation testing. This drop could be attributed to data overfitting. In contrast, the proposed method has the advantages of dropout as the regularization, which gives the network a considerable boost in performance. As a result, the overfitting issue that has been plaguing the other methods has minimal effect on the proposed method. Furthermore, the network is fed the data in raw format for processing, which is a distinct approach from previous methods. Due to this, the system can learn connections between the raw data values in a unique way. In the end, the proposed DNN had an 88.41% accuracy with 0.1 split validation, which is the new recorded accuracy for the PID dataset.

Table 3. Results of different methods compared

ID no.	Method name	Accuracy (%)	Remarks
1.	Smith et al.	76	Regression network
2.	Kayaer et al. Training set	82.99	DNN
3.	Kayaer et al. Test set	80.21	Same DNN
4.	Kayaer et al. Mean correct prediction	82.29	GRNN
5.	Proposed method	88.41	DNN, with dropout

4   Conclusion

In this paper, a reliable prediction system using a dropout method was proposed to reduce data overfitting in the predictive model which is used for forecasting the disease of diabetes. The proposed method employed a novel form of deep neural network for diabetes prognosis to increase the prediction accuracy. In the experiment, the proposed method achieved an accuracy of 88.41% over the PID Data Set. By diminishing the effect of overfitting in the proposed model, increased accuracy is achieved via experimentation. As a result, performance for predictive models for diabetes can now have better prediction scores or performance gains which can lead to future breakthroughs in health prognostication. In the future, a synergy effect of better diabetes prediction will be explored with real-time data of healthcare wearable devices.

Acknowledgements.   This work was supported by Electronics and Telecommunications Research Institute (ETRI) grant funded by the Korean government [17ZS1700, Development of smart HSE system for shipyard and onshore plant]. The authors also acknowledge department of Computer Science and Engineering, University of Asia Pacific for supporting this research in various ways.

References

1. Alberti, K.G.M.M., Zimmet, P.F.: Definition, diagnosis and classification of diabetes mellitus and its complications. Part 1: diagnosis and classification of diabetes mellitus. In: provisional report of a WHO consultation. Diabetic Med. **15**(7), 539–553 (1998)
2. National Diabetes Data Group: Classification and diagnosis of diabetes mellitus and other categories of glucose intolerance. Diabetes **28**(12), 1039–1057 (1979)

3. Vashist, S.K.: Non-invasive glucose monitoring technology in diabetes management: A review. Anal. Chim. Acta **750**, 16–27 (2012)
4. Potts, R.O., Tamada, A.J., Tierney, J.M.: Glucose monitoring by reverse iontophoresis. Diabetes Metab. Res. Rev. **18**, S49–S53 (2002)
5. Wentholt, I.M.E., Hoekstra, J.B.L., Zwart, A., DeVries, J.H.: Pendra goes Dutch: lessons for the CE mark in Europe. Diabetologia **48**(6), 1055–1058 (2005)
6. Harman-Boehm, I., Gal, A., Raykhman, A.M., Zahn, J.D., Naidis, E., Mayzel, Y.: Noninvasive glucose monitoring: a novel approach. J. Diabetes Sci. Technol. **3**(2), 253–260 (2009)
7. Bandodkar, A.J., Jia, W., Yardımcı, C., Wang, X., Ramirez, J., Wang, J.: Tattoo-based noninvasive glucose monitoring: a proof-of-concept study. Anal. Chem. **87**(1), 394–398 (2014)
8. Lee, H.J., Choi, T.K., Lee, Y.B., Cho, H.R., Ghaffari, R., Wang, L., Choi, H.J., Chung, T.D., Lu, N., Hyeon, T., Choi, S.H., Kim, D.H.: A graphene-based electrochemical device with thermoresponsive microneedles for diabetes monitoring and therapy. Nat. Nanotechnol. **11**(6), 566–572 (2016)
9. Zanon, M., Sparacino, G., Facchinetti, A., Talary, M.S., Mueller, M., Caduff, A., Cobelli, C.: Non-invasive continuous glucose monitoring with multi-sensor systems: a Monte Carlo-based methodology for assessing calibration robustness. Sensors **13**(6), 7279–7295 (2013)
10. Caduff, A., Zanon, M., Mueller, M., Zakharov, P., Feldman, Y., De Feo, O., Donath, M., Stahel, W.A., Talary, M.S.: The effect of a global, subject, and device-specific model on a noninvasive glucose monitoring multisensor system. J. Diabetes Sci. Technol. **9**(4), 865–872 (2015)
11. Park, Y.J., Seong, K.E., Jeong, S.Y., Kang, S.J.: Self-organizing wearable device platform for assisting and reminding humans in real time. Mobile Inform. Syst. **2016**, 15 (2016)
12. Smith, J.W., Everhart, J., Dickson, W., Knowler W., Johannes, R.: Using the adap learning algorithm to forecast the onset of diabetes mellitus. In: Proceedings of the annual symposium on computer application in medical care. American Medical Informatics Association, p. 261 (1988)
13. Kayaer, K., Yildirim, T.: Medical diagnosis on pima indian diabetes using general regression neural networks. In: Proceedings of the International Conference on Artificial Neural Networks and Neural Information Processing, pp. 181–184 (2003)
14. Ashiquzzaman, A., Tushar, A. K.: Handwritten arabic numeral recognition using deep learning neural networks. In: 2017 IEEE International Conference on Imaging, Vision & Pattern Recognition (icIVPR), pp. 1–4. IEEE (2017)
15. Dasgupta, J., Sikder, J., Mandal, D.: Modeling and optimization of polymer enhanced ultrafiltration using hybrid neuralgenetic algorithm based evolutionary approach. Appl. Soft Comput. **55**, 108–126 (2017)
16. Nielsen, M.A.: Neural networks and deep learning. http://neuralnetworksanddeeplearning.com.Accessed 29 May 2017
17. Srivastava, N., Hinton, G.E., Krizhevsky, A., Sutskever, I., Salakhutdinov, R.: Dropout: a simple way to prevent neural networks from overfitting. J. Mach. Learn. Res. **15**(1), 1929–1958 (2014)
18. Krizhevsky, A., Sutskever, I., and Hinton, G. E.: Imagenet classification with deep convolutional neural networks. In: Advances in Neural Information Processing Systems, pp. 1097–1105 (2012)
19. Nair, V., Hinton, G.E.: Rectified linear units improve restricted boltzmann machines. In: 27th International Conference on Machine Learning, pp. 807–814 (2010)

20. Glorot, X., Bordes, A., Bengio, Y.: Deep sparse rectifier neural networks. Aistats **15**(106), 275 (2011)
21. Heaton, J.: Introduction to neural networks with Java. Heaton Research, Inc. (2008)
22. Panchal, G., Ganatra, A., Kosta, Y., Panchal, D.: Review on methods of selecting number of hidden nodes in artificial neural network. Int. J. Comput. Theory Eng. **3**(2), 332–337 (2011)
23. Hinton, G.E., Srivastava, N., Krizhevsky, A., Sutskever, I., Salakhutdinov, R.R.: Improving neural networks by preventing co-adaptation of feature detectors. arXiv preprint arXiv: 1207.0580 (2012)
24. Warde-Farley, D., Goodfellow, I. J., Courville, A., Bengio, Y.: An empirical analysis of dropout in piecewise linear networks. arXiv preprint arXiv:1312.6197 (2013)
25. Lichman, M.: UCI machine learning repository (2013). http://archive.ics.uci.edu/ml Accessed 29 May 2017
26. National Institute of Diabetes and Digestive and Kidney Diseases. https://www.niddk.nih.gov/. Accessed 29 May 2017
27. Theano Development Team, Theano: a Python framework for fast computation of mathematical expressions. arXiv e-prints, vol. abs/1605.02688 (2016)
28. Chollet, F.: Keras https://github.com/fchollet/keras. Accessed 01 June 2017

An Improved SVM-T-RFE Based on Intensity-Dependent Normalization for Feature Selection in Gene Expression of Big-Data

Chayoung Kim[1] and Hye-young Kim[2(✉)]

[1] Kyonggi University, 154-42 Gwanggyosan-ro, Yeongtong-gu, Suwon, Gyeonggi, Korea
kimcha0@kgu.ac
[2] Hongik University, 2639 Sejong-ro, Jochiwon-eup, Sejong, Korea
hykim@hongik.ac.kr
http://nglab.kr

Abstract. Thanks to Next-Generation-Sequencing (NGS) revolutionary, high-throughput RNA sequencing data (RNA-seq) has become a highly sensitive and accurate method of measuring gene expression. Since RNA-seq generate a huge amount of data they have been struggling to overcome the lack of computational methods to exploit the enormous RNA-seq Big-Data. In most of cases, those methods have not been adequate for feature scaling scheme on RNA-seq Big-Data. So, RNA-seq encourages computational biologist to identify both novel and well-known features, although it have led to an increase in an adoption of previous methods and development of newly scalable data analysis ones. And it provides recognition of some deep learning methods which are scalable and adaptable for assuming and selecting the highly correlated genes for classification and prediction. However, some assumption of those methods have not been always correct and they have been considered unstable in terms of large-scale gene expression profiling. Therefore we propose improved feature selection technique of well-known support vector machine recursive feature elimination (SVM-RFE) with T-Statistics based on Intensity-dependent normalization, which uses log differential expression ratio (M vs A plot) for improving scalability. In each iteration of SVM-RFE, less dominated feature set with respect to relevance and redundancy is excluded from this set of features. In the proposed algorithm, the most relevant and less redundant feature is included in the final feature set, accomplishing comparable accuracy with a small subsets of Big-Data, such as NCBI-GEO. The proposed algorithm is compared with the existing one on several known data. It finds that the proposed algorithm have become convenient and quick than previous because it uses all functions in R package and have more improvement with regard to the time consuming in terms of Big-Data.

Keywords: Support Vector Machine Recursive Feature Elimination (SVM-RFE) · Intensity-dependent normalization (M vs A plot method) · T-Statistics · RNA-seq gene expression · Big-Data

© Springer Nature Singapore Pte Ltd. 2018
K.J. Kim et al. (eds.), *IT Convergence and Security 2017*,
Lecture Notes in Electrical Engineering 449,
DOI 10.1007/978-981-10-6451-7_6

1   Introduction

The Next-generation sequencing (NGS) such as RNA-seq as an alternative DNA micro-arrays have become an important part of genomics and systems biology [1] because it can become a highly sensitive and accurate method of measuring gene expression. They have been struggling to overcome the lack of computational methods to exploit the enormous data since RNA-seq are generating a huge amount of data. In response, the assistance of managing growing data volumes and those repositories has been realized. So Bioinformatics and IT professionals have been extremely to analyze huge volume of RNA-seq. However those methods have not been adequate for feature scaling scheme on RNA-seq Big-Data because technical variability, high noise levels and massive sample sizes also have been increasing. In terms of genomics experiments, there are some of big objectives which are to generate a few groups of relevant data for identifying differentially expressed genes and to build gene regulatory networks (GRN) for under-standing the underlying molecular mechanisms through inferring causality relationships [2]. In this paper, we focus on the object of identifying differentially expressed genes with the number of transcripts, which can differ significantly between samples in Big-Data. Gene expression profiling has been applicable for capturing the gene expression patterns in cellular responses to diseases, genetic perturbations and drug treatments [2]. The complex patterns of gene expression might evoke more scalable and adaptive tech-nique that substantially reduces technical variability and improves the quality of down-stream analyses. This imply those should be required samples have not been highly and differentially expressed features in accordance with reasonable time consumptions of normalization. So, RNA-seq encourages computational biologist to identify both novel and well-known features, although it have led to an increase in an adoption of previous methods and development of newly scalable data analysis ones. The well-known normalizations such as DEGseq [1] are based on total or effective counts and tend to perform poorly. Therefore, we propose the improved algorithm to overcome the lack of scalability with combining Support Vector Machine-Recursive Feature Elimination (SVM-RFE) with T-Statistics which can become a "filter-out" factor for improving scalability based on well-known normalization, DEGseq. We recognize that some deep learning methods such as SVM-RFE [3] are scalable and adaptable for assuming and selecting the highly correlated genes for classification and prediction. Guyon et al. [3] propose support vector machine recursive feature elimination (SVM-RFE) algorithm to recursively remove genes based on their weight vectors in the support vector machine (SVM) classifiers and classify the samples with SVM. Because feature selection can be regarded as a dimensionality reduction, some feature ranking algorithms exploit commonly used ranking matrices such as Signal-to-Noise and T-Statistics [4]. In this paper, T-Statistics as a "filter-out" factor in SVM-RFE can be named to remove the least possible bottom-ranked genes following to the previous method [4]. The T-Statistics could remove more than one gene at a single round. Moreover we use intensity-dependent normalization, which uses log differential expression ratio (Minus vs Add plot, MA-plot) as known DEGseq [1], as a preprocessor right before SVM-RFE with T-Statistics. Some literatures [4] develop alternative algorithms based on SVM-RFE, to overcome consuming a huge amount of training time and the problem of over-fitting persist and eliminating only one gene at each iteration. It can improve the accuracy of the resultant classifier and narrow down the potential set of cancerous genes. The proposed our algorithm can become more accuracy by exploiting DEGseq right before the previous SVM-RFE while reducing the size of the number of potentially distinguishable genes. If some algorithms [2] might use the proposed our algorithm as a preprocessor, it can be essential prerequisite for better time consumption with scalability and more accurate. And the proposed algorithm can become convenient and quick than previous, because it uses all functions in R package. It have more improvement with regard to the feature selection in terms of Big-Data. Also, we have computational results from the subsets of Big-Data, such as NCBI-GEO [5]. We compared the results of the proposed algorithm with those of the existing ones. The comparison was performed based on the number of ranked features. We can find that it can be accomplishing comparable scalability in Big-Data. This article is structured as follows: the next Sect. 2 describes the materials and methods that have been used for this work such as, the datasets, the algorithms like MA-plot-based normalization (DEGseq), SVM-RFE and T-Statistics. In Sect. 3, the algorithmic representation is demonstrated. In the next section, the result of the proposed algorithm have been analyzed with the existing ones. Finally, the conclusion describes the future directions.

2   Materials and Methods

2.1   Data

To compare our implementation results with a well-known result of a microarray-based technology, we downloaded leukemia [2] from their websites. It was analyzed by using R-package 'golubEsets'. This set was exploited with SVM-RFE based on MA-plot-based methods by R-package 'DEGseq'. Leukemia [2] was assayed using Affymetrix Hgu6800 chips. Also we are interested in RNA-seq like colon [6], which was assayed using Illumina HiSeq 2000. We have downloaded the colon dataset [6] from NCBI-GEO, GEO Series accession number GSE2109 in Gene Expression Omnibus (GEO, http://www.ncbi.nlm.nih.gov/geo/). RNA-seq data of 54 samples (normal colon, primary colorectal cancer (CRC), and liver metastasis) from 18 CRC patients are generated in [6], which are identified significant genes associated with aggressiveness of CRC for identifying a prognostic signature with diverse progression and heterogeneity. Only diverse statistical methods including generalized linear model likelihood ratio test have validated the results of two significantly activate regulators in [6]. Likewise the recent research [4], we try to the learning method, the well-known classifier, such as SVM-RFE for large-scale gene expression profiling. And for that the objective of this study is to identify the feature genes, although the number of input genes are too small, we downloaded Salt tolerance of rice (Oryza sativa) [7]. For GO term enrichment analysis on gene selected from our implementation results, we use 'QuickGO' (http://www.ebi.ac.uk/QuickGO-Beta/) [8].

2.2 SVM-RFE Algorithm

Guyon et al. [2] proposed a gene selection algorithm, Support Vector Machine-Recursive Feature Elimination (SVM-RFE). SVM is a classification algorithm. In terms of cancer classification, linear SVM has been already exploited as the basic classifier. For a linear kernel SVM, the weight vector, w can be calculated by $w = \sum_{i=1}^{n} \alpha_i x_i y_i$ (1), where i is the number of genes ranging from 1 to n, x_i is the gene expression vector of a sample i in the training set and y_i is the class label of i, $y_i \in [-1,1]$ and α_i is the *Lagrangian-Multiplier* estimated from the training set. Feature elimination algorithms work by iteratively removing one "least" gene at a time. In each loop, the remaining significant genes are ranked and it is possible the genes ranking can be modified at the result. As the well-known SVM-RFE as a recursive feature elimination algorithms achieved notable performance improvement with the weight vector $w = \|w\|^2$ (2). Figure 1 describes the SVM-RFE algorithm in detail.

Algorithm : SVM-RFE
1. Input: gene set, G={1,2,...n},
   Output: Ranked gene list for classification
2. Initialization Set S={ }
3. Do while if G is not empty
   Train SVM in G
   Compute gene ranking score by sum square of w=w² by equation (1) and (2)
   Sort the features by the ranking score
4. Return Ranked gene list

Fig. 1. SVM-RFE algorithm

2.3 Feature Ranking Algorithms for Gene Selection

Some feature ranking algorithms works in a way by ranking genes individually in terms of correlation-based metric. Because these ranked genes are selected to form the most informative gene subset, they can be regarded as a metric for eliminating more one gene at a time. Tang et al. [4] proposed feature ranking algorithm, a gene selection algorithm that extended the SVM-RFE algorithm by incorporating the T-Statistics. This method combined the statistical T-Statistics to predict higher accuracy and more significant genes. And they are aimed at training the algorithm in a much faster manner by eliminating many a genes at a time.

$$|(+) - (-)| \sqrt{i(+2)/n(+) + i(-)2/n(-)} - \text{T} - \text{Statistics} \qquad (3)$$

Combining Support Vector Machine-Recursive Feature Elimination (SVM-RFE) with T-Statistics which become a "filter-out" factor could overcome the lack scalability. There are some arguments about the good performance SVM-RFE is that it does not make handling features simultaneously. Many previous works makes the assumption that if only one gen is eliminated at each step, that is a smaller "filter-out", the final gene subset should be the best one. However, the research [4] shows the assumption is not always correct. So, we can use T-Statistics as a "filter-out".

2.4 MA-Plot-Based Methods for Normalization

The MA-plot, which is a statistical analysis method, which is having been widely used to detect and visualize intensity-dependent ratio of microarray data [1]. There are systems differences in gene expression data. The normalization of expressed gene adjusts the individual hybridization intensities in two-color (Red/Green) microarray assay to balance them appropriately reliable so that meaningful biological comparisons can be made. The intensity-based normalization which uses log differential expression ratio (Minus vs Add plot, $M = \log_2 C_1 - \log_2 C_2$ and $A = (\log_2 C_1 + \log_2 C_2)/2$, where C_1 and C_2 is the counts of reads obtained from two samples) in each element on the array for the Locally Weighted Linear Regression (LOWESS) [1] analysis can be exploited. Local variation as a function of intensity can be used to identify differentially expressed genes by calculating an intensity-dependent Z-score. In DEGseq [1] for this purpose, there is a plot which is color-coded depending on whether they are less than one standard deviation, between one and two standard deviations, or more than two standard deviations. That means the genes out of the line ($\log_2 C_1 = \log_2 C_2$) in the plot are identified for carrying out further analyses

3   The Proposed Algorithm

In this section, we propose SVM-T-RFE with T-Statistics with DEGseq algorithm for gene selection aims at eliminating more genes at a time recursively for an improved scalability in Fig. 1. The previous SVM-RFE has been trained in each iteration at eliminating a gene at a time, depending on different sets of genes. It is a state-of-the-art technique but has the flaws, which is consumption of the high amount of training time because elimination "worst" gene at a time. There are some researches that the implicit assumption about the SVM-RFE: The final ranked gene list could be better when the SVM-RFE removes one gene at a time. However, there are some researches that shows the assumption is not always correct when it simulated on the well-known AML/ALL [2]. So we conjunct the T-Statistics with the SVM-RFE for some different raking criteria as a larger "filter-out" factor for the good performance. In terms of the data size, gene expression Big-Data were recursively removed, making the algorithm faster enough to work with. In addition to this larger "filter-out", we might exploit the Intensity-based normalization which uses log differential expression ratio (M vs A plot by DEGseq) as a preprocessor before the SVM-T-RFE for the good scalability. The proposed SVM-T-RFE with DEGseq, which is a conjunction of SVM-RFE and T-Statistic (Welch's t-test statistic) based on DEGseq to aim at training the algorithm in a much faster manner by eliminating many genes at a time with a smaller input gene expression from Big-Data. So, the SVM-T-RFE with DEGseq provides the comparable time consumption compared with the previous SVM-RFE [2] and SVM-RFE with T-Statistics [4]. In the proposed algorithm, a new modified rank score is given in $w_{i2} = \cdot{}^*w_i + (1 - \cdot)^*t_i$, where

w_{i2} = a new rank of the ith gene, $\cdot$ = parameter determining the tradeoff between SVM weights, w_i = SVM weight vector for ith gene and T-Statistics range from 0 to 1, t_i = Welch's t-test of the ith gene. In MA-plot-based methods (DEGseq) of the proposed algorithm, for the number of input genes such as the preprocessor, the T-Statistics P-value or as $\log_2$ Fold-Change from the result of the DEGseq can be exploited as a threshold. We exploited the threshold in the AML/ALL and CRC RNA-seq. And for the generic comparison, in Salt tolerance of rice (Oryza sativa) [7], we compared with SVM-T-RFE without DEGseq (Fig. 2).

Algorithm: DEGseq-SVM-RFE with Welch's t-test
Input: Gene-Set, G = {1, 2, … n},
Output: Ordered Gene List for classification based on the ranking criterion, FRList

1. Initialization Gene-Set-Normal
2. Initialization Gene-Set-Cancer
3. Get the Output-Gene-List by DEGseq (MA-plot-based Methods)
4. Cut the Output-Gene-List with THRESHOLD such as $\log_2$ Fold-Change/P-value
5. Update the initialized Gene-Set with the Output-Gene-List
    {New-Gene-List} = {Gene-Set} − {Output-Gene-List}
6. Do while if New-Gene-List is not empty
    Train SVM in New-Gene-List
    Compute the Weight Vector (w_i) by eq (1) ($wi = \sum_{k=1}^{n} aiyixi$)
    Compute the T-Statistics (t_i) by eq (3)
$$(|(+) - (-)|\sqrt{i(+)2/n(+) + i(-)2/n(-)})$$
    If P-value by T-Statistics less than 0.05, then $w_{i2} = \cdot \, *w_i + (1 - \cdot)*t_i$
    Compute the Ranking Criterion, CR=$\|w_{i2}\|^2$ by eq (2)
    Sort the new Feature Ranks, New$_R$ based on CR, New$_R$ = sort (CR)
    Update the Feature Ranked List, FRList
        based on New$_R$ FRList=FRList+New-Gene(New$_R$)
    Eliminate the Features based on R,
        New-Gene-List = New-Gene-List − New-Gene (New$_R$)
7. Return the feature ranked list, FRList

Fig. 2. The proposed SVM-T-RFE with T-Statistics algorithm

4 The Proposed Algorithm Evaluation

Most of recent researches have been targeted to select a few perfect gene subset for providing a right validation accuracy: The smallest one is claimed to be the best. Most extracted perfect subsets have been analyzed by the evaluation methods which are meaningful in some senses. For example, with some small size of samples and highly correlated genes, "a few perfect subsets" could be extracted by an algorithm. However, the smallest perfect subsets cannot justify the gene selection algorithms outstanding compared to other algorithms Because of "Curse of Dimension", that is the scarcity of the samples in Big-Data. A small number of perfect subsets cannot be extracted in a comparable computation time. In the computation time, we try to select the subsets which explain the complex cancer regulation in AML/ALL [2] and CRC [6] or salt-resistance in Salt tolerance of rice [7]. Some results might not enough to uncover the cancer mechanisms. And those results could be our next movement for better under-standing of cancer mechanisms.. In our work, the proposed SVM-T-RFE with DEGseq is compared with the well-known SVM-RFE without DEGseq. For our DEGseq grounded on the p-values of the genes, the p-values less than equal to 0.05 are considered as statistically significant threshold. Before the iteration of SVM-T-RFE, DEGseq can be performed as a preprocessor.

The AML/ALL with 7,120 genes, CRC with 23,505 genes and Salt tolerance of rice with 200 genes have been prepared for comparison. After Degseq performed as a prepro-cessor, each one of samples can be made in AML/ALL with 425 genes, CRC with 761 genes for SVM-T-RFE. And we need Salt tolerance of rice [7] for meaningful biological comparisons such as when we use the whole genes without DEGseq, how many selected feature genes are in the almost same ranking. So, the whole 200 genes of Salt tolerance of rice have been taken into consideration. Table 1 shows that the feature genes of SVM-RFE and those of SVM-T-RFE without DEGseq of Salt tolerance of rice are almost same (simi-larity is 50% in the descending order of 20% of the top most subsets in 200 genes). This result indicates that the algorithm with a larger "filter-out" can improve the performance in terms of computational time in Big-Data with the comparative feature subsets. For the proof that the smaller the number of genes, the more practical computational time without scarifying the accuracy, we compare the results in the AML/ALL and in the CRC. Table 1 shows that the features genes of SVM-RFE and those of SVM-T-RFE with DEGseq of CRC are little bit of different with the smaller input genes (accuracy is 5% in the descending order of 10% with 761 input genes in 23,505 genes). That means the gene selection is not accurate if too few genes have been used an input. However, Table 1 shows that the feature genes of SVM-RFE and those of SVM-T-RFE with DEGseq of AML/ALL compared with the results of Golub [2] and [4] are quite comparative. In Table 1, genes in green for AML in Golub [2], ones in orange for ALL in Golub [2], one in red for ALL in Golub [2] and [4] (reliability is comparative as 10% in the descending order of 5% with 425 input genes in 7129). In results of AML/ALL, there is an improvement on the reliability/accuracy,

Table 1. Accuracy comparison on the 2 different algorithms on salt tolerance of rice, CRC, and AML/ALL

	Salt tolerance of rice		CRC(colorectal cancer)		AML/ALL	
	SVM-RFE without DEGseq	SVM-T-RFE without DEGseq	SVM-RFE with DEGseq	SVM-T-RFE with DEGseq	SVM-RFE with DEGseq	SVM-T-RFE with DEGseq
1	OsAffx.18823.1.S1_at	OsAffx.5284.1.S1_at	REG1A	SPINK4	M27891	D86967
2	OsAffx.5284.1.S1_at	Os.12300.1.S1_at	IFITM1	CEACAM7	M17733	
3	Os.49199.1.S1_at	OsAffx.23281.1.S1_at	MIR3936	COL3A1	V00787	M11722
4	Os.10838.1.S1_a_at	Os.11919.1.S1_at	CEACAM7	IFITM1	M19507	M23197
5	OsAffx.23281.1.S1_at	OsAffx.18823.1.S1_at	ACTG2	COL1A2	U05255	X80822
6	Os.26728.3.S1_at	Os.49199.1.S1_at	COL1A1	LCN2	Z19554	J04164
7	Os.11919.1.S1_at	Os.10838.1.S1_a_at	COL1A2	RN5-8S1	Z23090	
8	Os.12300.1.S1_at	Os.11037.1.S1_a_at	COL3A1	CCL20		X01677
9	Os.56929.1.S1_at	Os.26728.3.S1_at	DES	REG1A	M11147	M59465
10	Os.11037.1.S1_a_at	Os.56929.1.S1_at	LCN2	ACTG2	hum_alu	M34996
11			RN5-8S1	MIR3936	U68105	AFFX-HUMRGE/M10098
12			SPINK4	IGLL5	M27783	M31627
13			IGLL5	COL1A1	M23613	M89957
14			CCL20	DES		U23852
15					J04088	M31520
16					M11722	
17						X01703
18					M14328	U06155
19					L38941	J03592
20					L20941	X97267

comparing with those of CRC. We notice that it is difficult that decide accuracy and relia-bility vs. computational performance and that the assumption that too few small genes as an input are not always providing inferior results.

5   Conclusion

We have suggested the algorithm combining Support Vector Machine-Recursive Feature Elimination (SVM-RFE) with T-Statistics as a larger "filter-out" factor for more bottom-ranked genes at a time based on the one of well-known normalization, DEGseq for improving scalability. Our proposed algorithm tries to overcome consuming a huge amount of training time and eliminating only one gene at each iteration while narrowing down dominated feature set with respect to highly relevance. So, the proposed algorithm can be exploited DEGseq as a preprocessor right before the previous SVM-RFE for improving the accuracy and T-Statistics for reducing the size of the number of potentially distinguishable genes for better time consumption. In the performance evaluation of Salt tolerance of rice, we have the meaningful biological results that the algorithm can perform in a practical time with a larger "filter-out" factor. In our result of CRC, the smaller input genes scarify the accuracy of the gene selection. However, in our results of AML/ALL, we notice our assumption that too few small genes as an input are not always providing inferior results. We realize that improving the accuracy and reliability of the resultant classifier and narrow down the potential set of cancerous genes could be balanced with practical computational performance. And we can find that the proposed algorithm can be accomplishing compa-rable scalability in terms of Big-Data, comparing with the previous ones.

References

1. Wang, L., Feng, Z., Wang, X., Wang, X., Zhang, X.: DEGseq: an R package for identifying differentially expressed genes from RNA-seq data. Bioinformatics **26**(1), 136–138 (2010)
2. Golub, T.R., Slonim, D.K., Tamayo, P., Huard, C., Gaasenbeek, M., Mesirov, J.P., Coller, H., Loh, M.L., Downing, J.R., Caligiuri, M.A., Bloomfield, C.D., Lander, E.S.: Molecular classification of cancer: class discovery and class prediction by gene expression monitoring. Science **286**(5439), 531–537 (1999)
3. Guyon, I., Weston, J., Barnhill, S., Vapnik, V.: Gene selection for cancer classification using support vector machine. Mach. Learn. **46**, 389–422 (2002)
4. Tang, Y., Zhang, Y.-Q., Huang, Z.: Development of two-stage SVM-RFE gene selection strategy for microarray expression data analysis. IEEE ACM Trans. Comput. Biol. Bioinform. **4**(3), 365–381 (2007)
5. NCBI GEO. https://www.ncbi.nlm.nih.gov/geo/
6. Kim, S.-K., Kim, S.-Y., Kim, J.-H., Roh, S.-A., Cho, D.-H., Kim, Y.-S., Kim, J.-C.: A nineteen gene-based risk score classifier predicts prognosis of colorectal cancer patients. Mol. Oncol. **8**(8), 653–1666 (2014)
7. Wang, J., Chen, L., Wang, Y., Zhang, J., Liang, Y., Xu, D.: Computational systems biology study for understanding salt tolerance mechanism in rice. PLoS ONE **8**(6), e64929 (2013)
8. Quick GO. http://www.ebi.ac.uk/QuickGO-Beta/

Vehicle Counting System Based on Vehicle Type Classification Using Deep Learning Method

Suryanti Awang[✉] and Nik Mohamad Aizuddin Nik Azmi

Soft Computing & Intelligent System Research Group (SPINT), Faculty of Computer Systems & Software Engineering (FSKKP), Universiti Malaysia Pahang, Lebuhraya Tun Razak, 26300 Kuantan, Pahang, Malaysia
suryanti@ump.edu.my

Abstract. Vehicle counting system (VCS) is one of the technologies that able to fulfil the intelligence transportation system's (ITS) aim in providing a safe and efficient road and transportation infrastructure. This paper is aimed to provide a more accurate VCS based on vehicle type classification method rather than the current implementation in existing works that only count the vehicle as vehicle and non-vehicle. To fulfil the aim, we proposed to use deep learning method with convolutional neural network with layer skipping-strategy (CNNLS) framework to classify the vehicle into three classes namely car, taxi and truck. This VCS is motivated by current implementation of the traffic census in Malaysia where they record the vehicle based on certain vehicle classes. The biggest challenge in this paper is how to discriminate features of taxi and car since taxi has almost identical features as car. However, with our proposed method, we able to count based on correctly classified of the vehicle with the average accuracy of 90.5%. We tested our method using a frontal view of vehicle from the self-obtained database taken using mounted-camera at the selected federal road.

Keywords: Computational intelligence · Vehicle counting · Vehicle classification · Deep learning

1 Introduction

Intelligence transportation system (ITS) corresponds to the technology applied in road and transportation infrastructure to improve safety and efficiency in related applications for instance, expressway toll system, traffic census, traffic surveillance, etc. This paper will focus on the traffic census that many countries especially in Malaysia still implementing conventional methods for instance, manually observed by human, photography and sensor-based. One of the traffic census aims is to provide a good quality of transportation infrastructure based on the outcome from that census. Basically, the census in Malaysia will count the vehicle based on vehicle's class namely motorcycle, car, truck and public transportation. The outcome will be used as an input for further actions such as designing road pavement thickness, providing appropriate public transport, etc. Thus, an efficient approach of the

© Springer Nature Singapore Pte Ltd. 2018
K.J. Kim et al. (eds.), *IT Convergence and Security 2017*,
Lecture Notes in Electrical Engineering 449,
DOI 10.1007/978-981-10-6451-7_7

census is needed to achieve an accurate outcome. ITS offers various technologies to be applied in this issue. One of the technologies is Vehicle Counting System (VCS) where much related to observe the traffic by counting how many vehicles are using certain road at certain period of time or in a day.

VCS can be grouped into two categories; sensor based system and camera based system. We will focus on existing works of VCS for camera based system. There are a few works in VCS that implemented computer vision to count the vehicle. For example, a surveillance camera and traffic camera from top view are used to track and count vehicles in works done by [1, 2, 5, 6]. The vehicles are counted as general (vehicle or non-vehicle) without any specific classification. Simple background subtraction, color image-based adaptive background subtraction and vehicle box region are used respectively to differentiate the vehicle and background. These approaches are able to count the vehicles, however unable to give an accurate outcome as expected in the traffic census. [3] also counts the vehicle without classifying its class using a motion estimation with Taylor series approximation. Other than that, there is principle component pursuit (PCP) that is based on principle component analysis (PCA) used to extract features of vehicles from a satellite image, and vehicles are counted based on that features and does not consider the class of vehicle [4].

There is a VCS that is based on vehicle types proposed by [7]. They count the vehicles as small car, van and motorcycle using Gaussian Mixture Model (GMM) to detect the vehicle, Kalman filter for shadow removal and blob size to classify the size of vehicle. The drawback of this approach is the feature that extracted to be used in classifying the vehicle is too limited which is only rely on the size of blob. Thus, it will lead to misclassification when almost similar size but difference class of vehicle is classified. VCS also can be developed by using a combination of techniques in computer vision and machine learning. [8] proposed to implement this combination in VCS for traffic control analysis. They applied fast region-based convolutional neural network (FR-CNN) to detect the moving vehicles. However, this approach only able to detect and count the moving vehicles without able to do the counting based on vehicle types.

Based on the existing works, we found that the VCS can be improved with intelligent techniques in machine learning to classify the vehicle type. Therefore, we proposed to implement convolutional neural network with layer-skipping strategy (CNNLS) based on deep learning method. This framework is able to learn vehicle features in detail and the VCS is counted based on the classes of car, taxi and truck. The aim of this proposed approach is to provide accurate outcome for traffic census for further benefits. The explanation on the methodology of VCS with CNNLS is in the next section, followed by experiment and result section and end with conclusion section.

2   Methodology

There are two stages of methodology in this paper; first is VCS and second is vehicle type classification. We will explain the first methodology that is based on VCS process flow and later is the explanation on vehicle type classification based on CNNLS framework.

2.1   Vehicle Counting System (VCS)

We count the vehicle based on a mounted camera video. To detect the vehicle in the video, we use region-CNN (R-CNN) to detect the region of interest for moving vehicle. The R-CNN is used due to its ability to capture a moving object and separate the vehicle and background. Figure 1 shows the process flow of VCS. We set duration for vehicle counting is within 1 h. This is based on the practice by Public Works Department that held the census by hour and day. Thus, we decided to conduct it in hour for the initial observation. We use a counter for car, taxi and truck to record the number of each vehicle class. If there is a vehicle detected, vehicle type classification will classify it using CNNLS. If the vehicle is in a car class, the car counter will be increased by 1 and vice versa. However, if the vehicle is not in any of the class, therefore there is no increment in any counter. The process will be looped until the time limit is exceed. At the end, the total number of each class is counted based on the latest value in each counter.

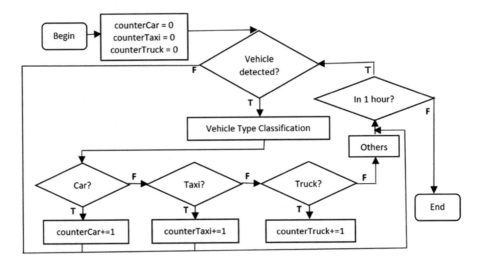

Fig. 1.   VCS process flow

2.2   Vehicle Type Classification - CNNLS

This section is about how CNNLS is implemented in classifying the vehicle type. Basically, before CNNLS is implemented, there are two processes namely image acquisition to acquire vehicle images and pre-processing to prepare the images prior feature extraction process. The CNNLS is used in the feature extraction to extract vehicle features to be classified in classification process. Generally, layer-skipping strategy of the CNNLS has capability to extract both local and global features of a vehicle. Figure 2 shows the general framework of CNNLS in this implementation.

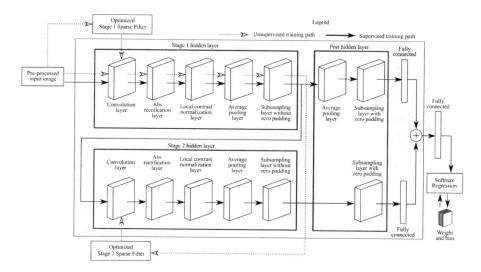

Fig. 2. CNNLS framework in vehicle type classification

There are 2 major phases in the vehicle type classification, namely, training and testing phase. Based on Fig. 2, there is a process to pre-process the images of training and testing separately prior to the CNNLS implementation. The CNNLS has two stages of hidden layer that will implement the feature extraction of the pre-processed images. For the training phase, note that there are two types of training namely unsupervised and supervised. The purpose of the unsupervised training is to generate two stages of optimized Sparse Filters, whereas the supervised training is to generate parameter weights and biases that will be used by the classifier in the classification process.

During the unsupervised training, a set of pre-processed images are firstly delivered into Sparse Filtering function to generate a set of optimized stage 1 sparse filters. Later, the set of pre-processed input images are delivered into CNNLS stage 1 hidden layer and convolved with the optimized stage 1 sparse filters. The output from the CNNLS stage 1 hidden layer is used as an input for the Sparse Filtering function to generate a set of optimized stage 2 sparse filters. Thus, the outcome of the unsupervised training is a set of optimized stage 1 sparse filters, and a set of optimized stage 2 sparse filters.

For the supervised training, the set of pre-processed images are delivered into CNNLS stage 1 hidden layer and will be convolved with the stage 1 sparse filters. The extracted features from this stage will be an input to the stage 2 hidden layer and the post hidden layer. In the stage 2 hidden layer, the input is convolved with the stage 2 sparse filters and the same layers are deployed to obtain the extracted features. This extracted features will be an input to the post hidden layer. The output from both stages which are the extracted features will be concatenated at the fully connected process into a single vector. This single vector feature is used to train the Softmax Regression classifier. Note that the testing set in testing phase has similar process with the supervised training except at the classifier where the trained weights and biases are used to calculate classification probabilistic in obtaining a classification result.

Image Acquisition and Pre-processing. The training and testing images that contain frontal-view of the vehicle are recorded using a surveillance camera. The pre-processing technique that has been used is a combination of existing works done by Baustita et al. [9], Dong et al. [10], and Pinto et al. [11]. There are five pre-processing methods used, which are resizing with maintained aspect ratio, converting from RGB color space to grayscale, histogram equalization (HE), normalizing to zero mean and unit variance, and local contrast normalization (LCN). The input image is first converted to grayscale. HE is performed on that grayscale image to normalize brightness and contrast of the image by using OpenCV 2.4 built-in library. After that, the image is resized with the maintained aspect ratio. The resized image is normalized to the zero mean and unit variance. LCN is applied as the final process to eliminate light illumination and image shading. This is to avoid a variation in image. The output in this layer is a set of LCN normalized features in 3D that is normalized with minimum 0.0 to maximum 1.0. This is to avoid an exponent underflow and overflow happens during convolutional layer in the stage 2 hidden layer.

Feature Extraction: CNNLS. Based on Fig. 2, each hidden layer consists of five components (layers) which are convolutional, AVR, LCN, average pooling and subsampling without zero padding. The output from each layer is the extracted features that will be an input features to another layer accordingly. The components in the post hidden layer for the stage 1 and stage 2 are different where for the stage 1, it contains the average pooling and subsampling with zero padding, while for the stage 2, it contains only the subsampling with zero padding.

Convolutional Layer. In the convolutional layer, the pre-processed image will be convolved with the optimized Sparse Filters. The purpose of convoluting the image is to extract the image features. For example, the pre-processed image with 256×174 pixel will be convolved with the 64 optimized Sparse Filters (9×9 pixel). From here, the convolved images are produced and a sigmoid activation function is applied on each convolved image. Later, the 64 extracted features with 248×166 pixel are obtained.

Absolute Value Rectification (AVR) Layer. AVR is inspired from biological system that human eyes do not perceived images in negative values. This layer applies absolute value operation on the extracted features from the previous layer and the output will have absolute value elements.

Local Contrast Normalization (LCN) Layer. LCN applied during CNNLS hidden layers is different compared to the LCN applied during the pre-processing process. Both LCNs has similar subtractive and divisive operations except that the input is a set of extracted features from the convolutional layer and different maximum value. The output from this algorithm is a set of LCN normalized extracted features.

Average Pooling Layer. The purpose of applying an average pooling on the extracted features is to ensure that the extracted features are robust to geometric distortion. Thus, the features become less sensitive to a variation in angle and size of a vehicle. The extracted features with size m by n pixels are convolved with an average filter with kernel size of t by t pixels. The purpose of the convolution is to produce an average features with size p by q. The size of the average features is smaller due to the border effect from the convolution operation where the border pixels is discarded.

Subsampling Layer without zero padding. The procedure for the subsampling layer without zero padding is similar to the procedure of resize with the maintained aspect ratio in the pre-processing process. The size of the features will be reduced from 240×158 pixel to 64×42 pixel based on 64 pixel of the resized size. However, the shape of the vehicle is maintained based on the aspect ratio that remain for example 79:120.

Subsampling Layer with zero padding. The subsampling layer with zero padding is applied on the extracted features to ensure every feature has the same size and has the aspect ratio of 1:1 before being delivered into Softmax Regression classifier. To describe the procedure, suppose that an input features with size of 158 by 240 pixels will be reduced to 64 pixels. Firstly, the input features is resized to 64 pixels with the maintained aspect ratio. Later, the shortest side of the resized input features will be padded with zero pixels to ensure the shortest side has the same length with the longest side.

Fully Connected. A fully connected vector is a vector that its elements are in 1-dimensional. For example, suppose that the features from the stage 1 hidden layer consists of element $\{1, 2, 3, \ldots, 9\}$ with size of 3×3 and the features from the stage 2 hidden layer consists of element $\{A, B, C, \ldots, I\}$ size of 3×3 as well. Each an image features will be firstly vectorized into a one-dimensional vector and then concatenated to form a single one-dimensional vector $\{1, 2, 3, \ldots, 9, A, B, C, \ldots, I\}$.

Classification using Softmax Regression. Classifying vehicles is performed by executing a Softmax Regression hypothesis function. The hypothesis is calculated for each vehicle classes to find the amount of probabilities that the extracted features belong to which class. There are two phases of the Softmax Regression implementation; supervised training and testing. The Softmax Regression for the supervised training is trained to produce optimum weights and biases. The weights and biases are prepared by minimizing both negative log-likelihood and MSE using gradient descent method. The optimization is executed for 10000 iterations to ensure the weights and biases are achieved an optimal convergence. Note that the Softmax Regression used in this research contains non-negative regularization parameter, λ, with the purpose to control the generalization performance of the Softmax Regression. A small amount of generalization could improve the classification accuracy because the classification will be more flexible where not too dependent on the training dataset. The value of the λ is preferred between 0.20 to 1.00. Setting λ to zero will disable the regularization. The optimum weight and bias later will be used in the testing phase where it will be loaded and the hypothesis is calculated instead of minimizing the negative log-likelihood. The testing is performed on each vehicle class dataset. Accuracy is calculated base d on number of vehicles that is correctly counted and classified to the total number of vehicles in the video.

3   Result and Discussion

We test our approach with a frontal view of single vehicle in single lane. The test is conducted based on mounted camera video taken in a selected road within 1 h. For the first phase, the vehicle classes are divide into car, taxi and truck. Table 1 shows the example of vehicle images in each class for each dataset. Note that the total samples for each vehicle class in the training dataset is 40 images, thus overall total is 120 images. While for the testing dataset, it is varying depends on how many vehicles captured in the video. Table 2 shows the result of a confusion matrix that consists of the correctly classified percentages and misclassification percentages.

Table 1. Example of vehicle images in each class for each dataset

Dataset	Car	Taxi	Truck
Training			
Testing			

Table 2. Confusion matrix

Predicted\Actual	Car (%)	Taxi (%)	Truck (%)
Car	92.5	4	3.5
Taxi	11	89	0
Truck	4	6	90

Looking at this table, the car and truck are able to be successfully counted and classified with the promising percentage of 92.5% and 90% respectively. However, the taxi is also successfully counted and classified with 89% and has been misclassified as the car with 11%. Yet, it is proved that using this approach we able to classified taxi as a different class of car. The percentage is quite low due to the almost similar features the taxi has to the car especially when the images are converted into grayscale. Other misclassification percentages are acceptable where below than 7%. Note that this CNNLS processed the images in grayscale. Figure 3 shows the output of the VCS based on each vehicle class where the number of each vehicle class is counted and recorded.

Fig. 3. The output of VCS based on each vehicle class

4 Conclusion

We proposed the implementation of vehicle counting system (VCS) based on vehicle type classification using deep learning method which is CNNLS to count the vehicle. The aim is to provide an accurate counting result for traffic census. The result thus can be used for further action by Public Works Department especially in Malaysia. Based on our approach, we able to automate the counting process by classifying the vehicle automatically into car, taxi and truck class. We proved that the taxi can be classified as different class from car which is none of the existing works have done this. In future, we will use color based image to improve the classification of taxi, multiple vehicle and multiple lane to support ITS aim in improving efficiency of modern cities.

References

1. Subaweh, M.B., Wibowo, E.P.: Implementation of pixel based adaptive segmenter method for tracking and counting vehicles in visual surveillance. In: 2016 International Conference on Informatics and Computing (ICIC), Mataram, Indonesia, pp. 1–5 (2016)
2. Li, S., Yu, H., Zhang, J., Yang, K., Bin, R.: Video-based traffic data collection system for multiple vehicle types. IET Intell. Transp. Syst. **8**(2), 164–174 (2014)
3. Prommool, P., Auephanwiriyakul, S., Theera-Umpon, N.: Vision-based automatic vehicle counting system using motion estimation with Taylor series approximation. In: 2016 6th IEEE International Conference on Control System, Computing and Engineering (ICCSCE), Penang, Malaysia, pp. 485–489 (2016)
4. Quesada, J., Rodriguez, P.: Automatic vehicle counting method based on principal component pursuit background modeling. In: 2016 IEEE International Conference on Image Processing (ICIP), Phoenix, AZ, pp. 3822–3826 (2016)
5. Kamkar, S., Safabakhsh, R.: Vehicle detection, counting and classification in various conditions. IET Intell. Transp. Syst. **10**(6), 406–413 (2016)
6. Seenouvong, N., Watchareeruetai, U., Nuthong, C., Khongsomboon, K., Ohnishi, N.: A computer vision based vehicle detection and counting system. In: 2016 8th International Conference on Knowledge and Smart Technology (KST), Chiangmai, pp. 234–238 (2016)
7. Valiere, P., Khoudour, L., Crouzil, A., Cong, D.N.T.: Robust vehicle counting with severe shadows and occlusions. In: 2015 International Conference on Image Processing Theory, Tools and Applications (IPTA), Orleans, pp. 99–104 (2015)
8. Zhang, Z., Liu, K., Gao, F., Li, X., Wang, G.: Vision-based vehicle detecting and counting for traffic flow analysis. In: 2016 International Joint Conference on Neural Networks (IJCNN), Vancouver, BC, pp. 2267–2273 (2016)
9. Bautista, C.M., Dy, C.A., Manalac, M.I., Orbe, R.A., Cordel, M.: Convolutional neural network for vehicle detection in low resolution traffic videos. In: 2016 IEEE Region 10 Symposium (TENSYMP), pp. 277–281, May 2016
10. Dong, Z., Pei, M., He, Y., Liu, T., Dong, Y., Jia, Y.: Vehicle type classification using unsupervised convolutional neural network. In: 2014 22nd International Conference on Pattern Recognition (ICPR), pp. 172–177. IEEE (2014)
11. Pinto, N., Cox, D.D., Di Carlo, J.J.: Why is real-world visual object recognition hard? PLoS Comput. Biol. **4**(1), 1–6 (2008)

Metadata Discovery of Heterogeneous Biomedical Datasets Using Token-Based Features

Jingran Wen[1], Ramkiran Gouripeddi[1,2], and Julio C. Facelli[1,2(✉)]

[1] Department of Biomedical Informatics, The University of Utah, Salt Lake City, UT 84108, USA
julio.facelli@utah.edu
[2] Center for Clinical and Translational Science, The University of Utah, Salt Lake City, UT 84108, USA

Abstract. Metadata discovery is the process of recognizing semantics and descriptors of data elements and datasets. This study uses a machine-learning approach to classify biomedical dataset characteristics for metadata discovery. Four common types of biomedical data sources were included in this study - genetic variant, protein structure, scientific publications, and general English corpus. Decision tree classification models were built using token-based features derived from these data files. These decision tree classification models are able to identify the four data sources with average F1 scores ranging from 0.935 to 1.000. This study demonstrates that biomedical data of different types have different distributions of token-based document structural features and that such structural features can be leveraged for metadata discovery.

Keywords: Metadata discovery · Text characterization · Data harmonization

1 Introduction

Sharing and reuse of biomedical data is critical to enhance research reproducibility and increase efficiency in translational biomedical sciences [1–3]. This requires biomedical data to be findable, accessible, interoperable and reusable according to the FAIR guiding principles [4]. The capture of sufficient metadata from heterogeneous data sources is a key requirement for successful data harmonization and integration [5]. Current approaches to metadata discovery are highly dependent on manual curation, which are expensive and time-consuming processes. Automatic or semiautomatic approaches for metadata discovery are necessary to enhance heterogeneous biomedical data integration.

We are developing a two-step metadata discovery architecture: (1) biomedical data type discovery, followed by (2) metadata discovery using methods specific for the data type and source identified in the first step. Our hypothesis here is that if we correctly identify different types of data sources by their intrinsic data file characteristics, we can then associate specific metadata discovery tools with each data source and type with a high level of confidence. A machine learning approach to accomplish the first step is described in this paper.

© Springer Nature Singapore Pte Ltd. 2018
K.J. Kim et al. (eds.), *IT Convergence and Security 2017*,
Lecture Notes in Electrical Engineering 449,
DOI 10.1007/978-981-10-6451-7_8

Biomedical and life science data are most commonly available in some type of text format. In this paper, we used different types of biomedical data sources in text format as examples of how to build machine-learning models for automatic identification of data files from specific sources and of specific types. Examples of such biomedical data files are described in Sect. 2.

Token-base features are higher-level characterizations of distributions of tokens in a given text document. Examples of token-based features include proportion of numerical tokens, proportion of capitalized tokens, and median length of tokens. These token-based features are a summarization of the structural distributions of a dataset. Principal advantages of using token-based features, over the word-based features, for the classification of textual data file are: (i) the dimensions of the feature vectors are relatively small; (ii) token-based methods are generalizable to any types of text document; and (iii) it is easier to get tokens as many steps of word processing, such as stemming and spelling correction, can be skipped when processing a text document for token extraction.

Token-based features have been used in text classification, but to the authors knowledge this method has not been used to distinguish different types of data files and data sources [6]. This study shows that the characteristics of a file as represented by token-based features are sufficiently different among the different data source types considered here and these features can be used to build classification models for data source identification.

2   Methods

2.1   Data Sources and Types

Four types of data from different sources were considered in this study: (1) protein structure; (2) genetic variant; (3) scientific publication; and (4) a general English corpus.

Protein structure files describe the three-dimensional arrangement of atoms in proteins or protein complexes. Both experimentally determined structures and computationally predicted structures were included in this study. The experimentally determined structures were extracted from the Research Collaboratory for Structural Bioinformatics (RCSB) Protein Data Bank (PDB) [7]. Computationally predicted structures were generated in the Facelli Lab (http://home.chpc.utah.edu/~facelli/) and included structures of ataxin-2 and ataxin-3 proteins predicted by I-TASSER [8] and structures of oncogene proteins modeled by Rosetta [9]. In this study, we considered protein structure files in PDB, macromolecular Crystallographic Information File (mmCIF) and Protein Data Bank Markup Language (PDBML) file formats. The PDB format is the standard representation of macromolecular structure data and is widely used by a variety of software tools. The mmCIF format is used by PDB to describe the information content of PDB entries. PDBML format provides an XML representation of the PBD data. Note that while the PDBML format uses XML modeling notations, for the purpose of this study, these files were treated as plain text. The angle brackets and xml tags were treated as tokens. The description and examples of these three formats can be found on the PDB website (http://www.wwpdb.org/documentation/file-format). We randomly selected and assigned 100 PDB, 100 CIF, 100 XML to the training set and 212 PDB, 255 CIF, 173 XML to the external evaluation set from the RSCB PDB source. We also randomly selected 100 PDB files from our I-TASSER collection and assigned them to the training set and 100 PDB from the Rosetta collection to the external validation set.

We used data from ClinVar [10] as an example of genetic variant data. ClinVar is a public available archive of information about relationship between genetic variances and human diseases [10]. We used both the ClinVar complete variant dataset (November 2012 and December 2013 full releases) and the summary data about variants (July 2016 release) in this study. The publically available complete ClinVar variance dataset is in XML format, but for the purpose of this study the files are treated as plain text documents. The variant summary data in ClinVar is in tab-separated-values (TSV) format with each line representing summary information of a variant. We randomly generated 100 XML and 100 TSV, with each file containing 1 to 2000 variants randomly selected from the ClinVar data, for training. Using the same sampling process, we randomly generated 100 XML and 100 TSV for external validations sets.

PubMed Central (PMC) is a free full-text archive of biomedical and life science papers, which are good examples of scientific publications in the biomedical field. We randomly selected 100 publications in Portable Document Format and assigned them to the training set after transforming them into plain text using Apache PDFBox (release 2.0.2). For the external test set, we used 393 scientific papers from the 2003 KDD Cup Competition (https://www.cs.cornell.edu/projects/kddcup/). Publications in the KDD data set include scientific papers in multiple fields of physics and related disciplines.

The Open American National Corpus (OANC) is an open and free electronic collection of American English, including texts of all genres and transcripts of spoken data (http://www.anc.org/). Google News articles (https://news.google.com/) are another good source representing general English. We included 129 randomly selected articles from OANC in the training set and 102 articles from Google News in the external validation set, all in plain text format.

All datasets used here were downloaded in July 2016. As described above, we used external evaluation datasets, having no overlap with the training datasets, to evaluate the classification performances and generalizability of the machine learning models.

2.2 Token-Based Features

Using the tokenization module in the Natural Language Toolkit 3.0 (NLTK, http://www.nltk.org/), we tokenized the text in each data file into tokens, such as words, symbols, numbers, and other meaningful elements by using white spaces as token separators. We categorized the tokens as Numerical, Word and Other tokens. Numerical tokens contain only numerical characters, while word tokens contains a mix of characters including numbers, letters and underscores. We categorized tokens that did not match these two categories as other tokens, which included punctuations characters and symbols. We calculated and normalized the counts for the three types of tokens to the total token counts within each data file. The proportion of numerical tokens with negative values, the word tokens that are capitalized and the word tokens that are all upper case letters were also calculated and normalized to their total count of tokens in their parent category. The median length of all the tokens in the file were also calculated. In total, eight token-based features were used: Normalized count of numerical tokens, Normalized count of negative numerical tokens, Normalized count of word tokens, Normalized count of UPPER-CASED tokens, Normalized count of Capitalized tokens, Normalized count of distinct word tokens, Normalized count of other tokens, and Median length of tokens.

2.3   Decision Tree Models Building and Evaluation

We built models using a decision tree classification method to distinguish protein structure, genetic variant, scientific paper, and general English corpus sources. We used the above-mentioned token-based features in the scikit-learn [11] machine learning package for building and evaluating machine learning models. We evaluated the performance of the decision tree models using both tenfold cross-validation and an external test dataset with precision, recall and F1 score as measures. Both tenfold cross-validation and external evaluation were repeated 100 times with different random seeds, and we report the average value of these 100 iterations.

3   Results

The total number of tokens was proportional to the size of each data file. Protein structure files in XML format have the largest number of tokens followed by ClinVar variant files in XML format, while general English corpus and scientific publications have fewer. The types of tokens in different data files have different distributions. Protein structure data files are dominated by numerical values, whereas word tokens dominate the general English corpus and scientific publications. The distribution of each token-based feature also varies across data files in different formats. For example, PDB and mmCIF format protein structure files have the highest proportion of numerical tokens, while XML format protein structure files have much smaller proportion of numerical values when compared with other types of protein structure files and genetic variant files. On evaluating the performance of the decision tree models using tenfold cross-validation approach, they were able to distinguish the biomedical data files as protein structure, genetic variance, scientific paper and general English corpus with F1 scores of 0.997, 0.997, 0.886 and 0.919, respectively (Table 1). The average precision and recall of classifying protein structure and genetic variance are above 0.99. Although the average

Table 1. Evaluation of the decision tree classification performance using 10-fold cross-validation.

Data type	Precision	Recall	F1-score
Protein structure	0.996	0.998	0.997
Genetic variant	0.995	1.000	0.997
Scientific paper	0.901	0.883	0.886
General English	0.930	0.914	0.919

precision and recall of classifying scientific papers and general English are lower than those of protein structure and genetic variance, these values are still around 0.9. A typical decision tree used for classification of biomedical data files is shown in Fig. 1.

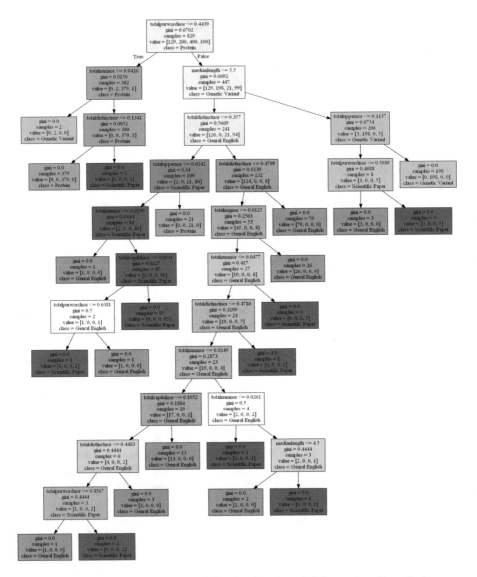

Fig. 1. Example of a decision tree used for classification of different data files. *Each square* represents a tree node. Decisions are listed on the first line of each decision node. For each decision, the child nodes on the left and right are based on the fact of being "true" and "false" respectively. The *Gini line* indicates the *Gini index* used to determine the splitting attributes. The *sample line* indicates the total number of instances in a node. The *value line* lists the number of instance in a node, from left to right English corpus, protein structure, genetic variant, and scientific paper.

When the performance of decision tree models were evaluated with external test datasets, the models were able to distinguish protein structure, genetic variance, scientific paper, and general English corpus with an F1 score of 1.000, 0.999, 0.980, and 0.935 respectively (Table 2). The average precision and recall of the models to distinguish protein structure files are both 1.000. The models also classified scientific literature from the KDD competition and Google News general English corpus with precisions and recalls of 0.979 and 0.982, and 0.949 and 0.921 respectively.

Table 2. Classification performance on the external test set.

Data type	Precision	Recall	F1-score
Protein structure	1.000	1.000	1.000
Genetic variant	0.997	1.000	0.999
Scientific paper	0.979	0.982	0.980
General English	0.949	0.921	0.935

4   Discussion

Data files embed in themselves different features that are inherent signatures. These include token-based, word-based, and sematic-based features [6]. Word-based features, such as word frequencies and TF-IDF scores [12], are widely used in document classification and text mining [12], but these sometimes require feature selection [13], as the dimensionality of the feature space can get extremely large. Moreover, word and semantic-base features are not agnostic to the domain specific content and require extensive training annotations for use in document classification problems. Token-based features considered here are high-level summaries of the structural content of the files and provide a simpler approach to dataset classification.

On average, decision tree models with token-based features are able to distinguish the four types of biomedical data, protein structure, genetic variance, scientific paper, and general English corpus, with F1 scores of 0.997, 0.997, 0.886, and 0.919 respectively, in the tenfold cross-validation. The classification performance of the models on the external evaluation dataset is comparable to the cross-validation results with average F1 scores of 1.000, 0.999, 0.980, and 0.935 on protein structure, genetic variant, scientific publication, and general English corpus, respectively. The scientific papers in the external evaluation dataset and the papers in the training dataset are from different resources and in different topics. Papers in the former dataset are from the KDD Competition and are in the physics domain, whereas papers in the latter dataset are from PMC open access archive and these papers are in biomedical and life science topics. However, the classification models still have a great prediction performance to distinguish the physics papers from other types of data with an average precision of 0.979 and recall of 0.982. These results indicate that it is robust to use token-based features for data file identification.

The varying distributions of token-based features among different data files and the common distribution of token-based features within the same type of data file necessitate the use of the machine learning methods for distinguishing them. This is evident from the decision tree in Fig. 1 in which each source is classified at multiple depths in the tree. However, at a depth of level four, the tree shows good performance in distinguishing the four data types.

While we used seven types of data files in this paper, our approach is highly scalable to very large numbers of data files as this approach requires minimum to no human annotation of the file types for developing training sets. All file type annotations can be automatically extracted based on their source in the training phase, for example, PDB files are labeled as such based on their source on the PDB website.

The results from this machine learning models are relatively high when compared to classification models in other domains, but our data file classification presented here is only the first step in the complete process of metadata discovery. The ultimate test for this approach would be the results from its use in integrated metadata discovery pipelines that utilize these classifications to programmatically assign specific metadata discovery tools and methods.

This study is not without limitations. Only four types of data are included. However, for these four types, datasets from a variety of sources and in different formats were included to assess the generality of the classification models. Using the file processing and machine leaning workflow built in this study it is easy to include a relatively large number of data types in the classification models. We are now considering including electronic medical records and environmental data in this model. In addition, the sample sizes for each dataset were relatively small, and further evaluation on larger sample sizes is required. We also plan to add an uncertainty quantification module to enable programmatic assignment of specific metadata discovery tools and methods.

5   Conclusion

Our approach shows that it is possible to automatically identify data files from specific sources and of specific types using only document structural token-based features. Our decision tree models performed well in distinguishing protein structure data, genetic-variant data, scientific publication, and general English corpus, and provide a promising way to facility metadata discovery. Therefore, it is reasonable to expect that it will be possible to develop and programmatically associate metadata extraction tools specific for each data source and type as next steps.

Acknowledgments. bioCADDIE is supported by the National Institutes of Health (NIH) through the NIH Big Data to Knowledge, Grant 1U24AI117966-01. OpenFurther has received support NCATS UL1TR001067, 3UL1RR025764-02S2, AHRQ R01 HS019862, DHHS 1D1BRH20425, U54EB021973, UU Research Foundation, NIBIB, NIH U54EB021973. Computer resources were provided by the University of Utah Center for High Performance Computing.

References

1. Federer, L.M., Lu, Y.L., Joubert, D.J., Welsh, J., Brandys, B.: Biomedical data sharing and reuse: attitudes and practices of clinical and scientific research staff. PLoS One **10**(6), e0129506 (2015)
2. Ross, J.S., Lehman, R., Gross, C.P.: The importance of clinical trial data sharing: toward more open science. Circ. Cardiovasc. Qual. Outcomes **5**(2), 238–240 (2012)
3. Gotzsche, P.C.: Why we need easy access to all data from all clinical trials and how to accomplish it. Trials **12**, 249 (2011)
4. Wilkinson, M.D., Dumontier, M., Aalbersberg, I.J., Appleton, G., Axton, M., Baak, A., et al.: The FAIR guiding principles for scientific data management and stewardship. Sci. Data **3**, 160018 (2016)
5. Gouripeddi, R., Schultz, N.D., Bradshaw, R.L., Madsen, R.P., Mo, Warner P.B., et al.: FURTHeR: an infrastructure for clinical, translational and comparative effectiveness research. In: American Medical Informatics Association 2013 Annual Symposium. Washington, DC (2013)
6. Brank, J., Mladenić, D., Grobelnik, M.: Feature construction in text mining. In: Sammut, C., Webb, G.I. (eds.) Encyclopedia of Machine Learning. Springer, Boston (2010)
7. Berman, H.M., Westbrook, J., Feng, Z., Gilliland, G., Bhat, T.N., Weissig, H., et al.: The protein data bank. Nucleic Acids Res. **28**(1), 235–242 (2000)
8. Roy, A., Kucukural, A., Zhang, Y.: I-TASSER: a unified platform for automated protein structure and function prediction. Nat. Protoc. **5**(4), 725–738 (2010)
9. Leaver-Fay, A., Tyka, M., Lewis, S.M., Lange, O.F., Thompson, J., Jacak, R., et al.: ROSETTA3: an object-oriented software suite for the simulation and design of macromolecules. Meth. Enzymol. **487**, 545–574 (2011)
10. Landrum, M.J., Lee, J.M., Benson, M., Brown, G., Chao, C., Chitipiralla, S., et al.: ClinVar: public archive of interpretations of clinically relevant variants. Nucleic Acids Res. **44**(D1), D862–D868 (2016)
11. Pedregosa, F., Varoquaux, G., Gramfort, A., Michel, V., Thirion, B., Grisel, O., et al.: Scikit-learn: machine learning in python. J. Mach. Learn. Res. **12**, 2825–2830 (2011)
12. Rajaraman, A., Ullman, J.D.: Mining of massive datasets. Data mining, pp. 1–17 (2011)
13. Mladenić, D.: Feature selection in text mining. In: Sammut, C., Webb, G.I. (eds.) Encyclopedia of Machine Learning. Springer, Boston (2010)

Heavy Rainfall Forecasting Model Using Artificial Neural Network for Flood Prone Area

Junaida Sulaiman[(✉)] and Siti Hajar Wahab

Soft Computing and Intelligent System Research Group (SPINT), Faculty of Computer Systems and Software Engineering, Universiti Malaysia Pahang, 26300 Kuantan, Pahang, Malaysia
junaida@ump.edu.my, sitihajar5152@gmail.com

Abstract. Interest in monitoring severe weather events is cautiously increasing because of the numerous disasters that happen in the recent years in many world countries. Although to predict the trend of precipitation is a difficult task, there are many approaches exist using time series analysis and machine learning techniques to provide an alternative way to reduce impact of flood cause by heavy precipitation event. This study applied an Artificial Neural Network (ANN) for prediction of heavy precipitation on monthly basis. For this purpose, precipitation data from 1965 to 2015 from local meteorological stations were collected and used in the study. Different combinations of past precipitation values were produced as forecasting inputs to evaluate the effectiveness of ANN approximation. The performance of the ANN model is compared to statistical technique called Auto Regression Integrated Moving Average (ARIMA). The performance of each approaches is evaluated using root mean square error (RMSE) and correlation coefficient (R^2). The results indicate that ANN model is reliable in anticipating above the risky level of heavy precipitation events.

Keywords: Computational intelligence · Time series forecasting · Neural network

1 Introduction

Heavy precipitation or rainfall is an extreme weather event that happens in tropical countries like Malaysia. It gives severe impact on social and economic aspects of affected areas. The most immediate impact of heavy precipitation is the prospect of flooding as streams and rivers in the region overflow main river banks. In Malaysia, high intensity of precipitation has been identified as the main factor of flood occurrences in states like Pahang, Kelantan and Terengganu [1, 2].

There are many data driven models like artificial neural network and auto regressive integrated moving average (ARIMA) implemented for weather forecasting. Conventional ARIMA method performs well with stationary data but artificial neural network works better with nonlinear type of data. ANN provides better approximation because its network is dynamic and works well with non-stationary data [3, 4].

© Springer Nature Singapore Pte Ltd. 2018
K.J. Kim et al. (eds.), *IT Convergence and Security 2017*,
Lecture Notes in Electrical Engineering 449,
DOI 10.1007/978-981-10-6451-7_9

Numerous studies have employed ANN in time series forecasting particularly on precipitation data [5–7]. These are due to its non-linear ability which is appropriate for seasonal precipitation. A study by Mislan et al. used backpropagation neural network (BPNN) to simulate heavy precipitation in Tenggarong station, East Kalimantan-Indonesia [8]. They concluded that BPNN can produce accurate prediction. ANN also been used in predicting rainfall using four years of hourly data from 75 rain gauge stations in Bangkok [9]. The combination data from different gauge stations produced better accuracy thus supported that ANN is effective when analyzing high number of input data [7]. Besides that, comparison between multiple linear regression (MLR) and ANN showed that the latter technique is significant in producing better forecast in long-term [10]. While, Abbot and Marohasy [11] discovered that lagged relationship in rainfall data greatly contribute to forecast accuracy of ANN model.

One of the variant of ANN is time delay neural network (TDNN). It is a dynamic model that works well with sequential data [12, 13]. Thus, the objective of this study is to implement time delay neural network (TDNN) in forecasting heavy precipitation and to compare its performance with ARIMA.

The paper is organized as follows: In Sect. 2, the materials and method implemented in the study is presented. In Sect. 3, the experimental results are discussed and finally conclusions are given in Sect. 4.

2   Materials and Methods

2.1   Study Area

Pahang is a large state in Peninsular Malaysia which is one of the flood-prone state in Malaysia. Two areas in the state are heavily affected with flood due to their closeness with the biggest river in Peninsular Malaysia, namely Sungai Pahang. The river becomes main basin for water that flows from upstream during heavy rainfalls.

Therefore, one of the best ways to cope with the flooding problem is to provide advance rainfall forecasting and flood warning. The Department of Irrigation and Drainage Malaysia (DID) has established the flood operations room DID for systematic and efficient management of operation and control of flood protection facilities. The state DID has 116 active rain gauges and 17 water level stations scattered throughout Pahang. Location of the rain gauges and water level stations are shown in Fig. 1. The highlighted words represent stations that were included in this study.

Historical rainfall data was collected from 116 rain gauge station for 50 years from 1965 to 2015. The data was divided into 80% training data and 20% testing data. Data from January 1965 to December 2005 and January 2006 to December 2015 were divided into training and testing sets, respectively. The focus areas in this study were prone to flood events and a total 20 stations along the main river is selected. Figure 2 shows the actual rainfall data from year 1965 to 2015. It can be observed that heavy rainfall exceeding 100 mm are consistent in every year for the past 50 years. Thus, this study is focusing on utilizing monthly heavy rainfall data.

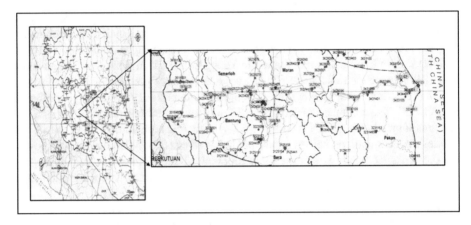

Fig. 1. Location of study area and the rainfall station

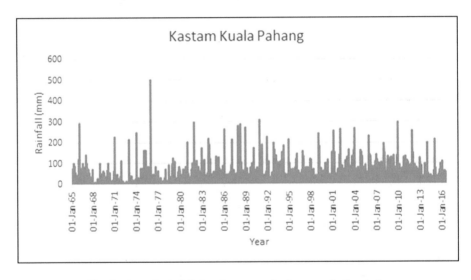

Fig. 2. Actual rainfall data from one of rain-gauge in the study.

2.2 Handling Missing Data

The collected data are prone to missing values due to reasons such as breakdown of equipment, site conditions and program maintenance. Therefore, it must be treated by using Expectation Maximization (EM) algorithm. This algorithm works by generating a set X of observed data, a set of unobserved latent data or missing values Z, and a vector of unknown parameters θ, along with a likelihood function $L(\theta;X,Z) = P(X,Z|\theta)$. The maximum likelihood estimate (MLE) of the unknown parameters is determined by the marginal likelihood of the observed data.

2.3 Data Gathering

The daily precipitation in millimeters (mm) were collected and ensemble in excel sheet and data pre-processing is implemented to prepare for TDNN model development. Pre-processing starts with transforming the daily data into maximum rainfall in each month for 50 years.

2.4 Data Normalization

The next process is to normalize the sample data into smaller intervals which are [0.1–0.8] using the min max technique [14] as shown in Eq. 1.

$$x' = \frac{x - \alpha}{b - \alpha} + 0.1 \tag{1}$$

Where; α is the minimum value, b is the maximum value, x is the data to be normalized, and x' is the data that have been normalized.

2.5 Development of Artificial Neural Networks (ANN)

ANN was inspired by biological neural networks. It contains neurons and connections that process information to find a relationship between inputs and outputs. A variant of ANN is used in this study called time delay neural network (TDNN). TDNN is an artificial neural network suitable for time series data [12, 13].

The general architecture of a TDNN is shown in Fig. 3. Here, the delay elements represented by the operator D. Therefore, the input rainfall R(t) to the neuron i can be defined as [15]:

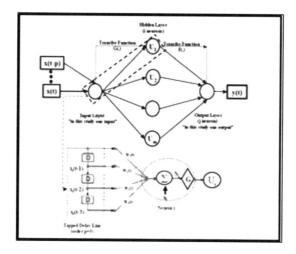

Fig. 3. Time delay neural network [15]

$$R_i(t) = \sum_{k=0}^{P} w_i(k)x(t - k) + b_i \tag{2}$$

From the equation, $w_i(k)$ is the synaptic weight for neuron i, and b_i is its bias. Then, the output of neuron U_i is obtained by processing $R_i(t)$ using non-linear activation function G named sigmoid function as shown in Eqs. 3 and 4, respectively:

$$U_i = G\left(\sum_{k=0}^{P} w_i'(k)x(t - k) + b_i'\right) \tag{3}$$

$$G(R_i(t)) = \frac{1}{1 + e^{-R(t)}} \tag{4}$$

The output of neuron j is the output of the TDNN with the following equation:

$$y_j(t) = F\left(\sum_{i=1}^{m} w_{ji}'' U_i + \alpha_j\right) \tag{5}$$

Where $F(\cdot)$ is the transfer activation function of the output neuron j, α_j is it bias and w_{ij} is the weight between the neurons of the hidden layer and the neuron of the output layer.

In developing the optimal forecasting model, several combinations of parameters and network architectures were tested to find the best representation of TDNN model. The architecture of TDNN used in this study are shown in Table 1. The first model (Model A) has structure of two nodes in the input layer, two hidden layers with five hidden nodes and one output node. The input to the model were monthly rainfall data (at time t) and past monthly rainfall with one month lag time (t − 1) while the output is the rainfall intensity of the next month (t + 1).

Table 1. Artificial neural network architecture

Model	Architecture	Training data
A	2-5-5-1	R_{t-1} and R_t
B	6-10-10-1	$R_{t-5}, R_{t-4}, R_{t-3}, R_{t-2}, R_{t-1}$ and R_t
C	12-10-10-1	$R_{t-11}, R_{t-10}, \ldots\ldots R_{t-2}, R_{t-1}$ and R_t
D	6-10-10-1	$R_{t1}, R_{t2}, R_{t3}, R_{t4}, R_{t5}$ and R_{t6}

The Model B has six inputs data and the hidden node is increased to 10 hidden nodes. In the Model C, only the input data are different between the Model B. The input data are increased to 12 that represent 12 months of past maximum precipitation. Lastly for the Model D, input variable was consisting of five nearby stations around the model station. The purpose of Model D is to see any relationship between precipitation data at nearby stations and future data at model station.

The process flow of time delay neural network forecasting model is given in Fig. 4. Basically, there are four phases in the model. The phase starts with gathering raw data as described in Sect. 2.1. Next, the data are processed using selected techniques for handling missing value and data normalization. The third phase is where the architecture of TDNN were developed and for this study, four TDNN with different architecture were developed.

Finally, the outputs of each architecture were compared for determining forecasting accuracy.

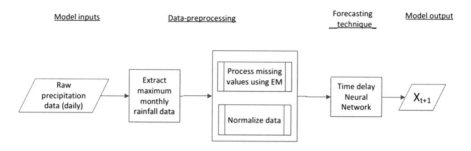

Fig. 4. Process flow of forecast model

Evaluation for Model A to Model D were based on the value of Root Mean Square Error (RMSE) and Correlation Coefficient (R^2). The formula for RMSE is as follows:

$$RMSE = \sqrt{\frac{1}{N} \sum_{i=1}^{N} (O_i - P_i)^2} \tag{6}$$

Where O_i is the observed value, P_i is the predicted value and N is the number of observations. An RMSE value close to 0 indicates higher predictive skill whereas an RMSE value close to 1 indicates poor predictive skill.

3   Results and Discussion

In this section, the results of the experiment are presented. The heavy precipitation is forecasted using (a) time delay neural network models and (b) auto regressive integrated moving average (ARIMA) model. The performance of both methods is compared to check the usefulness of time delay neural network approach.

The result of RMSE value from the forecasting model using ANN are compared to the ARIMA model. The best model for ARIMA was obtained as (2,0,1)(0,1,2) [12]. The testing result for four TDNN are shown in Table 2.

Table 2. Performance statistic of four TDNN using 2006–2015 testing data

Model	A	B	C	D
RMSE	0.075	0.073	0.06	0.06
R^2	0.165	0.208	0.469	0.008

Based on the results, Model C has the lowest RMSE and the highest correlation value. This showed that forecasting outputs for Model C were highly correlated to actual precipitation values and it outperformed other models by providing the best performance accuracy.

Model C has been selected for comparison with conventional time series model, ARIMA. Table 3 shows that the comparison of the performance of the ANN to the ARIMA model using RMSE indicator for testing data from 2006 to 2015. It is obviously from the Table 3, that the ANN model outperformed the ARIMA model with the lower value of RMSE with different of 0.01. This shown that TDNN is more superior than ARIMA in term of predictive ability when dealing with time series data.

Table 3. Comparison result between ANN and ARIMA.

Forecasting technique	RMSE	R^2
ANN	0.06	0.4689
ARIMA	0.07	0.6708

Figure 5 displays the time series of 1-month-ahead forecasting of ANN and ARIMA model and actual monthly precipitation for year 2007. The figure indicates that ANN shows a distinct forecasting improvement over the ARIMA model. It is clearly seen that forecasting result from ANN produced six expected outputs that are closed to the actual observed monthly precipitation as compared to ARIMA model that only have two expected outputs closed to the actual precipitation. In addition to that, the expected outputs of ANN are less risky than ARIMA in approximating the actual heavy precipitation.

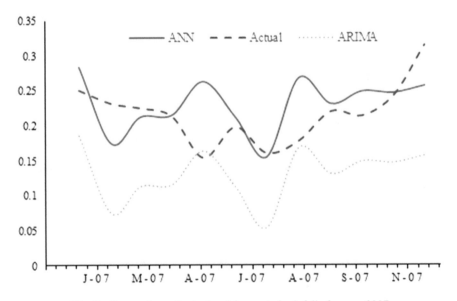

Fig. 5. Comparison of actual and forecasted rainfalls for year 2007

4   Conclusion

In this study, a time delay neural network was used to forecast heavy precipitation for one of the district in Malaysia which is prone to flood events. The forecasted outputs were compared with autoregressive integrated moving average (ARIMA) model for evaluating the performance for one month ahead forecasting. Based on the result, the TDNN outperformed ARIMA with slight advantage. But in term of forecasted output, TDNN has more closer outputs that ARIMA produced and less risky approximation.

References

1. Daud, A., Zakaria, R.A.A., Sahat, S., Ismail, N.F.A., Mohamad, N.F., Rosli, N.: The Study of Thunderstorm and Rainfall Occurrences over Pahang (in the Period 1998–2012). Malaysian Meteorological Department, Petaling Jaya (2015)
2. Daud, A., Mat Aji, S., Muhamad, N.: Synoptic and hydrological analysis of flood event over kelantan and terengganu, Kuala Lumpur, Malaysia (2011)
3. Latt, Z.Z., Wittenberg, H.: Improving flood forecasting in a developing country: a comparative study of stepwise multiple linear regression and artificial neural network. Water Resour. Manag. **28**, 2109–2128 (2014)
4. Shoaib, M., Shamseldin, A.Y., Melville, B.W., Khan, M.M.: A comparison between wavelet based static and dynamic neural network approaches for runoff prediction. J. Hydrol. **535**, 211–225 (2016)
5. Mekanik, F., Imteaz, M.: A multivariate artificial neural network Approach for rainfall forecasting: case study of Victoria, Australia. In: Proceedings of the World Congress on Engineering and Computer Science (2012)
6. Banihabib, M.E., Ahmadian, A., Jamali, F.S.: Hybrid DARIMA-NARX model for forecasting long-term daily inflow to Dez reservoir using the North Atlantic Oscillation (NAO) and rainfall data. GeoResJ. **13**, 9–16 (2017)
7. Kashiwao, T., Nakayama, K., Ando, S., Ikeda, K., Lee, M., Bahadori, A.: A neural network-based local rainfall prediction system using meteorological data on the Internet: a case study using data from the Japan Meteorological Agency. Appl. Soft Comput. J. **56**, 317–330 (2017)
8. Mislan, H., Hardwinarto, S., Sumaryono, A.M.: Rainfall monthly prediction based on artificial neural network: a case study in Tenggarong Station, East Kalimantan - Indonesia. Procedia Comput. Sci. **59**, 142–151 (2015)
9. Hung, N.Q., Babel, M.S., Weesakul, S., Tripathi, N.K.: An artificial neural network model for rainfall forecasting in Bangkok, Thailand. Hydrol. Earth Syst. Sci. **13**, 1413–1425 (2009)
10. Mekanik, F., Imteaz, M.A., Gato-Trinidad, S., Elmahdi, A.: Multiple regression and artificial neural network for long-term rainfall forecasting using large scale climate modes. J. Hydrol. **503**, 11–21 (2013)
11. Abbot, J., Marohasy, J.: Input selection and optimisation for monthly rainfall forecasting in Queensland, Australia, using artificial neural networks. Atmos. Res. **138**, 166–178 (2014)
12. Benmahdjoub, K., Ameur, Z., Boulifa, M.: Forecasting of rainfall using time delay neural network in Tizi-Ouzou (Algeria). Energy Procedia. **36**, 1138–1146 (2013)
13. Esposito, E., De Vito, S., Salvato, M., Bright, V., Jones, R.L., Popoola, O.: Dynamic neural network architectures for on field stochastic calibration of indicative low cost air quality sensing systems. Sensors Actuators B Chem. **231**, 701–713 (2016)

14. Awang, S., Sulaiman, J., Karimah, N., Noor, M.: Comparison of accuracy performance based on normalization techniques for the features fusion of face and online signature. In: International Conference on Computational Science and Engineering (ICCSE2016), Kota Kinabalu, Sabah (2016)
15. El-Shafie, A., Noureldin, A., Taha, M., Hussain, A., Mukhlisin, M.: Dynamic versus static neural network model for rainfall forecasting at Klang River Basin, Malaysia. Hydrol. Earth Syst. Sci. **16**, 1151–1169 (2012)

Communication and Signal Processing

I-Vector Extraction Using Speaker Relevancy for Short Duration Speaker Recognition

Woo Hyun Kang, Won Ik Cho, Se Young Jang, Hyeon Seung Lee, and Nam Soo Kim[✉]

Department of Electrical and Computer Engineering and INMC, Seoul National University, 1 Gwanak-ro, Gwanak-gu, Seoul 08826, Korea {whkang, wicho, syjang, hslee}@hi.snu.ac.kr, nkim@snu.ac.kr

Abstract. This paper presents a novel scheme for considering the frame-level speaker relevancy during i-vector extraction for speaker recognition. In the proposed system, the frame-level point-wise mutual information is utilized to directly modify the Baum-Welch statistics in order to extract a robust i-vector. Furthermore, a method for computing the frame-level speaker relevancy using deep neural network (DNN) analogous to the DNN used in robust automatic speech recognition (ASR) is proposed. The results show that the modified i-vectors obtained using the proposed methods outperformed the conventional i-vectors.

Keywords: Speaker recognition · i-vector · DNN

1 Introduction

Although the i-vector framework was successful in speaker recognition tasks dealing with long utterances such as the NIST SRE 2008 short2-short3 task [1, 2], which involves telephone speech utterances with average duration of 5 min, the uncertainty within the i-vector highly increases as the speech duration decreases [3]. This is mainly due to the lack of phonetic information provided by speech utterances with short durations [4]. Since a critical amount of speaker dependent information is inherent in the phonetic characteristics, the absence of phonetically informative frames can lead to degradation of speaker recognition performance [5]. Therefore, in real-life speaker recognition applications where the duration of the enrollment and test speech samples are not controlled, the classical i-vector framework may not be an optimal feature extraction technique [6].

In order to solve this problem, several methods were proposed to modify the probabilistic linear discriminant analysis (PLDA) scoring scheme to take account of the uncertainty caused by the short duration [3, 4, 7]. Furthermore, such unwanted variabilities caused by the short duration can be reduced via various normalization techniques such as the linear discriminant analysis (LDA) or within-class covariance normalization (WCCN) [8]. Despite their improvements in performance, these methods do not tackle the fundamental problem; the lack of speaker informative frames within the short speech sample.

© Springer Nature Singapore Pte Ltd. 2018
K.J. Kim et al. (eds.), *IT Convergence and Security 2017*,
Lecture Notes in Electrical Engineering 449,
DOI 10.1007/978-981-10-6451-7_10

In this paper, we focus on exploiting the frame-level speaker relevancy to extract a speaker recognition feature robust to the uncertainty caused by the scarce speaker dependent information within short duration speech samples. In order to emphasize the speaker relevancy within the speech sample when extracting the i-vector, we propose a novel scheme to modify the Baum-Welch statistics by weighting the frame-level posterior probabilities depending on the speaker information contained in the frames.

Moreover, motivated by the considerable gains in performance obtained by using deep neural networks (DNNs) for automatic speech recognition (ASR) [9], a DNN model is trained to extract the speaker relevancy more robustly and with less computational load. Experimental results on text-independent speaker recognition show that the proposed methods can enhance the performance in terms of classification error, equal error rate (EER), and decision cost function (DCF) measurements [2, 10].

2   Background and Related Work

2.1   The I-Vector Framework

I-vectors are now widely used to represent the idiosyncratic characteristics of the utterance in the field of speaker and language recognition [11], due to the fact that they can characterize various variabilities in a low dimensional vector [1]. Similar to the eigenvoice decomposition [12] or joint factor analysis (JFA) [13] technique, i-vector extraction can be understood as a factorization process decomposing the ideal GMM supervector as

$$\mathbf{m}_c = \mathbf{u}_c + \mathbf{T}_c \mathbf{w} \tag{1}$$

In (1), $\mathbf{T}_c$ is the factor loading submatrix corresponding to the cth GMM mixture component, which is a submatrix of the total variability matrix. $\mathbf{m}_c$ and $\mathbf{u}_c$ are the cth mixture component mean vector of the ideal GMM supervector dependent on a given speech utterance and the universal background model (UBM), respectively. Hence, the i-vector framework aims to find the optimal $\mathbf{w}$ and $\mathbf{T}_c$ to adapt the UBM to a given speech utterance. Analogous to the eigenvoice method, the total variability matrix is trained using the EM algorithm, but each utterance is assumed to be obtained from a different speaker. The 0th and 1st order Baum-Welch statistics are defined by

$$n_c(\mathbf{X}) = \sum_{l=1}^{L} \gamma_l(c), \tag{2}$$

$$\tilde{\mathbf{f}}_c(\mathbf{X}) = \sum_{l=1}^{L} \gamma_l(c)(\mathbf{x}_l - \mathbf{u}_c) \tag{3}$$

where for each frame l within utterance $\mathbf{X}$ with L frames, $\gamma_l(c)$ is the posterior probability that speech frame $\mathbf{x}_l$ is aligned to the cth Gaussian component of the UBM, $\mathbf{u}_c$ is the mean of the cth mixture component of the UBM, and $n_c(\mathbf{X})$ and $\tilde{\mathbf{f}}_c(\mathbf{X})$ are the 0th and the centralized 1st order Baum-Welch statistics, respectively. Once $\mathbf{T}_c$ for all the mixture components are trained, the i-vector of utterance $\mathbf{X}$ can be obtained as follows:

$$\mathbf{w}(\mathbf{X}) = \left(\mathbf{I} + \sum\nolimits_{c=1}^{C} n_c(\mathbf{X})\mathbf{T}_c^t\mathbf{\Sigma}_c^{-1}\mathbf{T}_c\right)^{-1} \sum\nolimits_{c=1}^{C} \mathbf{T}_c^t\mathbf{\Sigma}_c^{-1}\tilde{\mathbf{f}}_c(\mathbf{X}) \tag{4}$$

where C is the number of mixture components, $\mathbf{\Sigma}_c$ is the covariance matrix of the cth UBM mixture component, and superscript t indicates matrix transpose. Interested readers are encouraged to refer to [1] for further details of the i-vector framework.

2.2 Point-Wise Mutual Information

The point-wise mutual information (PMI) between two different events measures how much the knowledge of a single event reduces the uncertainty of the other event [14]. Thus speech frames with high PMI for the speaker identities are likely to contain rich information about the speaker. The PMI between speech frame $\mathbf{x}$ and speaker s is computed as follows:

$$pmi(\mathbf{x}; s) = \log\frac{p(\mathbf{x}|s)}{p(\mathbf{x})} = \log p(\mathbf{x}|s) - \log p(\mathbf{x}) \tag{5}$$

Given a Gaussian distribution dependent on speaker s and a speaker independent Gaussian distribution, the above equation can be approximated by

$$pmi(\mathbf{x}; s) \approx \log \mathcal{N}(\mathbf{x}|\Theta_s) - \log \mathcal{N}(\mathbf{x}|\Theta_{ind}) \tag{6}$$

where Θ_s and Θ_{ind} represent the Gaussian parameter of the speaker dependent and independent Gaussian distributions, respectively.

Although the value of PMI ranges from $-\infty$ to $\min(-\log p(\mathbf{x}), -\log p(s))$, it can be normalized between -1 and $+1$ for practical convenience [15]. The normalized PMI (NPMI) can be computed as below.

$$
\begin{aligned}
npmi(\mathbf{X}; s) &= \frac{pmi(\mathbf{X}; s)}{-\log p(\mathbf{X}, s)} \\
&= \frac{pmi(\mathbf{X}; s)}{-\log p(\mathbf{X}|s) - \log p(s)} \\
&\approx \frac{\log \mathcal{N}(\mathbf{X}|\Theta_s) - \log \mathcal{N}(\mathbf{X}|\Theta_{ind})}{-\log \mathcal{N}(\mathbf{X}|\Theta_s) - \log p(s)}.
\end{aligned} \tag{7}
$$

The value of NPMI is -1 if the two events never occur together, 0 if the two events are independent and $+1$ if the two events always occur together.

3 Baum-Welch Statistics Weighting Using Mutual Information

Given a pre-trained UBM, the classical i-vector framework computes the Baum-Welch statistics of the utterance with no discrimination among different frames. The approach we propose analyzes the relative importance of each speech frame in means of speaker recognition by measuring the NPMIs of each frame. This information can be used to emphasize the contribution of the frames with higher speaker relevancy while suppressing the ones with little speaker related information.

Let $\mathbf{X} = [\mathbf{x}_1,...,\mathbf{x}_l,...,\mathbf{x}_L]$ be the input speech utterance with L frames and $\{\Theta_1, ...,\Theta_s,...\Theta_S\}$ be a set of speaker dependent GMMs where Θ_s is the GMM parameters adapted to speaker s from a training set with S speakers. The speaker relevancy $\phi(\mathbf{x}_l|c)$ can be calculated as

$$\begin{aligned}\phi(\mathbf{X}_l|c) &= \frac{1}{S}\sum_{s=1}^{S} npmi(\mathbf{X}_l; s|c) \\ &\approx \frac{1}{S}\sum_{s=1}^{S} \frac{\log\mathcal{N}(x_l|\Theta_{s,c}) - \log\mathcal{N}(\mathbf{X}_l|\Theta_{UBM,c})}{\log\mathcal{N}(\mathbf{X}_l|\Theta_{s,c}) - \log p(s)}\end{aligned} \tag{8}$$

where $\Theta_{UBM,c}$ and $\Theta_{s,c}$ are the Gaussian parameters for the cth mixture component of the UBM and GMM dependent on speaker s, respectively. The speaker relevancy $\phi(\mathbf{x}_l|c)$ represents the average amount of speaker sensitive information that can be modeled by the c_{th} GMM mixture component contained in speech frame $\mathbf{x}_l$. Thus speech frames with high speaker relevancy can be considered relatively important compared to those with low speaker relevancy. Moreover, it would be safe to say that GMM mixture components with high speaker relevancy is more capable of modeling speaker-dependent patterns of the frame-level features.

The proposed system utilizes the speaker relevancy to weight the posterior probability $\gamma_l(c)$ when computing the Baum-Welch statistics. As mentioned in the earlier section, the i-vector framework adapts the UBM to the speech frames in a given utterance and Gaussian components with high number of aligned frames are more likely to contribute to the adaptation. Thus by modifying the 0_{th} and 1_{st} order Baum-Welch statistics, which represent the number and the mean of the speech frames aligned to the components respectively [16], the contribution of each frame can be weighted during the i-vector extraction process. Since the speaker relevancy formulated above is an average of NPMI, negative value can exist, hence it is not appropriate to directly use for weighting the probabilities. Therefore we define the weight α by

$$\alpha(\mathbf{x}_l, c) = \phi(\mathbf{x}_l|c) + 1, \tag{9}$$

which is simply the speaker relevancy with a slight offset added in order to make the value positive. The modified Baum-Welch statistics can be obtained as follows.

$$v_c(\mathbf{X}) = \sum_{l=1}^{L} \alpha(\mathbf{X}_l, c)\gamma_l(c), \tag{10}$$

$$\zeta_c(\mathbf{X}) = \sum_{l=1}^{L} \alpha(\mathbf{X}_l, c)\gamma_l(c)(\mathbf{X}_l - u_c) \tag{11}$$

where $v_c(\mathbf{X})$ and $\zeta_c(\mathbf{X})$ represents the modified $n_c(\mathbf{X})$ and $\tilde{\mathbf{f}}_c(\mathbf{X})$, respectively.

In the proposed scheme, the Baum-Welch statistics of the enrollment and test dataset are modified after training the total variability matrix. Once the total variability matrix is trained, i-vectors of the enrollment and test dataset are extracted using the modified Baum-Welch statistics and evaluated using the PLDA scoring scheme.

4 Weight Extraction Using DNN

In this section, we present an alternative method for extracting the speaker relevancy. Motivated by the robust nature and the high performance of the DNN in acoustic modeling for ASR [9], we propose a DNN-based method for extracting the frame-level weights.

A DNN is essentially a multi-layer perceptron (MLP) with two or more layers. DNN can represent the nonlinear relationship between two sets of vectors [17]. The DNN model we propose uses a stacked set of frame-level speech features as input analogous to the DNN used in most speech applications [18]. Given a DNN input $\mathbf{X}_\tau = [\mathbf{x}_{\tau-M}, \ldots, \mathbf{x}_\tau, \ldots \mathbf{x}_{\tau+M}]$, which is a stack of $2M + 1$ frames with $\mathbf{x}_\tau$ in the center, the target output of the c_{th} node of the sigmoidal output layer is the speaker relevancy for a single instance on the c_{th} Gaussian mixture component which is calculated by

$$o_{c,\tau} = \frac{1}{2}\alpha(\mathbf{x}_\tau, c). \tag{12}$$

In Eq. (11), the target output is set to be half of $\alpha(\mathbf{x}_\tau, c)$ because the value of the sigmoidal function ranges from 0 to 1, whereas the weight (8) ranges from 0 to 2. Thus for utterance $\mathbf{X}$ with L frames, a total of $L - 2M$ output vectors are generated by the DNN. The weight is computed by rescaling the DNN output as follows:

$$\alpha_{DNN}(\mathbf{x}_\tau, c) = 2\tilde{o}_{c,\tau}, \tag{13}$$

where $\tilde{o}_{c,\tau}$ represents the output from the cth node of the output layer given $\mathbf{X}_\tau$.

By using a DNN for extracting the weights, not only we can expect a better performance due to its robust attribute but can also be less time consuming since there is no need to compute the log-likelihood for each speaker dependent GMMs. Moreover, the usage of stacked frames as input enables the DNN to estimate the NPMI considering the contextual information.

Table 1. Speaker recognition performance of the baseline i-vector and the proposed methods.

	Class. Err. [%]	EER [%]	DCF08 [%]	DCF10 [%]
Baseline	12.62	3.36	2.01	0.07
Proposed	9.01	**2.82**	**1.75**	0.07
Proposed DNN	**8.89**	3.07	1.86	0.07

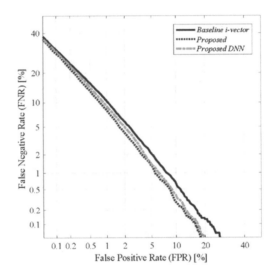

Fig. 1. DET curve comparison between the baseline i-vector system and the proposed methods.

5 Experiments

In order to evaluate the performance of the proposed algorithm under the uncertainty caused by short utterances, we conducted an experiment on speaker recognition using the TIDIGITS speech corpus [19]. The TIDIGITS corpus contains clean digit sequences and isolated digits spoken by 326 speakers, and a subset of 25096 samples was used for enrollment and another subset of 25096 samples was used for evaluation. For each frame, a Mel frequency cepstral coefficient (MFCC) feature was extracted using the SPro toolkit [20]. The UBM and the total variability matrix were trained with the TIMIT corpus [21], which consists of 6300 sentences (5.4 h) spoken by 630 speakers using the MSR identity toolbox [22]. For the baseline i-vector framework and the proposed algorithm, a UBM with 32 components was trained, and the dimension for the i-vectors was fixed to 200. Furthermore, LDA was applied to all the i-vectors before scoring with PLDA.

The proposed DNN model for speaker relevancy extraction consists of 2 hidden layers with 256 rectified linear (ReLU) activation nodes. The input of the DNN is a stacked vector of 11 frames (5 on each side of the current frame) of MFCC features. The output layer consists of 32 sigmoidal nodes, each corresponding to the speaker relevancy of an individual GMM component. The weights are randomly initialized and the parameters are fine-tuned using back propagation algorithm with a dropout regularization [23] (dropout rate 0.2).

A total of 4 types of performance measures were used in the following experiment: classification error (Class. Err.), EER [8], minimum NIST SRE 2008 DCF (DCF08) [2], minimum NIST SRE 2010 DCF (DCF10) [10].

To verify the performance of the i-vectors extracted using the proposed methods, we conducted a set of speaker recognition experiments on the TIDIGITS dataset. The experimented methods are as follows:

Baseline: standard 200 dimensional i-vector,

Proposed: i-vector extracted using the Baum-Welch statistics modified with $\alpha(\mathbf{x}_l, c)$ which is defined by Eq. (8),

Proposed DNN: i-vector extracted using the Baum-Welch statistics modified with $\alpha_{DNN}(\mathbf{x}_l, c)$ which is defined by Eq. (12).

Table 1 gives the results obtained by the experimented methods mentioned above. As shown in the results, the proposed methods (i.e. *Proposed* and *Proposed DNN*) outperformed the baseline i-vector framework (i.e. *Baseline*). *Proposed* achieved a relative improvement of 16.07% in terms of EER compared to *Baseline*. This may be due to the increased contribution of the frames with high speaker relevancy while extracting the i-vectors via Baum-Welch modification. The usage of DNN for generating the speaker relevancy (i.e. *Proposed DNN*) further improved the speaker classification performance, showing a relative improvement of 29.56% in terms of classification error compared to *Baseline*. Figure 1 shows the DET curves obtained from the three tested approaches.

6   Conclusion

In this paper, a novel scheme for considering frame-level speaker relevancy during the i-vector extraction is proposed. In order to extract a robust i-vector, the frame-level point-wise mutual information is utilized to directly modify the Baum-Welch statistics. Moreover, inspired by the DNN for ASR, a DNN based speaker relevancy extraction scheme is presented.

To investigate the performance of the i-vectors extracted using the proposed system in a short duration scenario, we conducted an experiment using the TIDIGITS dataset. We observed that the i-vector extracted via the proposed scheme outperforms the conventional i-vector.

In our future research, we will further develop the weighting criterion for modifying the Baum-Welch statistics to not only emphasize the speaker relevancy but to also take the speaker discriminability into account.

Acknowledgments. This research was supported by Projects for Research and Development of Police science and Technology under Center for Research and Development of Police science and Technology and Korean National Police Agency funded by the Ministry of Science, ICT and Future Planning (PA-J000001-2017-101), and by the National Research Foundation of Korea (NRF) grant funded by the Korean government (MEST) (NRF-2015R1A2A1A15054343).

References

1. Dehak, N.: Front-end factor analysis for speaker verification. IEEE Trans. Audio Speech Lang. Process. **19**(4), 788–798 (2011)
2. The NIST year 2008 speaker recognition evaluation plan (2008). http://www.itl.nist.gov/iad/mig//tests/sre/2008/
3. Saeidi, R., Alku, P.: Accounting for uncertainty of i-vectors in speaker recognition using uncertainty propagation and modified imputation, In: Proceedings INTERSPEECH, pp. 3546–3550 (2015)
4. Hasan, T., et al.: Duration mismatch compensation for i-vector based speaker recognition systems. In: Proceedings of IEEE International Conference Acoustic, Speech, and Signal Process, pp. 7663–7667 (2013)
5. Jung, C.S.: Selecting feature frames for automatic speaker recognition using mutual information. IEEE Trans. Audio Speech Lang. Process. **18**(6), 1332–1340 (2010)
6. Mandasari, M.I. et al.: Evaluation of i-vector speaker recognition systems for forensic application. In: Proceedings of INTERSPEECH, pp. 21–24 (2011)
7. Mandasari, M.I.: Quality measure functions for calibration of speaker recognition systems in various duration conditions. IEEE Trans. Audio Speech Lang. Process. **21**(11), 2425–2438 (2013)
8. Hansen, J., Hasan, T.: Speaker recognition by machines and humans. IEEE Signal Process. Mag. **32**(6), 74–99 (2015)
9. Hinton, G., et al.: Deep neural networks for acoustic modeling in speech recognition. IEEE Signal Process. Mag. **29**(6), 82–97 (2012)
10. The NIST year 2010 speaker recognition evaluation plan (2010). http://www.itl.nist.gov/iad/mig//tests/sre/2010/
11. Dehak, N. et al.: Language recognition via ivectors and dimensionality reduction, In: Proceedings of INTERSPEECH, pp. 857–860 (2011)
12. Kenny, P.: Eigenvoice modeling with sparse training data. IEEE Trans. Audio, Speech, Lang. Process. **13**(3), 345–354 (2005)
13. Dehak, N., et al.: Support vector machines and joint factor analysis for speaker verification. In: Proceedings of IEEE International Conference on Acoust, Speech, and Signal Process, pp. 4237–4240 (2009)
14. Hoang, H.H., et al.: A re-examination of lexical association measures, In: Proceedings of Workshop on Multiword Expressions Identification, Interpretation, Disambiguation and Application, pp. 31–39 (2009)
15. Bouma, G.: Normalized (pointwise) mutual information in collocation extraction, In: Proceedings of the Biennial GSCL Conference, pp. 31–40 (2009)
16. Kenny, P.: A small footprint i-vector extractor. In: Proceedings of Odyssey, pp. 1–25 (2012)
17. Salakhutdinov, R.: Learning deep generative models. Ann. Rev. Stat. Appl. **2**(1), 361–385 (2015)
18. Garcia-Romero, R., McCree, A.: Insights into deep neural networks for speaker recognition. In: Proceedings of INTERSPEECH, pp. 1141–1145 (2015)

19. Leonard, R.G.: A database for speaker-independent digit recognition. In: Proceedings of IEEE International Conference on Acoust, Speech, and Signal Process, pp. 328–331 (1984)
20. Gravier, G.: SPro: speech signal processing toolkit. Software. http://gforge.inria.fr/projects/spro
21. Lopes, C., Perdigao, F.: Phone recognition on the TIMIT database. Speech Technol. **1**, 285–302 (2011)
22. Sadjadi, S.O., et al.: MSR identity toolbox v1.0: a MATLAB toolbox for speaker recognition research. Speech Lang. Process. Tech. Comm. Newsl. **1(4)**, (2013)
23. Srivastava, N., et al.: Dropout: a simple way to prevent neural networks from overfitting. J. Mach. Learn. Res. **15**, 1929–1958 (2014)

A Recommended Replacement Algorithm for the Scalable Asynchronous Cache Consistency Scheme

Ramzi A. Haraty$^{(\boxtimes)}$ and Lama Hasan Nahas

Department of Computer Science and Mathematics,
Lebanese American University, Beirut, Lebanon
rharaty@lau.edu.lb, lama.nahas@lau.edu

Abstract. Acknowledging the widespread prevalence of mobile computing and the prominence of data caching in mobile networks has led to a myriad of efforts to alleviate their associated problems. These challenges are inherited from the mobile environment because of low bandwidth, limited battery power, and mobile disconnectedness. Many caching techniques were presented in literature; however, Scalable Asynchronous Cache Consistency Scheme (SACCS) that proved to be highly scalable with minimum database management overhead. SACCS was initially implemented with Least Recently Used (LRU) as a cache replacement policy. In this project we adopted the Extended LRU (E-LRU) as a cache replacement strategy to be applied in SACCS. A simulation of SACCS was done with the E-LRU, Least Recently Used, Most Recently Used, Most Frequently Used, and Least Frequently Used; and the comparative evaluation showed that SACCS with E-LRU is superior in terms of delay time, hit ratio, miss ratio and data downloaded per query.

Keywords: Mobile computing · Cache consistency · Replacement policy · Invalidation strategy · Hybrid algorithms

1 Introduction

Mobile computing allows greater ease of communication and versatility in using technology. Mobiles support a wide range of applications, allowing people to retrieve subtle information, such as stock prices, in real time. However, mobility comes hand-in-hand with several issues resulting from frequent disconnections and limited resources. Caching is storing desired data in local storage to improve data availability. It is an effective way of enhancing system performance. It improves the bandwidth utilization, while minimizing the query delay time and battery consumption. Nevertheless, maintaining cache consistency and a rigorous cache replacement policy are vital for successful caching. Cache consistency is usually actualized using either stateful or stateless approaches. The former is used for larger scale database systems, yet compromises nontrivial overhead in attempting to manage the server database. Meanwhile, according to the latter, namely the stateless approach, the mobile user's cache content is kept outside the awareness of the server. Yet, unlike the stateful

© Springer Nature Singapore Pte Ltd. 2018
K.J. Kim et al. (eds.), *IT Convergence and Security 2017*,
Lecture Notes in Electrical Engineering 449,
DOI 10.1007/978-981-10-6451-7_11

approach, it is not efficient when applied to large scale database systems, and is not capable of dealing with the user's mobility and disconnectedness [1].

A combination of both aforementioned approaches was proposed through what was termed "Scalable Asynchronous Cache Consistency Scheme". However, an effective replacement policy would substantially contribute to the success of the promising scheme. Based on the limited size of mobile's internal memory, a replacement strategy is needed to determine which data should be evicted when the mobile cache is full. The algorithm is a key factor, as it notably affects the performance of the whole scheme. Many replacement algorithms have been proposed in literature, and we're recommending an algorithm that proved to better suit SACCS.

The vast development in the computing realm allowed for massive advancement in real life applications, and there is still a wide range of improvement for future work. The client/server environment in today's wireless distributed networks gives more capabilities for mobile users. Nevertheless, it's a challenging environment due to bandwidth constraints and to the nature of mobile units that allows frequent disconnectedness from network and limited battery power. It allows the user to connect from different access points and to preserve their connection even when displaced [2].

In this work, we implemented an extended version of Least Recently Used replacement algorithm with SACCS. The E-LRU considers size and recency of usage when choosing the data to be replaced. The rest of the paper is organized as follows: Sect. 2 presents a literature review of cache consistency approaches and replacement policy strategies. Section 3 thoroughly explains the SACCS algorithm; Sect. 4 presents E-LRU, our recommended replacement policy for SACCS. Section 5 shows and compares the simulation results of the replacement policies integrated with SACCS. Finally, Sect. 6 concludes the study and suggests future work.

2 Literature Review

2.1 Data Caching Technique

Data caching is saving a copy of data in the client's side memory to avoid retrieving it again from source node soon. Data caching proved to be an extremely effective technique to preserve scarce resources. Its effect is further manifested in cases of small database and frequent queries, and when the communication cost is high [3]. Caching technology is extensively used in software and hardware, and is more crucial in a mobile environment in order to mitigate the usage of bandwidth, memory and energy. Cache consistency and replacement policy are two main pillars for successful cache management. Cache consistency techniques guarantee the validity of data stored in a mobile cache; while replacement policies determine the data item/s to be evicted from the local cache when it is too full to accommodate any new caches [4].

2.2 Cache Replacement Policy

Knowing the limited size of memory in mobile units, the subset dedicated for cache is obviously limited and valuable. Hence, when the cache is replete, a replacement policy is needed as a criterion to decide which data should be dropped from cache to make room for a new data item [5]. Cache replacement policies fall into three categories:

- Temporal locality expects recently used data items to be revisited shortly.
- Spatial locality anticipates that data nearby recently referenced data could be accessed in a future time.
- Semantic locality signifies that recently accessed data area is more likely to be accessed again soon.

Several factors play a role in deciding on the data items being replaced, the top of which are the following: access probability, recency of data access, access frequency, data size, fetching and communication cost, update rate, distance, etc. [6].

In a wireless environment, more parameters should be considered, such as connectivity, bandwidth and location. Many policies are based on a function that combines different parameters.

Some promising policies are modified to enhance their performance or to better fit in a different environment. Therefore policies have many variants; LRU for example uses temporal locality and has variants such as LRU-k, LRU-THOLD, E-LRU, etc. One of the LRU refined replacement policy is LRU-MIN which focuses on the size of data items being added to or removed from cache. It finds all cached data items of size equal to or greater than the size of the newly arrived data item, denoted as *Size-A*, then deletes the least recently used data among them. If all cached data items are smaller than *Size-A*, then LRU-MIN finds all cached data items of size equal to or greater than half of *Size-A*. In this case, the two least recently used data items will be deleted, and so on until enough space has been emptied [7]. Many other value function cache replacement strategies were presented in the literature, for instance SAIU, SAUD, Min SAUD, On Bound Selection, etc. [8].

2.2.1 Location-Based Cache Replacement

A wide range of cache replacement policies became more location oriented after the propagation of location-dependent information services via mobile devices. Effective parameters in these policies are distance, valid scope area and direction. Manhattan was one of the early introduced location aware policies. Farther Away Replacement (FAR) is a spatial locality approach that removes the farthest data item from the client's location. Predicted Region Based Replacement Policy (PRRP) is another location based policy. It relies on the size of data in the cache as well as the predicted region where the client will shortly arrive [9].

2.2.2 Coordinated Cache Replacement

Energy-efficient Coordinated cache Replacement Problem (ECORP) chose energy efficiency to be the main performance objective [10]. In ECORP and Dynamic ECORP DP adjacent nodes only store different data items in an attempt to better utilize the collective cache.

2.3   Invalidation Reports

Mobile data is prone to inconsistency since mobiles may reply to a query using local cache, while the requested data item is being updated in the server database. Broadcasting the updated data to all mobile units preserves consistency, yet it is costly and impractical. Invalidation Report (IR) technique proved to be more efficient since an IR may only contain the IDs and timestamps of the altered data; or the IR may contain a bit sequence that resembles the database, and assign bit value 1 to the updated data and 0 for the rest. Timestamp and bit sequence techniques significantly limit the overhead of transferring data items, since their size is lesser than the size of the actual data. Thus, IR approaches avoid congesting the network with unnecessary traffic. The role of IR is to invalidate data available in mobile units' caches. Whenever a mobile unit receives a data request from its user, it checks if data is available in its cache; and, if it is not available, the MU immediately sends a query to its MSS requesting the data. Otherwise, if the data was present in the cache, the mobile unit (MU) waits for the following IR to confirm the validity of the cache data so it can be sent to the user. If the IR indicates that the cache item is invalid, an original copy is imported from MSS and forwarded to the user. The IR interval length determines the mobile unit's pace of answering any query. The longer the IR interval gets the more delay that occurs in answering queries. Replicating IR m times within the IR interval was one approach to decrease the latency. Thus, the upper bound of the query latency is $1/m$ of the IR interval time. The additional IRs in this approach, called Updated Interval Reports, strictly contain the data items updated after the last IR [11]. Broadcasting IRs has many advantages, yet missing an IR threatens the cache consistency; unfortunately it is possible for an MU to lose an IR especially in the case of frequent disconnection. One antidote is to send back an acknowledgment for every IR received, but this will significantly increase the communication overhead [12].

2.3.1   Stateful Invalidation Schemes

It is rare to find stateful cache consistency maintenance approaches recommended for wireless networks. Kahol et al. [13] proposed an Asynchronous Stateful algorithm denoted AS, where the server keeps track of all data items in every MU, and sends IRs to MUs only when they are connected. In case of sleep, the IR is buffered at MSS to be resent later. However, MSS can't differentiate between a missing IR and a disconnected MU; therefore, it is necessary to allow a maximum number of retransmissions. A stateful approach for cooperative cache consistency called Greedy Walk-based Selective Push (GWSP) was proposed by Huang et al. [14]. In GWSP, the source node saves the Time to Live (TTL) and the request rate of every cached item in all caching nodes. This information is used to dynamically choose which caching node should receive the updated version of a data item. Whenever a cached data is updated in the source node, a greedy walk technique is used to deliver the updated data to the chosen caching nodes only.

3   The Scalable Asynchronous Cache Consistency Scheme

In this work, we consider the data communication system to consist of a wire-connected network that links many main servers to distributed Mobile Support Stations (MSS). The MSS is wirelessly connected to several mobile units. SACCS maintains cache consistency between the cache of MSS and the local cache of Mobile Units (MUs). SACCS assumes that the cache consistency between servers and MSSs is taken care of. Each data entry in the cache has three states: valid, uncertain and ID-only. Mobile user data cache can be used when requested if "valid". If the MU receives an IR that invalidates a valid or uncertain data item, only the data ID will remain in cache and the valid data will be downloaded later if needed. The state of "valid" cache entry will become "uncertain" when its TTL ends. Also, all "valid" data in the cache will become "uncertain" in case the mobile unit disconnects and reconnects to the network (sleep/wake situation). When an "uncertain" cache entry is requested, the MSS will either approve its validity or it will send the updated version to be downloaded [1].

4   Cache Replacement Policy for SACCS

Whenever the size of the cache reaches its limit, the replacement policy determines which data is to be evicted in order to allow new data into the cache. To make its decision, the Extended-LRU relies on size, frequency of references as well as recency. E-LRU sets a threshold for acceptable size of a cached data item. If the data item size is more than 50% of the cache size, the query requesting this data will be answered directly without importing the data item into the cache. E-LRU prioritizes the data item that is referenced only once for eviction. If more than one data item in the cache is referenced once, the data item least recently referenced among them will be deleted. In case all data in the cache is referenced more than once, the data item with the longest inter arrival time between its last two referrals will be the victim. In case more than one data item has the same maximum inter arrival time, the one with less TTL remaining will be evicted [6]. E-LRU is a replacement policy suggested for various systems. In this study, we recommended it with SACCS and compared it with well-known cache replacement policies.

5   Experimental Results

A simulation of SACCS was performed with five replacement policies: E-LRU, LRU, MRU, LFU and MFU. The simulation environment was unified by fixing the number of mobile clients, the size of cache, the number of data items, etc. Other parameters were randomly determined, such as: the size of data items, update rate, the bandwidth, the uplink and downlink message size, etc. The sleep/wake time and frequency were randomly chosen from a predetermined set. The simulation was run in eight time slots, varying from 50,000 microseconds to 400,000 in 50,000 intervals. The simulation environment assumes that the network architecture consists of one cell. The performance evaluation metrics consists of the hits/misses total, and the latency reflected by the average delay and the total delay time.

5.1   Hit and Miss Ratio

When the requested data item is found in the cache, the query is counted as one hit. On the other hand, when the query needs to download the requested data from the source node because it is not available in mobile client cache, this query is counted as one miss. Certainly, the algorithm with a higher number of hits and lower number of misses is more likely to outperform other algorithms. The E-LRU total hits reached 5,216 queries during the simulation period, where the second best replacement algorithm was LRU with 5,086 hits as depicted in Fig. 1.

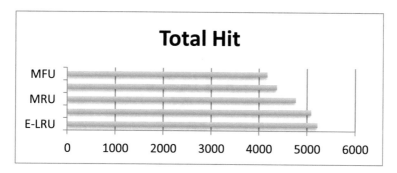

Fig. 1.   Total hit

The total missed queries summed up to 11527 with E-LRU, a number far less than other replacement policies: 11666, 11953, 12342 and 12519 in LRU, MRU, LFU and MFU respectively. This is shown in Fig. 2.

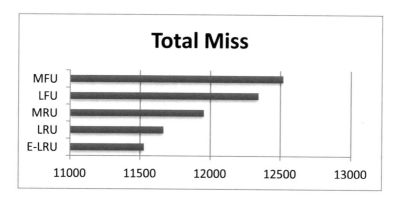

Fig. 2.   Total miss

E-LRU takes size into consideration when taking the replacement decision. It does not approve removing many items form cache to import one large item into cache. There is no data item that is permitted to occupy more than half of the size of MU cache. This sounds logical since it allows better exploitation of the cache memory. Size check is definitely an improvement over LRU. The impact of this extra check is clearly demonstrated in the high numbers of hits E-LRU achieved.

5.2 Latency

Another important performance metric is the latency. The latency is calculated by measuring the time the mobile client waits between issuing a query and receiving a response to it. In our simulation, the latency was considered negligible in case of a hit. The total delay over the whole simulation time summed up to 15248 with E-LRE, 15346 with LRU, 16823 with MRU, 19483 with LFU and the longest total delay was 21,751 with MFU (see Fig. 3). Also, the average delay reflected the same result.

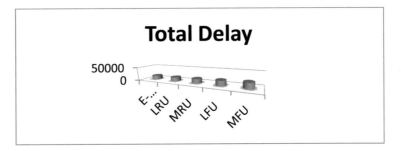

Fig. 3. Total delay

Also, the average delay, presented in Fig. 4, reflected the same results as the total delay. Query delay degrades dramatically upon improving data availability. Deleting data items referenced only once allows space for data items that are frequently referenced and more likely to be referenced again.

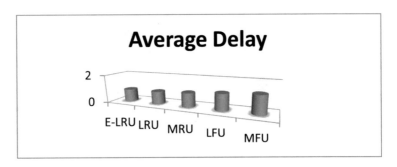

Fig. 4. Average delay

6 Conclusion

Mobile computing is imperative to the rapid advancement of technology, while in parallel, data caching is crucial to the efficiency and pragmatism of mobile computing. Ergo, dealing with the complications that arise from the dynamic environment of mobiles is inevitable. Given a rigorous highly scalable scheme as SACCS, finding a matching replacement policy is vital to its success. Initially, SACCS was implemented with the basic LRU. In this work, we recommended a cache replacement policy for SACCS that outperforms it. Simulating SACCS with various replacement policies shows the immense impact the replacement policies have on overall performance. The comparative evaluation shows that our recommended cache replacement policy, E-LRU, is superior to basic approaches (LRU, MRU, LFU and MFU) in terms of hit ratio, miss ratio, query delay and bytes downloaded per query.

There is no one perfect cache replacement policy that fits in every scenario. Therefore, selecting the right cache replacement algorithm out of the enormous variety presented in literature could always lead to better results. Moreover, comparing replacement algorithm against other advanced replacement policies will be more indicative than comparing with basic policies. In our future work, we would like to further improve SACCS by examining more replacement algorithms, and studying their performance in comparison with E-LRU and other promising policies.

References

1. Wang, Z., Das, S.K., Che, H., Kumar, M.: A scalable asynchronous cache consistency scheme (SACCS) for mobile environments. IEEE Trans. Parallel Distrib. Syst. **15**(11), 983–995 (2004)
2. Zeitunlian, A., Haraty, R.A.: An efficient cache replacement strategy for the hybrid cache consistency approach in a mobile environment. In: Proceedings of the International Conference on Computer, Electrical, and Systems Science, and Engineering (ICCESSE 2010), Rio De Janeiro, Brazil, March 2010
3. Barbara, D., Imielinksi, T.: Sleepers and workaholics: caching strategies in mobile environments. ACM SIGMOD Record **23**(2), 1–12 (1994)
4. Haraty, Ramzi A.: Innovative mobile e-Healthcare systems: a new rule-based cache replacement strategy using least profit values. Mobile Inf. Syst. (2016). doi:10.1155/2016/6141828
5. Jane, M.M., Nouh, F.Y., Nadarajan, R., Safar, M.: Network distance based cache replacement policy for location-dependent data in mobile environment. Paper presented at the Ninth International Conference on Mobile Data Management Workshops. IEEE, Beijing (2008)
6. Joy, P.T., Jacob, K.P.: Cache replacement strategies for mobile data caching. Int. J. Ad hoc Sens. Ubiquitous Comput. (IJASUC) **3**(4), 99 (2012)
7. Abrams, M., Standridge, C., Abdulla, G., Williams, S., Fox, E.: Caching proxies: limitations and potential. In: Paper presented at the Fourth International World Wide Web Conference: The Web Revolution, Boston, MA, USA (1995)

8. Hattab, E., Qawasmeh, S.: A survey of replacement policies for mobile web caching. In: Paper presented at the 2015 International Conference on Developments of E-Systems Engineering, pp. 41–46. IEEE, Dubai (2015)
9. Kumar, A., Misra, M., Sarje, A.: A new cache replacement policy for location dependent data in mobile environment. In: Paper presented at the 2006 IFIP International Conference on Wireless and Optical Communications Networks. IEEE, Bangalore (2006)
10. Li, W., Chan, E., Chen, D.: Energy-efficient cache replacement policies for cooperative caching in mobile ad hoc network. In: Paper presented at the Wireless Communications and Networking Conference, 2007, pp. 3347–3352. IEEE, Hong Kong (2007)
11. Haraty, R.A., Lana Turk, L.:. A comparative study of replacement algorithms used in the scalable asynchronous cache consistency scheme. In: Proceedings of the ISCA 19th International Conference on Computer Applications in Industry and Engineering, Las Vegas, USA (2006)
12. Haraty, Ramzi A., Zeitouny, Joe: Rule-based data mining cache replacement strategy. Int. J. Data Warehouse. Min. 9(1), 56–69 (2013). doi:10.4018/IJDWM. ISSN: 1548-3924, EISSN: 1548-3932
13. Kahol, A., Khurana, S., Gupta, S.K.S., Srimani, P.K.: A strategy to manage cache consistency in a disconnected distributed environment. IEEE Trans. Parallel Distrib. Syst. 12 (7), 686–700 (2001)
14. Huang, Y., Jin, B., Cao, J., Sun, G., Feng, Y.: A selective push algorithm for cooperative cache consistency maintenance over MANETs. In: Kuo, T., Sha, E., Guo, M., Yang, L.T. (eds.) Lecture Notes in Computer Science, vol. 4808, pp. 650–660. Springer, Berlin (2007)

Multiple Constraints Satisfaction-Based Reliable Localization for Mobile Underwater Sensor Networks

Guangyuan Wang[1], Yongji Ren[2(✉)], Xiaofeng Xu[3], and Xiaolei Liu[4]

[1] Department of Military Training,
Naval Aeronautical and Astronautical University, Yantai 264001, China
[2] Department of Command,
Naval Aeronautical and Astronautical University, Yantai 264001, China
lenglengqiuyu@sina.com
[3] Science and Technology on Communication
Information Security Control Laboratory, Jiaxing 314033, China
[4] Department of Electrical Engineering,
Yantai Vocational College, Yantai 264001, China

Abstract. In this paper, we proposed a novel Multiple Constraints Satisfaction Based Reliable Localization Algorithm for mobile Underwater Wireless Sensor Networks (UWSNs). In a typical application, e.g. the ocean battlefields, the localization process has been no doubt constantly restricted by several kinds of uncertainty, which lead to obvious degradation of localization reliability and accuracy, e.g. the confidence of reference information, the mobility of sensor nodes, the reliability of multi-hop localization link, etc. It implies a limit and an insufficiency of localization reliability. Thus, we transformed the reliable localization problem into a multiple Constraints Satisfaction Problem (CSP). In the CSP framework, we firstly integrated three kinds of constraints, i.e. confidence constraint, mobility constraint, and reliability constraint. Then game method has been utilized to deal with the CSP and determine the positions of underwater sensor nodes. Simulation results show that algorithm is effective and efficient.

Keywords: Underwater sensor networks · Localization · Constraint satisfaction

1 Introduction

In recent years there has been a rapidly growing interest in mobile Underwater Wireless Sensor Networks (UWSNs). UWSNs have significant advantages over traditional wired networks for sensing environments and monitoring targets in ocean battlefield, e.g. easy deployment, self-management, non-established infrastructure, etc. [1, 2].

For this kind of location-based applications, localization is an essential technology for UWSNs which significantly affects the network performance [3]. In complex ocean battlefield, the localization procedure has been no doubt restricted by several kinds of adverse challenges, e.g. the potential attacks, the unreliable reference nodes and multi-hop localization link, the non-ideal network conditions, the unknown mobility of

© Springer Nature Singapore Pte Ltd. 2018
K.J. Kim et al. (eds.), *IT Convergence and Security 2017*,
Lecture Notes in Electrical Engineering 449,
DOI 10.1007/978-981-10-6451-7_12

sensor nodes, etc. All the above adverse factors would lead to obvious degradation of localization reliability and accuracy [4, 5]. The substantial reason is that the localization performance has been restricted by multiple hybrid constraints simultaneously, i.e. the confidence constraint of reference information, the mobility constraint of underwater sensor nodes, and the reliability constraint of multi-hop localization link. Some localization approaches which mainly focused on a certain aspect independently have been studied. The state-of-art survey on localization has been proposed in [6], which classifying the algorithms based on static and mobile nodes. Although this is an excellent summary of recently proposed localization algorithms, it doesn't include the secure localization challenges. Mesmoudi et al. have discussed localization algorithms in [7], where the algorithms are primarily classified into range free and range based algorithms, each of which are further classified into full schemes and hybrid schemes. Though this work presents a comprehensive analysis of the algorithms, mobility and security issues are not covered extensively. Although many existing algorithms have been studied to deal with the uncertainty independently, the multiple constraints satisfaction problems pose new challenges and make it necessary to develop new reliable localization algorithms. Therefore, a novel multiple constraints satisfaction based localization algorithm has been proposed in this paper.

2 Multiple Constraints Satisfaction-Based Reliable Localization

In this paper, we assume that the underwater sensor nodes are deployed in ocean battlefield, that is, there are potential malicious attackers in the hostile circumstance. The attackers intentionally attack the UWSNs anchor nodes. Therefore, besides the constraints of mobility and reliability, the localization process has been restricted by constraints of confidence. To solve this problem, we transformed the reliable localization problem into a multiple Constraints Satisfaction Problem [8].

2.1 Constraint Satisfaction Problem Framework

In this paper, we mainly consider three kinds of constraints, i.e. confidence constraint, mobility constraint, and reliability constraint. Thus, a simple constraint satisfaction problem of localization, i.e. the determination problem of reliable localization for sensor nodes, is defined by a set of 3 constraints $C = \{c_1, c_2, c_3\}$, with a set of interval domains $\{[x_1], [x_2], [x_3]\}$. To each constraint, we have to build a contractor.

The first one $C1$ contracts the confidence with respect to the reference information of anchor nodes. Recall that a malicious node can claim fake locations to the other benign nodes. The contractor $C2$ is related to the reliability with respect to the reliability of multi-hop localization link. It makes possible to contract the corresponding anchor node as well as the ordinary node. The contractor $C3$ associated with the mobility of sensor nodes. It provides contractions for the to-be-localized ordinary node as well as the corresponding reference nodes.

Solving the CSP problem in an interval analysis framework actually is finding the intersection that contains all possible solutions. The detailed procedure is as follows.

2.2 Confidence Constraint

In the game scenario, the benign anchor node's strategy is to prevent the running states of protective modality information being approximated by the potential malicious attackers. The to-be-localized node's strategy is to make its running states of reference information get closer to the corresponding states of benign nodes. The malicious attacker's strategy is to disseminate fake locations to the benign nodes.

Assume that the UWSNs are composed by N nodes and there are m localization groups acting on it. For the to-be-localized node N_k with the localization group V_k, the running cost function is given by

$$L^k\left(t, x, u^k\right) = \sum_{i \in V_k} c_i(u_i) - \sum_{i \in V_k} \sum_{j \in V_{k', k' \neq k}} \left[a_{i,j} e^{-\theta_{i,j}^k} - a_{i,j} e^{-\omega_{i,j}^k} \right] \tag{1}$$

where c_i is control cost function of node N_i, $a_{i,j} e^{-\theta_{i,j}^k}$ are attack profit running functions of node N_i to node N_j and the team $a_{i,j} e^{-\omega_{i,j}^k}$ are the corresponding information loss running functions. According to the strategies, all the localization groups wish to minimize their respective cost functions. The cost function of group k is given by

$$J^k\left(t, x, u^k\right) = \int_t^{t_f} L^k\left(t, x, u^k\right) dt + \Psi^k\left(x_{t_f}^k\right), \quad 1 \leq k \leq m \tag{2}$$

where u^k is the control vector and Ψ^k is terminal cost functions of group k. Let group k's admissible control set be U^k. The admissible control combination $(\hat{u}^1, \hat{u}^2, \cdots, \hat{u}^m)$ is said to be a Nash equilibrium solution if it satisfies the inequalities

$$J^k\left(0, x, u^k\right) \leq J^k(0, x, \langle \hat{u}|k \rangle), \quad 1 \leq k \leq m \tag{3}$$

2.3 Reliability Constraint

On the basis of confidence game-paly, in this phase, we will find out which anchor nodes and corresponding multi-hop links should be employed so that the utilization in the UWSNs localization is reliable.

Step 1: Assume that an ordinary node initiates an inviting request to its neighbor or multi-hop anchor nodes, namely set X. If the members in set X are free, they respond the joining ACK to ordinary node x. Otherwise, they respond the abandoning ACK to x. Then the loose game domain is created, and the anchors that send the joining ACK will become the game players.

Step 2: The ordinary node x announces the localization reference information to all the players in the game domain. Each player in the game domain receiving the announcement calculates its payoff J_k. As a game player, there are two actions < keep, reject > to enforce for the ordinary node. At the first time of the play, all players make their action based on their payoffs, i.e. if $J_k > 0$, broadcasting a 'keep' message to all players, or else broadcasting 'reject' message.

Step 3: Each player in the game domain will increasingly receive action messages from other game domain players. Once there is at least one player's payoff that is larger than it has, it will broadcast action messages with 'reject' if its previous message is 'keep'. As the game repeats, there will be only one player with 'keep' remaining. And the ordinary node will adapt to it.

Step 4: The ordinary node x re-estimates whether it is also needs other anchor nodes. If it still has to choose anchors, go to step 2, otherwise it broadcasts the message to all players to dismiss the game domain.

2.4   Mobility Constraint

The contractor C3 associated with the mobility of the to-be-localized node as well as the corresponding anchor nodes. A mobility constraint, i.e. the determination problem of location domain, is defined by a set of k constraints, $f_1, f_2, \cdots, f_k$. Each constraint is actually a discriminant relation $f_i : d_{ui} - c_{ui} \leq \|W_u - W_i\|_2 \leq d_{ui} + c_{ui}$ linking the corresponding anchor nodes coordinates as well as the to-be-localized node.

This constraint raises an interesting problem. The known mobility of sensor node may change with time due to the complex underwater environment. In such a case, choosing one fixed error distribution is unreliable. The constraint intersection will risk overbounding or underbounding, where the former produces a loose intersection, whereas the latter results in a void admissible space. The complicated intersection of the feasible solutions makes the exact computation infeasible.

2.5   Location Estimation

The distance from X_u to X_i can be denoted by $\zeta_{u1}^I = \left[\zeta_{u1}^{I-}, \zeta_{u1}^{I+}\right]$, where ζ_{u1}^{I-} and ζ_{u1}^{I+} are the corresponding minimal and maximal bounds. The set of the intervals regarding to X_u actually is the set of the constraints $f_1, f_2, \cdots, f_k$. Solving the CSP problem in an interval analysis approach consists of finding the intersection that contains all possible solutions. Consider two intervals ζ_{u1}^I and ζ_{u2}^I, their intersection can be computed by

$$\zeta_{u1}^I \cap \zeta_{u2}^I = \left[max\{\zeta_{u1}^{I-}, \zeta_{u2}^{I-}\}, \ min\{\zeta_{u1}^{I+}, \zeta_{u2}^{I+}\}\right] \tag{4}$$

Regarding the coordinates of all sub-boxes' centers as samples of X_u, we can get a sample set $F_u = \{\Theta_1, \Theta_2, \cdots, \Theta_n\}$, and the centre of Θ_n can be found by $\zeta_n^* = \left(\zeta_n^- + \zeta_n^+\right)/2$. Then the optimum point estimate, i.e. the desired coordinates of ordinary node X_u, can be obtained by

$$\hat{W}_u = \underset{W_u}{\arg\min} \sum_{i=1}^{k} \left(\left\|\zeta_n^* - W_i\right\|_2 - d_{ui}\right)^2 \tag{5}$$
$$subject \ to \ W_u \in F(w)$$

3   Performance Evaluation

In our simulation experiments, 200 nodes with adjustable transmission range R are randomly distributed in a large-scale three dimensional region with a size of $2000 \times 2000 \times 200$ (see Fig. 1). We control the density and connectivity of the network by changing the transmission range while keeping the area of deployment the same. Different effective anchor percentages are considered in our simulation. In addition, we simulated the DV-distance localization scheme for comparison.

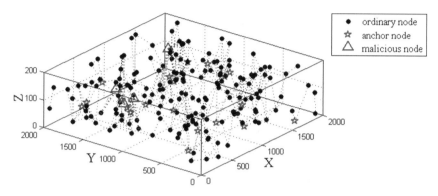

Fig. 1. Topology of UWSNs

Figure 2 plots the relationship between the average localization error (ALE) and network connectivity. It should be noted that our scheme can achieve relatively high localization accuracy even with low network connectivity than the DV-distance method. We can also observe that our scheme can maintain steady ALE below $0.6R$ varying with the network connectivity when the percentage of malicious node is 10%. The performance indicates that our scheme can achieve better localization accuracy even restricted by malicious and non-ideal network conditions.

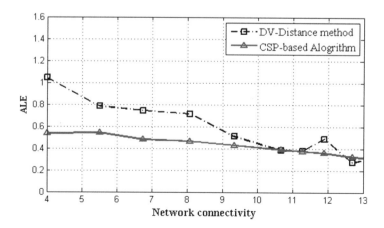

Fig. 2. Average localization error vs. network connectivity

Figure 3 plots the relationship between the percentage of high-error nodes and network connectivity, with the anchor percentage is 10% and malicious nodes percentage is 5%. For our scheme, we can see that the percentage of high-error nodes (larger than $0.5R$) is well below 40%, with the network connectivity varying from 4 to 14. However, the percentage of high-error nodes of DV-distance is always higher than us, especially in the condition of low network connectivity. This suggests that in malicious networks, the traditional method will suffer the high localization error under the influence of malicious nodes, but our scheme can achieve reliable and high localization accuracy performance.

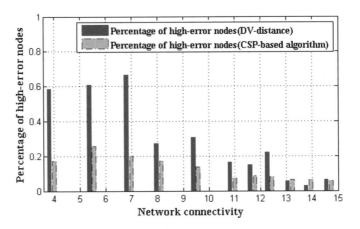

Fig. 3. Percentage of high-error nodes vs. network connectivity

4   Conclusion

In this paper, we proposed a novel Multiple Constraints Satisfaction Based Reliable Localization Algorithm for mobile UWSNs. We transformed the reliable localization problem into a multiple Constraints Satisfaction Problem. In the CSP framework, we mainly integrated three kinds of constraints, i.e. confidence constraint, mobility constraint, and reliability constraint. The advantage of our framework is that both the malicious anchor nodes and the true anchor nodes can be treated as information uncertainty and casted into information game process. Then, the localization issues can be tackled in the constraint satisfaction problem framework. Simulation results show that our algorithm is an effective and efficient approach to mobile UWSNs.

Acknowledgment. The authors are grateful to the anonymous reviewers for their industrious work and insightful comments. This work was supported by the National Natural Science Foundation of China (grant no. 61501488).

References

1. Erol-Kantarci, M., Mouftah, H.T., Oktug, S.: A survey of architectures and localization techniques for underwater acoustic sensor networks. IEEE Commun. Surv. Tutor. **13**(3), 487–502 (2011)
2. Tan, H., Diamant, R., Seah, W.K.G., Waldmeyer, M.: A survey of techniques and challenges in underwater localization. Ocean Eng. **38**(14–15), 1663–1676 (2011)
3. Mao, G., Fidan, B., Anderson, B.: Wireless sensor network localization techniques. Comput. Netw. **51**, 2529–2553 (2007)
4. Franceschini, F., Galetto, M., Maisano, D., Mastrogiacomo, L.: A review of localization algorithms for distributed wireless sensor networks in manufacturing. Int. J. Comput. Integr. Manuf. **22**, 698–716 (2009)
5. Diamant, R., Lampe, L.: Underwater localization with time-synchronization and propagation speed uncertainties. IEEE Trans. Mob. Comput. **12**(7), 1257–1269 (2013)
6. Han, G., Xu, H., Duong, T.Q., Jiang, J., Hara, T.: Localization algorithms of wireless sensor networks: a survey. Telecommun. Syst. **52**(4), 2419–2436 (2013)
7. Mesmoudi, A., Feham, M., Labraoui, N.: Wireless sensor networks localization algorithms: a comprehensive survey (2013). arXiv preprint: arXiv:1312.4082
8. Jaulin, L., Kieffer, M., Didrit, O., Walter, E.: Applied Interval Analysis, with Examples in Parameter and State Estimation, Robust Control and Robotics. Springer, London (2001)

A Design of Kernel-Level Remote Memory Extension System

Shinyoung Ahn[1], Eunji Lim[1], Wan Choi[1], Sungwon Kang[2], and Hyuncheol Kim[3(✉)]

[1] ETRI, Yuseong-Gu, Daejeon, 34129, South Korea
syahn@etri.re.kr
[2] KAIST, Yuseong-Gu, Daejeon, 34141, South Korea
[3] Namseoul University, Cheonan, South Korea
hckim@nsu.ac.kr

Abstract. Big data scientists are struggling with a memory capacity of the computer system because of the expansion speed of memory capacity has not kept up with the increasing requirement of large memory applications. Even though large memory application vitally requires big memory system, big memory machine has been too expensive for many researchers and students. By the way, as very high-speed networking technologies such as Infiniband EDR(100 Gbps) have been developed, approaches to utilize remote memory has been considered as a cost effective way to run large memory applications in the HPC cluster environment. For the general users of HPC cluster system who want to run large memory application with administrator's support, we suggest a kernel-level remote memory extension system. We designed a remote memory extension system which mapped remote memory pages to the virtual address space of the large memory application process. The system includes three components such as remote memory consumer, Integrated Memory Manager, and memory provider. We developed a kernel-level remote memory extension device and achieved 4 × improvement of the latency of page fault handling on remote memory.

Keywords: Remote memory · Large memory application · Memory extension · Remote memory library · Page fault handling · Remote memory consumer integrated memory manager · Memory provider

1 Introduction

Enterprises, academy, researcher are struggling with a memory capacity of the computer system because of the expansion speed of memory capacity has not kept up with the increasing requirement of large memory applications. Large memory applications such as IMDB(in-memory database), IMDG(in-memory data grid), denovo assembly application in the human genome sequencing area, business application acceleration, big data analytics, and large-scale scientific calculation are increasing exponentially. Such applications need cost-effective memory scale-out system which provides big memory because buying big memory system is a too expensive way.

© Springer Nature Singapore Pte Ltd. 2018
K.J. Kim et al. (eds.), *IT Convergence and Security 2017*,
Lecture Notes in Electrical Engineering 449,
DOI 10.1007/978-981-10-6451-7_13

As time passes, networking technologies have been improved. Modern networking technologies in LAN/SAN area such as Infiniband, Quadrics, and Myrinet have very low latency and very high bandwidth. Especially, Infiniband EDR(Enhanced Data Rate) shows a few microseconds level end-to-end latency and achieves 100 Gbps data transfer rate. This performance can be compared with a bandwidth of PCI-E(Gen3 x16, 128 Gbps) used as internal system bus of the computer. These technologies also support Remote Direct Memory Access(RDMA) operation mode in which CPU don't need to coordinate transferring between local memory and remote memory. The RDMA feature reduces the number of memory copy between user-level and kernel level. The feature improves the access latency and the bandwidth to the remote memory dramatically. The access latency to remote memory is a few orders of magnitudes faster than that of local disk. Therefore, there have been many trials to use remote memory like local memory, local block, and local file system.

The concept of approaches utilizing remote memory as local memory is adding an additional layer between main memory and local disk in the memory hierarchy. This makes sense because the latency of memory is faster 10,000~100,000 times rather than HDD's. If we can achieve a few microseconds as the access latency of remote memory which is faster 100~1,000 times than HDD, it would be the very reasonable solution. In HPC cluster computing environment, the benefit of utilizing remote memory is that we can execute large memory application without additional HW cost by harvesting idle memory on remote nodes. In this paper, we don't talk about memory harvesting issues. We assume that we can use abundant idle memory.

The majority of previous approaches to extending virtual address space to remote memory have preferred an implicit method which conceals the fact that users actually use remote memory when they run their large memory applications. This method requires no special re-compiling or linking with the special library but requires the modification of kernel's memory management core.

In this paper, we designed a kernel-level remote memory extension system with user-level APIs by adding a brand-new component, Remote Memory Extension Device to our existing user-level solution [19–21]. The system provides application developer APIs for remote memory allocation and access. The remainder of this paper is organized as follows. In Sect. 2, we summarize and discuss the related work. In Sect. 2, we describe the details of kernel-level remote memory extension system. Finally, in Sect. 4, we conclude by talking about our current and future work.

2   Related Works

Comer suggested remote paging mechanism based on the remote memory model which consists of several client machines and one or more dedicated remote memory server [1]. In the remote paging mechanism, clients use remote memory rather than local disk as a backing storage. Clients move memory pages to the remote memory server when the client machines exhaust their local memory [1].

Since Comer's suggestion, there have been many research works on the utilization of remote memory. Such research works have been approached from different perspectives because of the variety of research goals and network characteristics [2]. Various approaches tried to use remote memory just like a local memory [1, 3–8], or as a cache for local or distributed storage [2, 9–15], or as a storage [2, 16].

Anderson et al. classified the approaches to utilize remote memory like local memory based on the implementation layer/methods: (1) explicit program management, (2) user-level, (3) device-driver (4) kernel modification and (5) Network interface [4]. Explicit program management requires the programmer to coordinate all data movement to/from the remote memory [5]. User-level implementation method requires the programmer modify their code using a new malloc for remote memory allocation [4]. Device-driver method replaces the swap device with a new device which sends the pages to remote memory [8, 17]. Kernel modification method modifies the virtual memory(VM) subsystem. It shows the highest performance but very lower portability [1, 3]. Network Interface method is HW-level method which replaces memory controller with some sort of chip. This is most unportable and hardest to implement.

User-level approach more portable than device driver approach and doesn't require root privilege or administrator's support. However, device driver approach doesn't require user-code modification and kernel source modification [4, 8]. Even though device-driver method has been the majority of research, the method requires root privilege. If the administrator of cluster system does not support to install the device driver for remote memory extension, then most of the non-privileged user can't use remote memory. To decrease the communication cost of TCP/IP networking, RDMA enabled network technologies such as Infiniband, Quadrics, and Mirinet became to be used [8, 18].

Our approach is mixing (2) user-level and (3) device driver method to improve the performance of page fault handling. However, our new device can't be used as a swap device.

3   Kernel-Level Remote Memory Extension System

Figure 1 shows an architecture of kernel-level remote memory extension system and behaviors. The remote memory extension system requires three types of computing node: remote memory consumer node, memory manager node, and memory provider node. Each node runs one or more of components of the remote memory extension system: Remote Memory Consumer, Remote Memory Extension Device, Integrated Memory Manager, and Memory Provider.

The Memory Provider in the memory provider node allocates some part of its own idle memory as a granted memory. After allocation, it configures the granted memory readable or writable via RDMA operation by register the virtual address of allocated memory to Infiniband HCA(Host Channel Adaptor). Then, it registers the granted memory to Integrated Memory Manager and eventually provides the granted memory block to Remote Memory Consumer by the administration of the Integrated Memory Manager [19].

The Integrated Memory Manager configures and builds a remote memory(RM) pool composed of the granted memory blocks which the Memory Providers

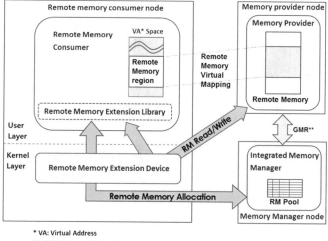

* VA: Virtual Address
** GMR: Granted Memory Registration

Fig. 1. Remote memory extension system and behaviors

registered. It also assigns the granted memory block to Remote Memory Consumer. The Integrated Memory Manager and Memory Provider can be grouped and named as remote memory extension servers [20, 21].

The Remote Memory Consumer link a library(Remote Memory Extension Library) which enables it to use a remote memory service from the remote memory servers. The Remote Memory Consumer requests allocation of a remote memory from the remote memory pool on the Integrated Memory Manager through the Remote Memory Extension Library, which maps the allocated remote memory blocks in its virtual address space(refer Fig. 1. Remote Memory Virtual Mapping). The Remote Memory Consumer can use the remote memory like its own local memory because the Remote Memory Extension Device hides the details of accessing the granted remote memory.

We added a new component, Remote Memory Extension Device, to existing user-level remote memory extension system [21]. It is a new virtual device which executes remote memory allocation, read/write, and de-allocation instead of user-level Remote Memory Library [21]. The Remote Memory Extension Library in Fig. 1 just provides Application Programing Interface(API) to interact with it.

3.1 The Mechanism of Remote Memory Access

To use remote memory, first, the remote memory should be mapped on virtual address space of Remote Memory Consumer process as shown in the Fig. 2. Second, page fault handling for remote memory region is required whenever consumer process try to access the remote memory region.

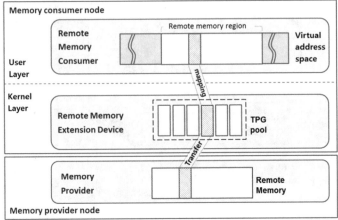

Fig. 2. A mechanism of remote memory access

Figure 2 shows a mechanism of remote memory access. As explained above, Remote Memory Consumer process can use remote memory with the help of Remote Memory Extension Device and Remote Memory Provider. Remote Memory Providers grant the access to their own memory by Remote Memory Consumer. Remote Memory Extension Device handles the kernel-level page fault for the remote memory region. It reads real data from remote memory page to the temporary page(TPG) of consumer node via RDMA Read and dynamically maps the TPG to virtual address space of consumer process. In fact, the physical memory pages of Remote Memory Extension Device are allocated to Remote Memory Consumers. The role of TPG is similar to that of page cache in the kernel. The difference between TPG and page cache is that page cache use free memory as much as it can use, but the size of TPG pool is restricted because we need to minimize the overhead to serve remote memory extension to consumer process. Because the TPGs are very limited, it should be recycled efficiently. Therefore, recycled TPGs may be mapped to different addresses of the remote memory region.

3.2 User-Level API

Table 1 shows user-level APIs for remote memory extension. The rminit() make a connection to Integrated Memory Manager and register the Remote Memory Consumer

Table 1. User-level APIs for remote memory extension.

APIs	Description
int rminit();	Initialize remote memory extension library
int rmterminate();	Terminate remote memory extension library
void *rmalloc(size_t size);	Allocate remote memory
void rmfree(void *ptr);	Deallocate the allocated remote memory

to the Integrated Memory Manager. On the contrary, the rmterminate() deregister the Remote Memory Consumers and disconnect the connection to Integrated Memory Manager. After initialization, the programmer can allocate remote memory via rmalloc() and may release the memory via rmfree() after using remote memory.

3.3 Page Fault Handling Method of Remote Memory

In general, the behavior of page fault handling is that kernel allocates physical memory page and map the page to the virtual address of the user process by updating page table of the process. It is not allowed for the user-level code to allocate physical memory page and update the page table of the process. Therefore, for the user-level remote memory extension system, we suggested a tricky way use signal handling mechanism of segmentation fault signal(SIGSEGV) by setting the remote memory region not readable and not writeable. By doing so, we could handle the page fault at user-level by changing the action taken by the user process on receipt of SIGSEGV signal [19–21]. In this paper, we design to use a new Linux device, Remote Memory Extension Device, which is not a physical device but virtual devices. To support the allocation of remote memory in the device driver for Remote Memory Extension Device, we developed mmap() function as file operations of the device driver. We also specify a page fault handling function for each memory allocation(by setting vm_area_struct->vm_ops->fault()) in the mmap() function. Linux kernel calls the specified page fault handler when a page fault occurs in the remote memory region. The page fault handler function selects a TPG from TPG pool and read real data from remote page to the TPG, and then return control to Linux kernel. Linux kernel maps the TPG to the virtual address of the user process by calling kernel API to update page tables.

3.4 Temporal Page Management

The TPG is very limited resource, it is very important to recycle TPG efficiently. The system manage TPG by using four double linked lists: (1) Unused, (2) Active, (3) Inactive dirty, and (4) Inactive clean. The Unused list holds all the unallocated TPGs. The Active list holds all the TPGs allocated to consumer thread and used by consumer thread. The Inactive dirty list holds all the modified TPGs out of inactivated TPGs. The Inactive clean list holds all the clean or unmodified TPGs out of inactivated TPGs. The TPG lifecycle is managed through 6 operations. When a page fault occurs in the remote memory region, the page fault handler acquires a TPG from the Unused list(1.TPG Allocation). The TPG_manager inactivates some TPGs of the Active list(2.TPG Inactivation) for recycling. The inactivated TPGs are selected from the head of the Active list via LRU policy. It saves the data of the modified TPGs to the remote page(3.TPG Laundering). After laundering, it periodically moves all the TPGs in the Inactive clean list to the tail of the Unused list(4.TPG Reclaiming). When Remote Memory Consumer revisit the already mapped pages, it move a revisited TPG of Inactive dirty to the Active list(5.TPG Reactivation) or moves the revisited TPG to the tail of the Active list(TPG Reordering) because the Active list should be ordered by least recently used(LRU) time [20].

4   Conclusion

In this paper, we designed and partially implemented a kernel-level Remote Memory Extension System. The system provides user-level API for Remote Memory Consumer applications, kernel-level page fault handling, dynamic TPG management functions, reading/writing data from/to the remote memory, and prefetching functions.

At this writing, we do not finish all implementation of Remote Memory Extension Device because we did not finish the kernel-level Infiniband communication module to support communication with other memory provider and Integrated Memory Manager, but we finished developing kernel-level page fault handling mechanism.

Our previous study on user-level remote memory extension system referred to the need of kernel-level page fault handling because of the latency of user-level page fault handling is about 3.5~3.6 us for random access pattern for reading remote memory region [21]. We had expected that kernel-level page fault handler might decrease the page fault handling time to about 1~2 us. By developing the kernel-level device driver for Remote Memory Extension Device, we achieved 4 × improvement of the latency of page fault handling: 0.89 us per page. We experimented this with same machine and condition. If we finish developing communication module, we might get a cost-effective solution for large memory application. Now, we also have a mission to optimize the Infiniband RDMA operations. It would improve the expected average read latency to under 10 us.

Our system may not yet be a good solution for latency sensitive application which requires large memory. Our suggestion might be a good candidate for someone who needs large data sharing among cluster machines because remote memory can be very fast shared memory.

Acknowledgments. This work was supported by Institute for Information & Communications Technology Promotion(IITP) grant funded by the Korea government(MSIP) (No. 2016-0-00087, Development of HPC System for Accelerating Large-scale Deep Learning).

References

1. Comer, D.: A new design for distributed systems: the remote memory model. In: Proc. the USENIX Summer Conference. Anaheim, California, pp. 127–135 (1990)
2. Roussev, V., Richard III, G.G., Tingstrom, D.: dRamDisk: efficient RAM sharing on a commodity cluster. In: Proceedings of 25th IEEE International Performance, Computing, and Communications Conference (IPCCC 2006), Phoenix, Arizona, pp. 193–198 (2006)
3. Iftode, L., Petersen K., Li, K.: Memory servers for multicomputers. In: Proceedings of IEEE COMPCON 1993 Conference (1993)
4. Anderson, E., Neefe, J.: An Exploration of Network RAM. Technical report CSD-98-1000, UC Berkley (1998)
5. Koussih, S., Acharya, A., Setia, S.: Dodo: a user-level system for exploiting idle memory in workstation clusters. In: Proceedings of 8th IEEE International Symposium on High-Performance Distributed Computing (1999)
6. McDonald, I.: Remote paging in a single address space operating system supporting quality of service. Technical report, Department of Computer Science, University of Glasgow (1999)

7. Jeegou, Y.: Implementation of page management in Mome, a User-Level DSM. In: Proceedings of 3rd IEEE International Symposium on Cluster Computing and the Grid (CCGRID 2003), Tokyo, Japan (2003)

8. Liang, S., Noronha, R., Panda, D.K.: Swapping to remote memory over infiniband: an approach using a high performance network block device. In: Proceedings of IEEE Cluster Computing (2005)

9. Dahlin, M., Wang, R., Anderson, T.E., Patterson, D.A.: Cooperative caching: Using remote client memory to improve file system performance. In: Proceedings of Operating Systems Design and Implementation (1994)

10. Zhao, M.: Dynamic policy disk caching for storage networking, IBM Research report. Publication (2006)

11. Liu, J., Huang, W., Abali, B., Panda, D.K.: High performance VMM-bypass I/O in virtual machines. In: Proceedings of the Annual Conference on USENIX 2006 Annual Technical Conference (USENIX ATC 2006) (2006)

12. Chen, H., Wang, X., Wang Z., Wen, X., Jin, X., Luo, Y., Li, X.: REMOCA: hypervisor remote disk cache. In: Proceedings of IEEE International Symposium on Parallel and Distributed Processing with Applications, pp. 162–169 (2009)

13. Oleszkiewicz, J., Ziao, L., Liu, Y.: Parallel network RAM: effectively utilizing global cluster memory for large data-intensive parallel programs. In: Proceedings of IEEE 2004 International Conference on Parallel Processing (ICPP 2004) (2004)

14. Wang, L., Zhan, J., Shi, W.: In cloud, can scientific communities benefit from the economies of scale? IEEE Trans. Parallel Distrib. Syst. 23(2), 296–303 (2011)

15. Hanj, H., Lee, Y.C., Shin, W., Jung, H., Yeom, H.Y., Zomaya, A.Y.: Cashing in on the cache in the cloud. IEEE Trans. Parallel Distrib. Syst. 23(8), 1387–1399 (2012)

16. Ousterhout, J., Agrawal, P., Erickson, D., Kozyrakis, C., Leverich, J., Mazi`eres, D., Mitra, S., Narayanan, A., Parulkar, G., Rosenblum, M, Rumble, S.M., Stratmann, E., Stutsman, R.: The case for RAMClouds: scalable high-performance storage entirely in DRAM. SIGOPS Oper. Syst. Rev. 43, 92–105 (2010)

17. Newhall, T., Amato, D., Pshenichkin, A.: Reliable adaptable network ram. In: Proceedings of IEEE Cluster 2008 (2008)

18. Liu, J., Wu, J., Panda, D.K.: High performance RDMA-based MPI implementation over infiniBand. Int. J. Parallel Prog. 32(3), 167–198 (2004)

19. Ahn, S., Cha, G., Kim, Y., Lim, E.: Exploring feasibility of user-level memory extension to remote node. In: Proceedings of the 6th International Conference on Internet (2014)

20. Ahn, S., Cha, G., Kim, Y., Lim, E.: Design and implementation of user-level remote memory extension library. In: Proceedings of 18th International Conference on Advanced Communication Technology, pp. 716–721 (2015)

21. Ahn, S., Cha, G., Kim, Y., Lim, E.: A study on remote memory extension system. In: Proceedings of 18th International Conference on Advanced Communication Technology (2016)

A Comparison of Model Validation Techniques for Audio-Visual Speech Recognition

Thum Wei Seong[1], Mohd Zamri Ibrahim[1](✉),
Nurul Wahidah Binti Arshad[1], and D.J. Mulvaney[2]

[1] Faculty of Electrical and Electronic Engineering,
University Malaysia Pahang, 26600 Pekan, Pahang, Malaysia
weiseong91@hotmail.com, {zamri,wahidah}@ump.edu.my
[2] School of Electronic, Electrical and Systems Engineering,
Loughborough University, Loughborough LE11 3TU, UK
d.j.mulvaney@lboro.ac.uk

Abstract. This paper implements and compares the performance of a number of techniques proposed for improving the accuracy of Automatic Speech Recognition (ASR) systems. As ASR that uses only speech can be contaminated by environmental noise, in some applications it may improve performance to employ Audio-Visual Speech Recognition (AVSR), in which recognition uses both audio information and mouth movements obtained from a video recording of the speaker's face region. In this paper, model validation techniques, namely the holdout method, leave-one-out cross validation and bootstrap validation, are implemented to validate the performance of an AVSR system as well as to provide a comparison of the performance of the validation techniques themselves. A new speech data corpus is used, namely the Loughborough University Audio-Visual (LUNA-V) dataset that contains 10 speakers with five sets of samples uttered by each speaker. The database is divided into training and testing sets and processed in manners suitable for the validation techniques under investigation. The performance is evaluated using a range of different signal-to-noise ratio values using a variety of noise types obtained from the NOISEX-92 dataset.

Keywords: Audio-visual speech recognition · Hidden markov models · HTK toolkit · Holdout validation · Leave-one-out cross validation · Bootstrap validation

1 Introduction

This work adopts an established audio-visual speech recognition (AVSR) system that uses a range of modern techniques for feature extraction, frond-end processing, model integration, classification approaches and validation methods. Although, it would initially appear that combining two modalities (audio and visual) is likely to result in better overall system performance, many AVSR researchers have found this not to be the case in practice and this is at least partly due to poor selection of a validation technique to apply to dataset samples. Although a large body of literature exists that confirms that researchers are aware of the need to identify a suitable validation

© Springer Nature Singapore Pte Ltd. 2018
K.J. Kim et al. (eds.), *IT Convergence and Security 2017*,
Lecture Notes in Electrical Engineering 449,
DOI 10.1007/978-981-10-6451-7_14

technique that provides the most consistent and accurate estimation, no consensus has been reached. For example, one recent study found that the bootstrap validation approach was the best [1], while another claimed that leave-one-out cross validation (LOOCV) achieves the most accurate classification results [2].

In this paper, a comparison of three validation techniques (holdout, LOOCV and bootstrap) for an AVSR system is carried out. Section 2 concentrates explains the model validation techniques, Sect. 3 presents the methodology to be adopted to analyze the AVSR system and performance results of the different types of validation techniques are addressed in Sect. 4. The conclusions are discussed in Sect. 5.

2   Model Validation Techniques

This section describes the most popular validation methods for estimating AVSR recognition performance, namely the holdout method, LOOCV and bootstrap validation [3].

2.1   Holdout Method

The holdout method can be considered to be one of the most basic validation methods for result estimation. Its operation involves simply dividing the sample set into two; the first is used as a training set and the second is used as a test set, see Fig. 1. The bootstrap method performs well if the training set contains no corrupted data, but in practice corrupted data are often hard to detect among a large set of samples and, if they are not removed, poor performance results when evaluated using the testing set. Despite such drawbacks, there remains a number of applications to which the approach is well suited and there is a considerable body of research that has exploited this method [4, 5].

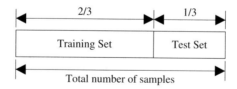

Fig. 1. Example of holdout validation distribution ratio

2.2   Leave One Out Cross Validation (LOOCV)

LOOCV is an extreme case of k-fold cross validation, where k represents the total number of samples. In k-fold cross validation, the validation process is carried out k times. In LOOCV, $k-1$ samples are used for training purposes and only a single sample is used for testing, see Fig. 2. According to Kocaguneli and Menzies [2], this technique has been shown to have low bias and is able to overcome the drawback of the holdout method in having poor performance in the presence of corrupted data. However, Kocaguneli and Menzies also found that there is no definitive solution regarding whether the holdout approach or the LOOCV method perform the better as the training and test sets used during validation are very different [2].

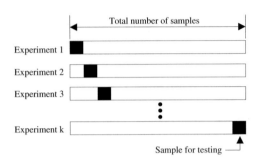

Fig. 2. Illustration diagram of leave-one-out cross validation

2.3   Bootstrap Validation

In the bootstrap model validation technique, assuming that there are N samples in the data set, then a number of samples are selected at random and are used for training, while those not selected are used for testing. The process is carried out M times and the final performance estimation is obtained by averaging the M sets of results.

Table 1 shows an example of replacement process in which the set of selected samples is given by X_1, X_2, X_3, X_4 and X_5, assuming $N = 5$ in this case. For example, in experiment set 2, then once X_2 and X_4 are selected as the test set, the training set contains X_1, X_3 and X_5, but, as two of the entries are repeated, the actual entries in the training set become X_1, X_3, X_3, X_5 and X_5. This process is carried out M times and the final validation outcome is averaged from all the experiment sets.

Table 1. Example of bootstrap validation, where the samples available are X_1, X_2, X_3, X_4, X_5

Experimental set number	Training set	Test set
1	X_1, X_2, X_3, X_4, X_5	X_4
2	X_1, X_3, X_3, X_5, X_5	X_2, X_4
3	X_1, X_1, X_2, X_2, X_4	X_3, X_5
.	.	.
.	.	.
.	.	.
M	X_1, X_3, X_3, X_3, X_3	X_2, X_4, X_5

3   Methodology

This AVSR implementation has been carried out in previous research and the contribution of this paper is principally the results of a comparison of cross validation techniques. In the previous work, Matlab R2015a [6] together with the OpenCV open source image processing library [7] was used for simulation and testing and the hidden Markov model toolkit (HTK) was used to generate and manipulate the nine states of a hidden Markov model [8]. HTK originates from the Machines Intelligence Laboratory at Cambridge University's Engineering Department [9].

A system diagram of the AVSR processes used in this work are shown in Fig. 3.

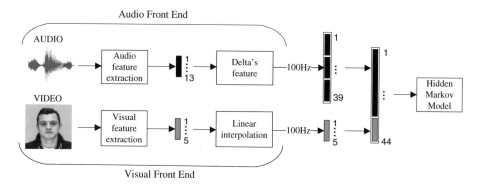

Fig. 3. AVSR processes carried out in this work

3.1   Visual Feature Extraction

The visual feature extraction techniques followed the steps from previous research [10]. It was also shown that this extraction technique is robust to head rotation and illumination changes [11]. The process is as follows. Firstly, visual information from the speaker is extracted in the form of geometrical-based features. A Viola-Jones face detection algorithm [12] was applied in which face and then mouth detection processes were carried out, as can be seen in Fig. 4. An HSV color filter was applied to differentiate the lip region [13], then border following [14] and finally convex hull techniques were used to extract the actual shape of speaker's lips.

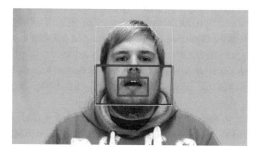

Fig. 4. Example of face and mouth detection

3.2 Audio Feature Extraction

In the literature, the mel-frequency cepstral coefficient (MFCC) and the linear prediction coefficient (LPC) are currently popular audio feature extraction techniques [15],
although a recent investigation has suggested that MFCC may be the better approach to
providing human speech features [16].

In this work, the HTK library was employed for MFCC feature extraction and a
feature vector of 39 dimensions was obtained. The vector includes dynamic feature
(delta-MFCCs and delta-delta MFCCs) as these were shown in previous work to
improve the performance of speech recognition systems [17].

4 Experimental Results

This experiments were conducted using the newly developed database known as the
Loughborough University audio-visual (LUNA-V) speech data corpus [10]. Compared
to other existing databases, the video recordings have a relatively high resolution of
1280×720 pixels, making more detailed information available to the recognition
process and so perhaps enabling improvements in the performance of AVSR systems
[18]. The database has contributions from 10 speakers (9 male and 1 female) with each
speaker providing five separate samples of uttering the English digits from 'zero' to
'nine'. Varies types of noise were applied at a number of different signal-to-noise ratios
(SNRs) in order to test the robustness of the AVSR system.

For each of the holdout, LOOCV and bootstrap validation techniques, a range of
noise types with SNR values in the interval 25 dB to − 10 dB relative to the speech
signals were added. In the results presented here, NOISEX-92 was used to supply the
noise signals and the types of noise used are known as 'white', 'babble' and 'factory1'
in the database archive.

The results of the 'white' noise experiments are shown in Table 2. White noise
contains contributions for all frequencies in the audible sound range and is known to
have a more profound effect on the perceived audibility of certain words, including 'six'
which is not strongly sounded. Furthermore, the word is often difficult for AVSR
systems to detect as its production requires only minimal lip movements. As can be seen
in Table 2, LOOCV achieved better accuracy in the AVSR tests than other two validation techniques when operating in the SNR range from 20 dB to 0 dB and holdout
only performed well on clean audio and when the strength of the noise signal was greater
than that of the speech. Bootstrap consistently performed the worst of the three methods.

Table 3 shows the recognition results when the speech signals were corrupted by
'babble' noise, which was captured from 100 people talking in a canteen. The digit
'seven' was found to be the word most adversely affected in the recognition results.
Apart from at very low noise levels where its performance was only slightly worse than
holdout, LOOCV achieved the best performance. Again, bootstrap performed the worst
of the three methods.

In Table 4, the noise used to contaminate the audio signal was 'factory1' noise,
recorded in the proximity of plate-cutting and electrical equipment. Again, except in
cases where the noise content was very low or very high, LOOCV achieved the greatest

Table 2. Word accuracy of the validation techniques when 'white noise' is added to the speech signals. Figures in bold type show the technique producing the best result at each SNR value.

SNR (dB)	Holdout (%)	LOOCV (%)	Bootstrap (%)
Clean	**100.0**	99.4	98.1
25	95.5	**97.6**	94.2
20	94.0	**94.6**	90.0
15	84.0	**86.2**	80.7
10	72.0	**73.2**	69.0
5	58.5	**60.4**	56.5
0	46.0	**50.0**	47.0
−5	**44.0**	41.4	40.2
−10	**37.0**	36.2	35.6

Table 3. Word accuracy of the validation techniques when 'babble noise' is added to the speech signals. Figures in bold type show the technique producing the best result at each SNR value

SNR (dB)	Holdout (%)	LOOCV (%)	Bootstrap (%)
Clean	**100**	99.4	98.1
25	**99.5**	99.2	97.4
20	**99.0**	**99.0**	96.5
15	96.5	**97.0**	93.7
10	90.5	**91.0**	87.1
5	79.0	**81.2**	77.7
0	64.0	**67.4**	63.4
−5	49.0	**50.6**	48.8
−10	43.0	**43.5**	42.1

Table 4. Word accuracy of the validation techniques when 'factory1 noise' is added to the speech signals. Figures in bold type show the technique producing the best result at each SNR value.

SNR (dB)	Holdout (%)	LOOCV (%)	Bootstrap (%)
Clean	**100**	99.4	98.1
25	**99.5**	99.2	97.2
20	97.5	**98.8**	95.8
15	92.5	**96.0**	92.2
10	86.5	**89.2**	83.9
5	75.0	**78.2**	73.1
0	59.0	**62.2**	58.8
−5	45.0	**48.6**	46.2
−10	**41.5**	39.4	39.3

accuracy compared to the holdout and bootstrap validation methods. The performance of the bootstrap validation method was again somewhat worse across the full range of SNR values.

Overall, the bootstrap methods exhibited the worst accuracy across the full range of SNR values. The holdout method performed particularly well when there was no noise contamination and when little noise was present. It is known that the holdout method is particularly susceptible to the presence of corrupted samples and if any were present during training, this could have led to a biased result. Furthermore, from previous work, the holdout method is known to be more sensitive to the quantity of data used in training, and if the number of values used was insufficient this may have also affected the accuracy available from the predictive model [19].

5   Conclusion

This paper has presented a comparison of the speech recognition results generated by a range of validation techniques when tested on the word accuracy of an AVSR operating in noisy environments. The work used an existing AVSR system that attempted to recognize English digits using a combination of speech and high-definition video sequences from the LUNA-V data corpus. Based on the experiment results, the LOOCV technique achieved a slightly better performance compared to the holdout and bootstrap validation methods.

Acknowledgments. This work was supported by Universiti Malaysia Pahang and funded by the Ministry of Higher Education Malaysia under FRGS Grant RDU160108.

References

1. Kokkinidis, K., Panagi, A., Manitsaris, A.: Finding the optimum training solution for Byzantine music recognition—a Max/Msp approach, pp. 6–9 (2016)
2. Kocaguneli, E., Menzies, T.: Software effort models should be assessed via leave-one-out validation. J. Syst. Softw. **86**(7), 1879–1890 (2013)
3. Kohavi, R.: A study of cross-validation and bootstrap for accuracy estimation and model selection. Int. Jt. Conf. Artif. Intell. **14**(12), 1137–1143 (1995)
4. Receveur, S., Scheler, D., Fingscheidt, T.: A Turbo-decoding weighted forward-backward algorithm for multimodal speech recognition, pp. 179–192. Springer, Berlin (2014)
5. Ibrahim, M.Z., Mulvaney, D.J., Abas, M.F.: Feature-fusion based audio-visual speech recognition using lip geometry features in noisy enviroment. ARPN J. Eng. Appl. Sci. **10**(23), 17521–17527 (2015)
6. Matlab, M.: R2015a. http://www.mathworks.com/products/matlab
7. Bradski, G., Kaehler, A.: Learning OpenCV: Computer Vision with the OpenCV Library. O'Reilly Media Inc., Newton (2008)
8. Young, S., Evermann, G., Gales, M., Hain, T., Kershaw, D., Liu, X., Moore, G., Odell, J., Ollason, D., Povey, D., et al.: The HTK book (for HTK version 3.4). Camb. Univ. Dep. Eng. **2**(2), 2–3 (2006)

9. Pawar, G.S., Morade, S.S.: Isolated english language digit recognition using hidden markov model toolkit. Int. J. Adv. Res. Comput. Sci. Softw. Eng. **4**(6), 781–784 (2014)
10. Ibrahim, M.Z.: A Novel Lip Geometry Approach for Audio-Visual Speech Recognition. Loughborough University, Loughborough (2014)
11. Ibrahim, M.Z., Mulvaney, D.J.: Robust geometrical-based lip-reading using hidden Markov models. In: IEEE EuroCon 2013, pp. 2011–2016, July 2013
12. Viola, P., Jones, M.: Rapid object detection using a boosted cascade of simple features. Comput. Vis. Pattern Recognit. **1**, I–511–I–518 (2001)
13. Kakumanu, P., Makrogiannis, S., Bourbakis, N.: A survey of skin-color modeling and detection methods. Pattern Recognit. **40**(3), 1106–1122 (2007)
14. Li, H., Greenspan, M.: Model-based segmentation and recognition of dynamic gestures in continuous video streams. Pattern Recognit. **44**(8), 1614–1628 (2011)
15. Chauhan, K., Sharma, S.: A review on feature extraction techniques for CBIR system. Signal Image Process. An Int. J. **3**(6), 1–14 (2012)
16. Tripathy, S., Baranwal, N., Nandi, G.C.: A MFCC based Hindi speech recognition technique using HTK Toolkit. In: 2013 IEEE 2nd International Conference on Image Information Processing, IEEE ICIIP 2013, pp. 539–544, January 2016
17. Wahid, N.S.A., Saad, P., Hariharan, M.: Automatic infant cry pattern classification for a multiclass problem. **8**(9), 45–52 (2016)
18. Chitu, G., Rothkrantz, L.J.M.: Building a data corpus for audio-visual speech recognition. **1**, Movellan 1995 (2007)
19. Tantithamthavorn, S., Mcintosh, A., Hassan, E., Matsumoto, K.: An empirical comparison of model validation techniques for defect prediction models. IEEE Trans. Softw. Eng. **5589**, 1–16 (2016)

Multi-focus Image Fusion Based on Non-subsampled Shearlet Transform and Sparse Representation

Weiguo Wan[1] and Hyo Jong Lee[1,2](✉)

[1] Division of Computer Science and Engineering,
Chonbuk National University, Jeonju, Korea
`wanwgplus@gmail.com`, `hlee@chonbuk.ac.kr`
[2] Center for Advanced Image and Information Technology,
Chonbuk National University, Jeonju, Korea

Abstract. To overcome the artifact phenomenon caused by the incomplete registration of the source images, a new multi-focus image fusion approach is proposed based on sparse representation and non-subsampled shearlet transform (NSST). Firstly, the source images are decomposed to low- and high-frequency coefficients by NSST. The sparse representation is then adopted to fuse the low-frequency coefficients. For the high-frequency coefficients, a maximum sum-modified-Laplacian (SML) rule is put forward to merge them. Finally, the resultant image is obtained by the inverse NSST on the fused coefficients. Experimental results indicate that the proposed method can achieve satisfied effect compared with various existing image fusion methods.

Keywords: Multi-focus image fusion · Non-subsample shearlet transform · Sparse representation · Sum-modified-Laplacian

1 Introduction

Image fusion aims at merging the multiple images with complementary and redundant information to one image which can better interpret the current scene than single source image and more suitable for people and machine perception [1]. Multi-focus image fusion is a main branch of image fusion.

In recent years, many image fusion methods have been proposed, in which the multiscale analysis based method is one of the most popular fusion methods due to its multi-resolution and multi-direction characters. Classical multiscale transforms mainly include pyramid transform and wavelet transform. However, these transforms are limited on the direction selection, and not good at representing the singular features in images. In order to better represent high order singular features, more effective multiresolution geometric analysis tools, such as curvelet transform [2] and contourlet transform [3] were proposed. These transforms are anisotropic and have good directional selectivity. However, these transform lack of shift-invariant, the fused results may be affected by the noise or mis-registration of source images. To overcome this disadvantage, Cunha et al. [4] proposed non-subsampled contourlet transform (NSCT)

© Springer Nature Singapore Pte Ltd. 2018
K.J. Kim et al. (eds.), *IT Convergence and Security 2017*,
Lecture Notes in Electrical Engineering 449,
DOI 10.1007/978-981-10-6451-7_15

which is an improved version of contourlet transform. But the computational complex of NSCT is high, lead to the fusion process take too much time.

The shearlet transform [5] is a new multi-scale transform method; compared with the contourlet transform and NSCT, the shearlet transform does not have restrictions on the number of directions, and its inverse transform only requires a summation of the shearing filters rather than an inversion of the directional filter banks. Thus, the implementation of the shearlet transform is computationally more efficient. However, akin to the traditional multiresolution transform, the shearlet transform is not shift-invariant. Therefore, the non-subsampled shearlet transform (NSST) was proposed, as this transform can obtain more information from the source images, and it reduces the pseudo-Gibbs phenomenon effectively. Moreover, the computational complexity of the NSST is lower than that of the NSCT; therefore, the NSST is more suitable for the image fusion process [6]. A sound multiscale transform based image fusion method not only relies on the transform approaches, but also depends on the coefficient selection rules.

Sparse representation [7] is a hotspots research in machine learning and computer vision fields in recent years, it decomposes and represents the original signal with nonzero coefficients as few as possible using over-complete dictionary. Motivated by this, we propose a novel multi-focus image fusion method based on NSST and sparse representation. After NSST multiscale decomposing, the obtained low-frequency coefficients of the source images are usually not sparse, fuse them with sparse representation can effectively extract the saliency information of the low-frequency sub-band images. As to the sparse high-frequency coefficients, in view of the correlation of the neighbor information, the local sum-modified-Laplacian (SML) is adopted as the selection rule. Several experiments are tested to compare the proposed algorithm with several existing fusion methods. The fused results demonstrate the superiority of the proposed method.

2 Relevant Work

2.1 NSST Image Decomposition

Although, the theory of NSST is described in [5] in detail, the main idea of NSST is briefly explained for reader. NSST is composed of two phases including multi-scale decomposition and multi-directional decomposition. In the multi-scale decomposition process, the non-subsampled Laplacian pyramid transform (NSLP) is utilized, thus it has superior performance in terms of shift-invariance. After j level scale decomposition, an image can be decomposed into $j + 1$ sub-bands with the same size of the source image in which one sub-band image is the low frequency component and other m images are the high frequency sub-band images. In the multi-directional decomposition process, the realization is via improved shearlet filters. These filters are formed by avoiding the subsampling to satisfy the property of shift-invariance. Shearlet filters allow the direction decomposition l with stages in high frequency images from NSLP at each level and produce $2^l + 2$ directional sub-images with the same size as the source image. The NSST is a fully shift-invariant, multi-scale and multi-directional expansion.

2.2   Sparse Representation

Sparse representation addresses the essential sparsity of signals according to the characteristic of human visual system. It is a very powerful image modeling technique which has been successfully used in many image processing applications, such as image denoising and super-resolution. The basic idea behind SR is to assume that a natural signal can be well approximated by a sparse linear combination of atoms with respect to a dictionary, i.e., $x \approx D\alpha$, where $x \in R^n$ is the signal, $\alpha \in R^m$ is the sparse coefficients vector, and $D \in R^{n \times m}(n < m)$ is an over-complete dictionary, which contains m prototype signals be referred to atoms. The goal of sparse representation is to calculate the sparsest α for an input signal x with a given dictionary D through the following optimization problem:

$$\min_{\alpha} ||\alpha||_0 \ \ s.t. \ \ ||y - D\alpha||_2^2 \le \varepsilon^2 \tag{1}$$

where $\varepsilon > 0$ is the error tolerance and the sparsity of vector α is often measured by its $l_0 - norm$, which counts the number of nonzero entries. Some pursuit algorithms have been proposed to solve this NP-hard problem, such as basis pursuit (BP) and orthogonal matching pursuit (OMP).

3   NSST Based Multi-focus Image Fusion Framework

The NSST maps the standard shearing filters from the false polarization grid system into a Cartesian coordinate system. It greatly reduces the computational complexity and preserves excellent multiscale analysis ability simultaneously. Based on the effectiveness and the flexibility of the NSST, a multi-focus image fusion method with sparse representation and SML is proposed in the present study. The low-frequency coefficients of the source images after NSST decomposition can be regarded as the approximate version of the source images. Due to the accurate image feature representation ability, the sparse representation is adopted as fusion rule of low-frequency coefficients. Meanwhile, the high-frequency coefficients reflect the edge and texture information of the source images. In this paper, we use the maximum SML approach to select the high-frequency coefficients.

The schematic diagram of the proposed fusion framework is shown in Fig. 1. The detail implement process can mainly divide to five steps as following:

(1) Decompose the source images A and B to obtain the different scales and directions sub-band coefficients $\{L_{j_0}^A, H_{j,l}^A\}$ and $\{L_{j_0}^B, H_{j,l}^B\}$ by using NSST, where $L_{j_0}^A$ and $L_{j_0}^B$ are low-frequency coefficients, $H_{j,l}^A$ and $H_{j,l}^B$ are the jth scale, lth direction high-frequency coefficients.

(2) Use the initial low-frequency fused coefficients as training sample to adaptively structure the over-complete dictionary D.

(3) Divide the low-frequency sub-images into small patches by sliding window, and then transform these patches to sparse coefficients using sparse representation with the obtained over-completed dictionary. Utilize the maximum absolute value to

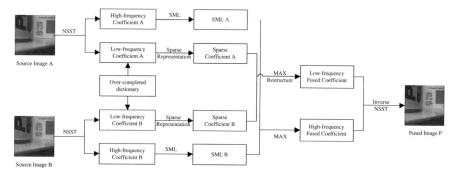

Fig. 1. The schematic diagram of the proposed fusion method

fuse the sparse coefficients, and restructure the fused sparse coefficients to get the fused low-frequency sub-band coefficients.

(4) Fuse the high-frequency coefficients based on maximum SML rule:

$$SML_{j,l}(x, y) = \sum_{m=-M}^{M} \sum_{n=-N}^{N} ML_{j,l}(x+m, y+n) \qquad (2)$$

where $(2M + 1)(2N + 1)$ is the size of the local window. $ML_{j,l}(x, y)$ is the discrete Laplacian operator which can be defined as:

$$ML_{j,l}(x, y) = |2H_{j,l}(x, y) - H_{j,l}(x - s, y) - H_{j,l}(x + s, y)| \\ + |2H_{j,l}(x, y) - H_{j,l}(x, y - s) - H_{j,l}(x, y + s)| \qquad (3)$$

where s is the distance of the coefficients. After obtaining the SML maps of the high-coefficients, the fusion coefficients can be selected as following rule:

$$H_{j,l}^F(x, y) = \begin{cases} H_{j,l}^A(x, y) & SML_{j,l}^A(x, y) \geq SML_{j,l}^B(x, y) \\ H_{j,l}^B(x, y) & SML_{j,l}^A(x, y) < SML_{j,l}^B(x, y) \end{cases} \qquad (4)$$

(5) Restructure to obtain the final fused image by inverse NSST transform on the fused low- and high-frequency coefficients.

4   Results

To verify the performance of the proposed method, one group of multi-focus images and seven existing fusion methods were used for comparison experiment. Comparison methods includes traditional gradient pyramid (GP) and discrete wavelet transform (DWT) methods; recently proposed curvelet transform, contourlet transform, and NSCT methods; Reference [6] NSST method and Reference [8] pulse coupled neural network (PCNN) based NSST method. In the proposed method, the four-level NSST

Fig. 2. Reference image and source images on Cameraman dataset

decomposition with [30 30 36 36] shearing filter matrix and [3 3 4 4] direction parameters were applied. The pyramid filter is the 'maxflat' wavelet. The dictionary size of the sparse representation is 64×256, the error tolerance $\varepsilon = 1.15$.

Figure 2 shows the reference image and source multi-focus images. The fused results by different methods are displayed in Fig. 3(a). From the visual contrast, it can be seen that the fused image of GP method has low contrast and inferior visual effect. Due to the lack of the shift-invariance, the fused images of the DWT, curvelet, and contourlet methods display artifact around the edge area. The fused images of the NSCT and NSST based methods are better in the quality; it means the non-sample process can overcome the pseudo-Gibbs phenomena and improve the fusion perfor-mance. To get the clearer comparison, this paper provides the residual maps between the fused images by different fusion methods and the top-focused source image in

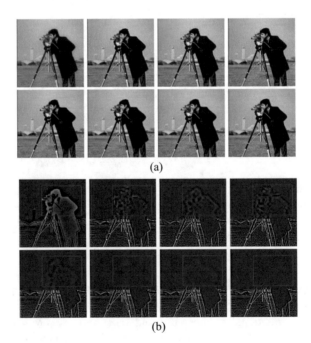

Fig. 3 The fused images (a) and their relevant residues maps (b) of different fusion methods (GP, DWT, curvelet, contourlet, NSCT, Reference [6], Reference [8], and the proposed method in sequence) on Cameraman dataset

Fig. 3(b). In the residual map, the less residual feature means more information in focused area of the source images transfer to the fused image. The residual pixels of the proposed method are least which indicates that the proposed method obtains the most information in the focused area.

In addition, to assess the fusion performance objectively, four image quality evaluation indexes, RMSE [9], SSIM [10], MI [11], and $Q^{AB/F}$ [12], were used to measure the fused images. The assessment results are shown in Fig. 4. It can be seen that the proposed method has least RMSE value and largest SSIM value. It means that the fused image of the proposed method is closest as the reference image compared with other methods. Moreover, the largest MI and $Q^{AB/F}$ values reflect the superiority of the proposed method.

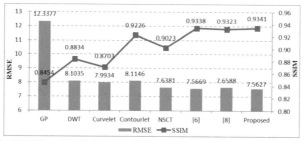

(a) RMSE and SSIM comparisons between different fused results

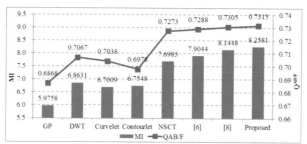

(b) MI and $Q^{AB/F}$ comparisons between different fused results

Fig. 4 The objective evaluation of the fusion results by different methods

5   Conclusion

In this paper, a new multi-focus image fusion method based on NSST and sparse representation was proposed. In allusion to the different sparsity of the low- and high-frequency coefficients, we utilized the pertinence section rules to fuse sub-band coefficients. For the low-frequency coefficients, a sparse representation based fusion rule was proposed. For the high-frequency coefficients, the SML algorithm was adopted. Experimental results indicate that the proposed method can extract the useful information from the source images more effective and can obtain superior performance regarding both the visual quality and the objective measurements.

Acknowledgments. This work was supported by the Brain Korea 21 PLUS Project, National Research Foundation of Korea. This research was also supported by the MSIT (Ministry of Science and ICT), Korea, under the ITRC (Information Technology Research Center) support program (IITP-2017-2015-0-00378) supervised by the IITP (Institute for Information & communications Technology Promotion). This research was also supported by Basic Science Research Program through the Nation-al Research Foundation of Korea (NRF) funded by the Ministry of Education (GR 2016R1D1A3B03931911)

References

1. Wan, W., Lee, H.: Multi-Focus image fusion based on focused regions detection. In: IEEE International Conference on Computational Science and Computational Intelligence (CSCI), pp. 814–817 (2016)
2. Candès, J., Donoho, L.: Recovering edges in ill-posed inverse problems: optimality of curvelet frames. Ann. Stat. **30**(3), 784–842 (2000)
3. Do, N.: The contourlet transform: an efficient directional multiresolution image representation. IEEE Trans. Image Process. **14**(12), 2091–2106 (2005)
4. Cunha, L., Da, J., Do, N.: The nonsubsampled contourlet transform: theory, design, and applications. IEEE Trans. Image Process. **15**(10), 3089–3101 (2006)
5. Easley, G., Labate, D., Lim, W.: Sparse directional image representations using the discrete shearlet transform. Apply Comput. Harmon. Anal. **25**, 25–46 (2008)
6. Cao, Y., Li, S. Hu, J.: Multi-focus image fusion by nonsubsampled shearlet transform. In: IEEE International Conference on Image & Graphics, pp. 17–21 (2011)
7. Aharon, M., Elad, M., Bruckstein, A.: K-SVD: An algorithm for designing overcomplete dictionaries for sparse representation. IEEE Trans. Sig. Process. **54**(11), 4311–4322 (2006)
8. Kong, W., Zhang, L., Lei, Y.: Novel fusion method for visible light and infrared images based on NSST-SF-PCNN. Infrared Phys. Technol. **65**(7), 103–112 (2014)
9. Yang, Y., Tong, S., Huang, S.: Image fusion based on fast discrete Curvelet transform. J. Image Graph. **02**, 219–228 (2015)
10. Wang, Z., Bovik, A., Sheikh, H., et al.: Image quality assessment: from error visibility to structural similarity. IEEE Trans. Image Process. **13**(4), 600–612 (2004)
11. Huang, W., Jing, Z.: Evaluation of focus measures in multi-focus image fusion. Pattern Recogn. Lett. **28**(4), 493–500 (2007)
12. Xydeas, C., Petrović, V.: Objective image fusion performance measure. Electron. Lett. **36** (4), 308–309 (2000)

Implementation of Large-Scale Network Flow Collection System and Flow Analysis in KREONET

Chanjin Park[1], Wonhyuk Lee[1], and Hyuncheol Kim[2(✉)]

[1] KREONET Operation and Service Division of Supercomputing,
Korea Institute of Science and Technology Information, Daejeon, South Korea
{pcj0722,livezone}@kisti.re.kr
[2] Deptartment of Computer Science,
Namseoul University, Cheonan, South Korea
hckim@nsu.ac.kr

Abstract. There have been many attempts to collect and analyze network flows for qualitative analysis of networks. However, the rapid increase in traffic and the size of the data made it difficult to collect and analyze flows. Especially, it is difficult to collect and analyze flows in Internet Service Provider (ISP) such as the Korea Research Environment Open NET-work (KREONET). This study attempted to examine how to build a network flow analysis system on a large-scale network such as KREONET in an effective fashion and review analysis results.

Keywords: KREONET · Network flow · Network flow collection · Network flow analysis

1 Instruction

For efficient network management and operation, the qualitative analysis of traffic, as well as analysis on traffic volume has become more important. For this, a network flow should be analyzed. However, it is not easy to collect and analyze a flow on a large-scale network such as the KREONET [1]. This study investigates the architecture and system implementation for the effective development of a flow collection & analysis system on the KREONET and reviews the matters to be considered during the implementation and some analysis results.

1.1 Korea Research Environment Open NETwork (KREONET)

The KREONET is a national R&D network operated by the Korea Institute of Science and Technology Information (KISTI). It operates a 600G backbone and provides high-performance network services to science & technology engineers. This R&D network operates 17 domestic centers in the Republic of Korea and 4 abroad centers as shown in Fig. 1. and sends tens of petabytes of research data annually. In 2016, more than 70 petabytes of science traffic were transmitted through KREONET.

© Springer Nature Singapore Pte Ltd. 2018
K.J. Kim et al. (eds.), *IT Convergence and Security 2017*,
Lecture Notes in Electrical Engineering 449,
DOI 10.1007/978-981-10-6451-7_16

Fig. 1. Map of KREONET 2017

1.2   Network Flow

A network flow is a sequence of packets from a source computer to a destination, which allows users to figure out how much traffic was generated by whom for whom, using what application. Router vendors provide flow information via netflow (Cisco), Jflowd (Juniper) and cflowd (Alcatel-Lucent). As stated in Fig. 2, they provide the following data: source IP address, destination IP address, source port, destination port and traffic volume.

srcIf	srcIPadd	DstIf	DstIPadd	Protocol	SrcPort	DstPort	Byte/Pkt	Pkts
Fa1/0	36.85.32.9	Fa0/0	158.43.192.1	17	2353	53	59	1
Fa0/0	158.43.192.1	Fa1/0	36.85.32.9	17	53	2353	134	1

Fig. 2. Network flow example (Netflow)

The netflow provides detailed information about who originated the traffic, who the destination is, and which applications are using the traffic. With this information, it is possible to identify users who are generating excessive traffic on the line and solve the problem.

2   Implementation of the Network Flow Collection System

2.1   Architecture

The KREONET's network flow collection and analysis system are structured in four stages like Fig. 3.: (i) flow collection, (ii) storing collected data in the database after encrypting [5] if necessary, preprocessing may go through depending on purposes, (iii) analysis of flow information in the database, (iv) providing the information to users.

Fig. 3. Network flow collection system architecture

2.2   Implementation

For system development, the commercial program 'logsee' was used. As shown in Fig. 4, the system consists of an analysis server including 1 web, two collection servers, and 1 database server. The flow information such as netflow and cflow is collected from 'N' routers via 2 collection servers. Since about 100 GB flow data are collected per day, it was impossible to collect the data with a single collection server only. The collection server keeps the raw data for about 5 days. Those which are stored for more than five days are being sent to the database server. Data analysis is faster in the collection server instead of getting the data from the database server. Therefore, data are stored in the collection server as many as possible. For communication with each server, a network switch supporting 1 Gbps was used. With the architecture illustrated in the Fig. 4, it was able to collect and analyze large-scale network flow information in the KREONET. However, the volume of data which are stored about 100 GB a day is ballooned up to 3 TB just in a month. As a result, it takes a lot of time to analyze such a huge amount of data.

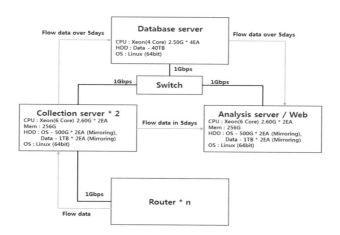

Fig. 4. Network flow collection system configuration

To solve this problem, a preprocessing step was added. It would be discussed in detail in Sect. 2.4.

2.3   Router Configuration for Flow Collection

The Fig. 5. reveals the collection of flow information in the KREONET2, an international network of the KREONET. For the effective collection of network flow, the KREONET has collected the flow coming from the network edge router to all ports linked with external networks via an ingress filter [3]. Under the same mechanism, it is able to collect all flow data which pass through the KREONET2. If the flow data both coming in and going out are collected, flow redundancy may occur. For example, if the flow comes into Daejeon and goes out to Hong Kong on the KREONET2, data are collected in Daejeon and later in Hong Kong again. Therefore, a redundancy problem can take place.

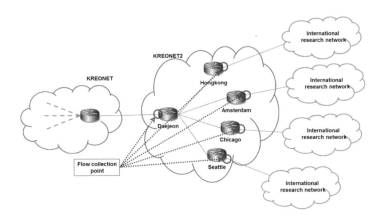

Fig. 5. Flow collection point in KREONE2

2.4   Flow Preprocessing to Increase Processing Speed

Since it is hard to analyze large-scale network flow information, this study decided to define the target information and implement preprocessing. In other words, it estimated statistics on the following data: (a) traffic volume of each group, (b) top application 100 of each group, (c) top talker and connection 100 of each group.

Then, raw data were stored after going through preprocessing (minute data, hour data, and day data) as illustrated in the Fig. 6 [2, 4]. During the preprocessing, minute data needs to be considered when generating the data. For example, if minute data is created using one minute ago, minute data is generated without proper accumulation of raw data, so accuracy can not be guaranteed. Considering the transmission time of the flow information from the router to the collection server and the processing time at the collection server, the system generates minute data with data 20 min before.

At statistical analysis on the said data ('a' thru 'c'), the preprocessed data, not raw ones, are analyzed. Therefore, analysis has become much faster. For example, the

Raw data

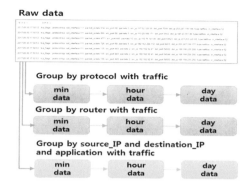

Fig. 6. Flow raw data preprocessing

analysis time was reduced from more than 30 min to less than 30 s in getting statistics on the data collected for about a month.

It should also be considered how much storage time is required, because creating min data, hour data, and day data occur duplication in terms of storage. In this system, the min data stores 2 weeks, the hour data stores 2 months, and the day data stores 1 year. However, there should be a further study on how long the said data (min data, hour data and day data) should be stored.

3   Statistical Analysis of Network Flows in KREONET2

In this section, among the network flow data collected through this system, those on the KREONET2, collected during May 2017 were analyzed. The Logsee used for the analysis is able to analyze data, using a query similar to SQL.

3.1   Application Analysis

According to analysis on top 5 applications on the KREONET2 as of May 2017, SSH was the highest with over 50% in terms of traffic volume. The reason why SSH revealed such as heavy traffic is that SFTP is used via an SSH port. Then, HTTP (29.28%) and SMTP (4.63) followed. Overall, many users have used well-known ports (Fig. 7).

3.2   Talker Analysis

Here, talker represents a pair of source IP and destination. In the said information, the last values were deleted to make specific IP information unclassifiable. According to analysis on each IP using the KISA's WHOIS service, the Korea Astronomy, and Space Science Institute, KISTI and KAIST generated a lot of traffic during collaboration with foreign research institutes via the KREONET. It appears that they usually are takers, which means that they get data from such foreign organizations, not givers (Fig. 8).

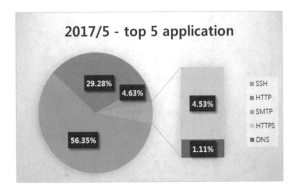

Fig. 7. Top application in KREONET2 (2017/05)

Rank	source IP	Destination IP	Traffica
1	150.203.153.—	210.98.54.—	50,961 Mbytes
2	171.64.103.—	210.98.54.—	29,303 Mbytes
3	192.146.239.—	150.197.30.—	25,950 Mbytes
4	130.237.11.—	143.248.100.—	18,649 Mbytes
5	136.172.30.—	210.98.49.—	13,908 Mbytes
6	131.154.130.—	134.75.124.—	11,951 Mbytes
7	171.67.205.—	143.248.31.—	11,940 Mbytes
8	90.147.132.—	203.250.156.—	10,008 Mbytes
9	131.154.130.—	134.75.124.—	9,868 Mbytes
10	193.61.196.—	210.125.45.—	9,410 Mbytes

Fig. 8. Top talker 10 in KREONET2 (2017/05)

4   Conclusion

This study investigated the design and implementation of a flow analysis system on a large-scale network such as the KREONET and reviewed an effective flow collection method and preprocessing for statistical analysis. As a result, it was able to reduce analysis time considerably. However, it is needed to define the results during flow collection and implement preprocessing accordingly. In this sense, this study is less flexible. There should be further studies on the data structure to improve the usefulness of the flow analysis system and the characteristics of the KREONET's network traffic after collecting flow information for a long period of time.

References

1. Liao, W., Fu, Z.: Cloud platform for flow-based analysis of large-scale network traffic. In: ICSSC, pp. 259–262 (2013)
2. RRDtool (2017). https://oss.oetiker.ch/rrdtool/doc/rrdtool.en.html
3. Hofstede, R., Celeda, P., Trammell, B., Drago, I., Sadre, R., Sperotto, A., Pras, A.: Flow monitoring explained: from packet capture to data analysis with netFlow and IPFIX. IEEE Commun. Surv. Tutorials **16**(4), 2037–2064 (2014)
4. Balantrapu, C., Potluri, A., Das, N.: A novel approach to netflow monitoring in data center networks. In: 2014 Sixth International Conference on Communication Systems and Networks (COMSNETS), pp. 1–4 (2014). doi:10.1109/COMSNETS.2014.6734934
5. Villani, A., Riboni, D., Vitali, D.: Obsidian: a scalable and efficient framework for netFlow obfuscation. In: INFOCOM, pp. 14–19 (2013)

Computer Vision and Applications

A Novel BP Neural Network Based System for Face Detection

Shuhui Cao[1], Zhihao Yu[1], Xiao Lin[1], Linhua Jiang[1(✉)],
and Dongfang Zhao[2]

[1] Shanghai Key Lab of Modern Optical Systems,
University of Shanghai for Science and Technology,
No. 516 JunGong Road, Shanghai 200093, People's Republic of China
honorsir@yandex.com
[2] Department of Computer Science and Engineering,
University of Nevada, Reno, NV 89557, USA

Abstract. We describe a new neural network, which can improve the performance of face detection system. In this paper, we propose a system that combines the Gabor feature and momentum factor back propagation algorithm for face detection. First, the Gabor feature of the training set is extracted and is inputted to the momentum factor of Back Propagation neural network for training. Then, using the trained system detects whether the face targets exist in the input image, and marking the target with the window. In order to enhance the training effect of the traditional Back Propagation neural network, the momentum factor is added to the Back Propagation algorithm, which can effectively slow down the trend of the network training in the shock and avoid the algorithm drop into the local minimum. Furthermore, the added momentum factor can adaptively adjust each layer weight of the Back Propagation neural network. Extensive experimental results demonstrate that our solution is effective and also competitive, compared to the classic and also state-of-the-art face detection models.

Keywords: Face detection · Back Propagation · Momentum term · MFBP

1 Introduction

Face detection is a well-studied problem in computer vision. In the past decades, massive efforts have been made on face detection [1–3]. The first essential step of face detection is to mark the target face with the window in the image. With the expansion of the application of face detection, it gradually developed into an independent research topic and received the attention of researchers.

Generally, Face detection can be classified into two categories: one is the face detection in a static image, including color image, it can detect a single face or multi-face [4, 5]. Another is the face detection in the dynamic image [6], can be known as target tracking. Our research is to detect multi-face in color image. The process of face detection is actually a comprehensive judgment of the face pattern features [7, 8]. The input face image contains abundant pattern features, these features can be divided

© Springer Nature Singapore Pte Ltd. 2018
K.J. Kim et al. (eds.), *IT Convergence and Security 2017*,
Lecture Notes in Electrical Engineering 449,
DOI 10.1007/978-981-10-6451-7_17

into two classes by color properties: one is the skin-color characteristics, another is the gray feature. Our paper is to combine gray features and the neural network as our face detection system. Due to the complexity of the input images, recent researches tend to use abundant faces training samples to construct a classifier and detect the faces in the test set. Remark that the neural network method is to implicitly describe the statistical properties of the model in the structure and parameters of the network, and modeled the target face, which is difficult to describe in the machine.

In this paper, we describe a new neural network to detect the faces in the input image. The momentum factor is added to the Back Propagation algorithm [9, 10], and then combine with Gabor features as our face detection system. Our system can effectively improve the BP algorithm with long convergence time, it also slows down the trend of the network training in the shock and avoids the algorithm into the local minimum.

2   Related Work

2.1   Neural Network Based on Face Detection

Early in 1994 Vaillant et al. [11] applied neural networks for face detection. In their algorithm, they consider using the trained convolutional neural network to detect whether it exist faces in the input image. In 1996, Rowley et al. [12] proposed a retinally connected neural to improve the performance frontal face detection. In 2002, Garcia et al. [13] proposed a neural network to detect semi-frontal human faces. In 2006, Osadchy et al. [14] applied a convolutional network for simultaneous face detection and pose estimation. In 2013, Girshick et al. [15] applied R-CNN method to detect the face, it takes class-specific classifiers to recognize the object category of the proposals. In 2015, Li et al. [16] proposed a convolutional neural network cascade for face detection. In 2016, Tao et al. [17] proposed a robust face detection using local CNN and SVM based on kernel combination. In 2016, Xia et al. [18] proposed a face occlusion detection using deep convolutional neural networks.

2.2   Face Features

The pattern features of the face include skin-color features and gray features. The skin color model has been used to describe skin color features, which is related to chrominance space including r-g, CIE-xy, TSL, HSV, YIQ etc. For the performance of Gaussian model and Mixed Gaussian model in different chrominance space, the latter can well describe the distribution of skin-color features region in a few cases. At present, gray features are widely used in statistical learning of face detection methods. Nguyen [19] utilized a wavelet transform to extract face of the multi-resolution features as a basis for classification. Sahoolizadeh [20] proposed the hybrid approaches for face recognition based on combining Gabor wavelet with ANN feature classifier which achieves 93% recognition rate on ORL data set. Mohammad Abadi [20] proposed an approach based on ANN and Gabor wavelets to detect the desirable number of faces in a fixed photo with a gray background. Compared with these facial features, our work classifier adds the Gabor feature to detect the faces which can achieve a better performance.

3 Proposed Approach

3.1 Gabor Filter

Gabor transform is a windowed Fourier transform. The Gabor function can extract the relevant features in different scales and directions in the frequency domain [20]. A Gabor kernel can extract a feature of an image on a certain frequency.

In this paper, we select five scales and eight directions to extract the image Gabor feature and convolve the input image with 5*8 Gabor cores in Fig. 1, and generate the image feature of 40 different scales at different frequencies.

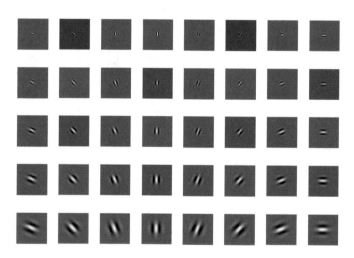

Fig. 1. A schematic of the 40 Gabor filters in this paper. The rows represent eight different angles, and the columns represent five different scales.

The input image as the input signal f_{in} is concerted into a frequency domain signal $\hat{f}_{in}$ using Fourier transform computed as follows.

$$\hat{f}_{in}(\sigma_x, \sigma_y) = \iint f_{in}(x, y)e^{-i2\pi(\sigma_x x + \sigma_y y)}dxdy \qquad (1)$$

(x, y) represents coordinate in the spatial domain. Then the result of the spatial signal is used to multiply the Fourier transform of Gabor core that can obtain the result image which is filtered by Gabor filter.

Using the convolution theorem formula as follows,

$$\text{Gabor} * f_{in} = \hat{Gabor} \cdot f_{in} \tag{2}$$

the Gabor core and the input signal are convoluted, and the response of the input signal near a certain neighborhood is obtained as shown in Fig. 2(b) and Fig. 2(a) is the original image of the input.

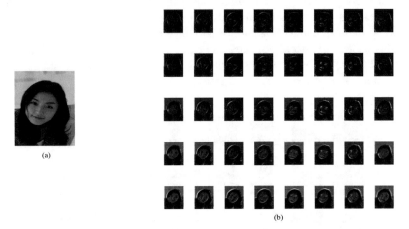

Fig. 2. (a) is the original image of the input. In (b), we show 40 filtered face images which include five scales and eight directions in different frequency bands extracted from the original image features.

Two-dimensional complex wave represented by formula (3) is multiplied by the two-dimensional Gaussian function that computed in formula (4) to obtain the Gabor kernel in formula (5).

$$s(x, y) = \exp(i(2\pi(u_0 x + v_0 y)) + P) \tag{3}$$

Where the initial phase P has little effect on the Gabor and can be omitted. In this formula, we set P = 0.1.

$$\omega(x, y, \delta_x, \delta_y) = K \exp\left(-\frac{\pi(x - x_0)_r^2}{\delta_x^2} + \frac{(y - y_0)_r^2}{\delta_y^2}\right) \tag{4}$$

δ_x and δ_y control the Gaussian function in the x-axis and y-axis on the "distribution" situation.

$$\begin{aligned} \text{Gabor}(x_0, y_0, \theta, \delta_x, \delta_y, u_0, v_0) &= s(x, y)\omega(x, y, \delta_x, \delta_y) \\ &= K \exp\left(-\frac{\pi(x - x_0)_r^2}{\delta_x^2} + \frac{(y - y_0)_r^2}{\delta_y^2}\right) \exp(2\pi i(u_0 x + v_0 y)) \end{aligned} \tag{5}$$

(x_0, y_0) is the center of the Gaussian kernel, θ is the direction of rotation of the Gaussian kernel, (δ_x, δ_y) is the scale of the Gaussian kernel in the x, y direction, (u_0, v_0) is the frequency domain coordinates and K is ratio of the magnitude of the Gaussian kernel. We set $K = 2, u_0 = 0.5, v_0 = 0.5, \delta_x = 0.8, \delta_y = 0.8$ and $\theta = \pi/4$.

3.2 MFBP

In this paper, the commonly used BP neural network is used as the basis for this experiment. Meanwhile, in order to overcome the fault of BP algorithm, the momentum term can be used to improve the convergence speed and avoid falling into a local minimum of the training network. The overall steps of the neural network of our face detection are shown in Fig. 3 which briefly demonstrate the workflow of the neural network and will introduce the momentum factor Back Propagation neural network, dubbed as MFBP, for short. The method is simulated in MATLAB.

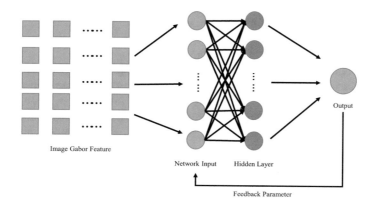

Fig. 3. The process of BP (Back Propagation) algorithm is mended by appending momentum factors which are the feedback parameter. The training performance of the conventional BP neural network is improved by adding feedback parameter when forwarding feedback error signal.

The extracted image Gabor features are used as input to the neural network which is a fully connected network structure. As shown in Fig. 3, the network has two layers, one is the hidden layer and another is the output layer.

The transfer function of hidden neurons is as follows,

$$y = f(\text{net}) = \frac{1}{1 + e^{-net}} \tag{6}$$

In this function, the net is the input data which comes from the input layer. Each input is treated as x_i, if giving n input datum, $1 \leq i \leq n$, then,

$$net = x_1 w_1 + x_2 w_2 + \ldots + x_n w_n \tag{7}$$

where w_i is the initialization weight of the neural network. In addition, in the BP neural network, the number of nodes in the input and output layers is known and the amount of nodes in the hidden layer h is decided by formula as follows.

$$h = \sqrt{m+n} + A \tag{8}$$

where m and n are the number of nodes in the input and output layers respectively and A is an adjustable constant between 1 and 10.

The process of forwarding propagation which can be calculated by formula as follows.

$$S_j = \sum_{i=0}^{m-1} w_{ij} x_i + b_j \tag{9}$$

Where $x_j = f(S_j)$, net $= S_j$ and we set $b_j = 0.5$.

The heart of the training network is the back-propagating error arithmetic. The initial weight w can be used in the formula of the forward propagation of the Back Propagation network and it is generated by random initialization. Then, the actual output may produce a large error, compared with the ideal output. To constantly adjust the w, the error function be obtained in formula (10), where y_i represents the actual output.

$$E(w, b) = \frac{1}{2} \sum_{j=0}^{n-1} (d_j - y_j)^2 \tag{10}$$

The samples entered are operated in the following computational steps during the training.

1. The error term for each cell of each layer is calculated from the output layer in reverse.

$$error_i = y_j(1 - y_j)(d_j - y_j) \tag{11}$$

2. Calculate the error of the nodes of the hidden layer.

$$error_h = y_h(1 - y_h)error_h \tag{12}$$

3. Update each weight.

$$w_{ik} = w_{ik} + \mu \cdot error_k \cdot x_{ik} \tag{13}$$

Where $\Delta w_{ik} = \mu \cdot error_k \cdot x_{ik}$ is the law of weight update, x_{ik} represents the input value and w_{ik} is the corresponding weight between nodes i and k.

These are the most basic mathematical ideas in the neural network which can train the positive examples and negative examples, having relatively long time consumption. The following is an improvement of the neural network.

$$\Delta w_{ik}(n) = \mu \delta_k x_{ik} + \alpha \Delta w_{ik}(n-1) \tag{14}$$

Where $0 < = \alpha < 1$ is efficient of the momentum. In our algorithm, we set $\alpha = 0.65$ and $\mu = 0.6$. The above formula which make the weight of the n time iteration partly depending on the weight of the n − 1 time iteration, modifying the law of weight update. Adding impulse items to a certain extent, to increase the effect of the search step, which can be faster convergence. On the other hand, since the multilayer network tends to cause the loss function to converge to the local minimum, the impulse term can, to some extent, cross some narrow local minimum to achieve a smaller place.

4 Experimental Result

For training, The Face Detection Data Set and Benchmark and the face databases at CMU and Harvard which contain approximately 5000 positive examples, the frontal faces with slight variations in pose angle and simple background. We used 250 images of scenery for collecting negative examples in bootstrap manner, selecting 2500 non-face images from sub-images of mentioned above images.

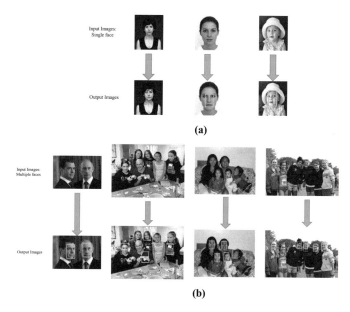

(a)

(b)

Fig. 4. The figure shows some results by face image detection using the MFBP network with the Partheenpan Dataset. In (a), we select the input image with a frontal face. In (b), each input image contains multiple faces. In the (a) and (b), the output images are marked with a rectangle with the detected face.

In Fig. 4(a), each image contains a frontal view of the face with a simple background and the face of each output image is marked out precisely. In Fig. 4(b), false detections are present in the second and the forth. We used the trained network to test some common face database, Partheenpan Dataset, The Annotated Faces in the Wild (AFW), The Face Detection Data Set and Benchmark and CMU Dataset.

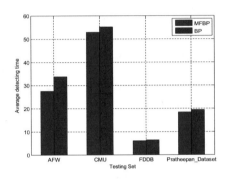

Fig. 5. Comparison of average consumption time on MFBP and BP

In Fig. 5, we do a comparison, compared with the average detection time of the MFBP algorithm and the average time of the BP algorithm. Because each of the AFW and CMU images may have multiple faces and the image background is more complex. As the graph shows, the improvement in the detection time performance is more obvious, for these two sets.

In Table 1 we compare our method with other two face detection methods Ada-Boost and SVM, it shows the detection rate and false detect rate of four data sets on three algorithms. From the Table 1, we can clearly see that our method is superior to the other two methods on the experimental results on simple test pictures on FDDB.

Table 1. Comparison of the detecting performance on testing set

Testing set	Face detect algorithm	Missed faces	Detect rate	False detects	False detect rate
AFW	MFBP	65(451)	85.6%	39	7.95%
	AdaBoost	57(451)	87.4%	33	6.81%
	SVM	68(451)	84.9%	42	8.52%
CMU	MFBP	126(507)	75.1%	43	7.8%
	AdaBoost	113(507)	77.7%	56	9.95%
	SVM	140(507)	72.4%	64	11.2%
FDDB	MFBP	381(5062)	92.5%	179	3.4%
	AdaBoost	456(5062)	90.1%	223	4.2%
	SVM	512(5062)	89.9%	276	5.2%
Pratheepan_Dataset	MFBP	46(352)	86.9%	28	7.3%
	AdaBoost	54(352)	84.7%	37	9.5%
	SVM	63(352)	82.1%	39	9.9%

However, the test images contain more face and the background are more complex on AFW, CMU and Pratheepan_Dataset testing sets, the false detect rate is relatively higher. Thus, our algorithm has also the limitation on some images.

5 Conclusion

In this paper, we proposed a new Back Propagation neural network based face detection method. The central idea of our method is to add the momentum factor to traditional Back Propagation algorithm and obtained the momentum factor Back Propagation neural network. Then our new Back Propagation neural network combines with the Gabor features as our face detection system. We have accelerated the convergence time of our proposed system and obtained an acceptable number of false detections by testing the test sets. Experimental results demonstrate that our proposed solution is competitive and attractive, compared against classical and our state-of-the-art algorithms.

Acknowledgement. The research was partly supported by the program for Professor of Special Appointment (Eastern Scholar) at Shanghai Institutions of Higher Learning, USST incubation project (15HJPY-MS02), National Natural Science Foundation of China (No. U1304616, No. 61502220).

References

1. Bellil, W., Brahim, H., Amar, C.B.: Gappy wavelet neural network for 3D occluded faces: detection and recognition. Multimed. Tools Appl. **75**(1), 1–16 (2016)
2. Zhang, K., Zhang, Z., Li, Z., et al.: Joint face detection and alignment using multitask cascaded convolutional networks. IEEE Signal Process. Lett. **23**(99), 1499–1503 (2016)
3. Zhan, S., Tao, Q.Q., Li, X.H.: Face detection using representation learning. Neurocomputing **187**, 19–26 (2016)
4. Jain, V., Learned-Miller, E.: FDDB: a benchmark for face detection in unconstrained settings. Technical report, University of Massachusetts, Amherst (2010)
5. Mahmoodi, M.R., Sayedi, S.M.: A face detection method based on kernel probability map. Pergamon Press, Inc., Tarrytown (2015)
6. Arceda, V.E.M., Fabián, K.M.F., Laura, P.C.L., et al.: Fast face detection in violent video scenes. Electron. Notes Theor. Comput. Sci. **329**, 5–26 (2016)
7. Pavani, S.K., Delgado-Gomez, D., Frangi, A.F.: Gaussian weak classifiers based on co-occurring Haar-like features for face detection. Pattern Anal. Appl. **17**(2), 431–439 (2014)
8. Kamaruzaman, F., Shafie, A.A.: Recognizing faces with normalized local Gabor features and Spiking Neuron Patterns. Pattern Recogn. **53**, 102–115 (2016)
9. Jia, W., Wei, Q.Y.: BP-neural network for plate number recognition. Int. J. Digit. Crime Forensics **8**(3), 34–45 (2016)
10. Ramirezquintana, J.A., Chaconmurguia, M.I., Chaconhinojos, J.F.: Artificial neural image processing applications: a survey. Eng. Lett. **20**(1), 68–81 (2012)
11. Vaillant, R., Monrocq, C., Cun, Y.L.: Original approach for the localisation of objects in images. IEE Proc. Vis. Image Signal Process. **141**(4), 245–250 (1994)

12. Rowley, H.A., Baluja, S., Kanade, T.: Neural network-based face detection. In: Conference on Computer Vision and Pattern Recognition, pp. 203–208. DBLP (1996)
13. Garcia, C., Delakis, M.: A neural architecture for fast and robust face detection. In: Proceedings of the International Conference on Pattern Recognition, vol. 2, pp. 44–47. IEEE (2002)
14. Osadchy, M., Cun, Y.L., Miller, M.L.: Synergistic face detection and pose estimation with energy-based models. J. Mach. Learn. Res. **8**(1), 1197–1215 (2006)
15. Girshick, R., Donahue, J., Darrell, T., et al.: Rich feature hierarchies for accurate object detection and semantic segmentation, 580–587 (2014)
16. Li, H., Lin, Z., Shen, X., et al.: A convolutional neural network cascade for face detection, 5325–5334 (2015)
17. Tao, Q.Q., Zhan, S., Li, X.H., et al.: Robust face detection using local CNN and SVM based on kernel combination. Neurocomputing **211**, 98–105 (2016)
18. Xia, Y., Zhang, B., Coenen, F.: Face occlusion detection using deep convolutional neural networks. Int. J. Pattern Recogn. Artif. Intell. **30**(9), 401–408 (2016)
19. Zhu, Y., Schwartz, S., Orchard, M.: Fast face detection using subspace discriminant wavelet features. In: Proceedings of the IEEE Conference on Computer Vision and Pattern Recognition, vol. 1, pp. 636–642. IEEE Xplore (2000)
20. Abadi, M., Khoudeir, M., Marchand, S.: Gabor filter-based texture features to archaeological ceramic materials characterization. Image and signal processing, pp. 333–342. Springer, Heidelberg (2012)
21. Wang, X.X., Shi, B.E.: GPU implemention of fast Gabor filters. In: IEEE International Symposium on Circuits and Systems, pp. 373–376. IEEE Xplore (2010)

A Distributed CBIR System Based on Improved SURF on Apache Spark

Tingting Huang[1], Zhihao Yu[1], Xiao Lin[1], Linhua Jiang[1(✉)], and Dongfang Zhao[2]

[1] Shanghai Key Lab of Modern Optical Systems,
University of Shanghai for Science and Technology,
No. 516 JunGong Road, Shanghai 200093, People's Republic of China
honorsir@yandex.com
[2] Department of Computer Science and Engineering,
University of Nevada, Reno, NV 89557, USA

Abstract. This paper investigates the problem of image retrieval in abundant volume of image data. We propose an improved Content Based Image Retrieval (CBIR) system based on Apache Spark, a lightning-fast engine of cluster computing for large-scale data processing, to overcome the shortcomings in retrieval speed and accuracy. Specifically, binary descriptors, which consume less memory and accelerate the retrieval speed, are built through uniform sampling patterns in Binary Robust Invariant Scalable Keypoints (BRISK) to represent images instead of floating-number descriptors in the original SURF. Then we eliminate the mismatched point pairs with Random Sample Consensus (RANSAC) in the pre-matching point pairs to further improve the accuracy of the retrieval. Experimental results show that the proposed system significantly improves both the retrieval speed and accuracy compared to traditional CBIR systems.

Keywords: Image retrieval · CBIR · Binary descriptors · Spark

1 Introduction

Image is an important and easy-to-access information carrier. With the booming of multimedia technology and Internet, large volume of image data are generated at an extremely high rate. How to retrieve a desired image rapidly and accurately among massive image databases is becoming an urgent problem. In particular, it is currently a hot topic in Content Based Image Retrieval (CBIR [1]) about how to effectively retrieve pertinent images in large-scale image databases. CBIR indexes images by using low-level features that are extracted from the image and are used to represent image in database such as shape, color, texture, and spatial layout. In the image feature extraction phase, Scale-Invariant Feature Transform (SIFT [2]) and Speeded-Up Robust Features (SURF [3]) are two popular algorithms. SIFT utilizes image pyramid to detect feature points at different scales and builds keypoint descriptor through calculating a set of orientation histograms around one particular keypoint. Although SIFT delivers high performance, establishing image pyramid is usually time consuming. SURF is characterized by the use of integral images, which greatly accelerates the

© Springer Nature Singapore Pte Ltd. 2018
K.J. Kim et al. (eds.), *IT Convergence and Security 2017*,
Lecture Notes in Electrical Engineering 449,
DOI 10.1007/978-981-10-6451-7_18

calculation time. Generally, SURF exhibits better robustness and is faster than SIFT, although it takes considerable amount of time to build the feature descriptor. In order to further shorten the computing time, researchers have proposed some new binary feature descriptors such as Oriented FAST and Rotated BRIEF(ORB) descriptor [4] and the BRISK descriptor [5]. CBIR is widely used in information retrieval [6], medical diagnosis, crime prevention, among many others. Nevertheless, in image databases at the scale of terabytes or petabytes, traditional CBIR retrieval speed will the throttled by the process of image retrieval. To that end, we propose a distributed CBIR system based on the SURF-BRISK algorithm and the popular big data system Spark.

2   Related Background

2.1   Feature Detection of SURF

Different from SIFT, the feature point detection of the SURF is based on the Hessian matrix. SURF locates the feature point by finding the local maximum of the Hessian matrix determinant and uses the integral image to dramatically accelerate the computing speed. On the σ scale, Hessian matrix [7] of image's point $X = (x, y)$ is defined as follows:

$$H(x, \sigma) = \begin{pmatrix} L_{xx}(x, \sigma) \, L_{xy}(x, \sigma) \\ L_{xy}(x, \sigma) \, L_{yy}(x, \sigma) \end{pmatrix} \tag{1}$$

Here $L_{xx}(x, \sigma)$ is the convolution of the Gaussian second order differential and the image $I(x, y)$ at point $X = (x, y).L_{xy}(x, \sigma)$ and $L_{yy}(x, \sigma)$ is similar to $L_{xx}(x, \sigma)$.

In order to increase the computation speed, Bay and other people proposed using a box filter as an approximation of the Gaussian second order partial derivatives and utilizing the integral image to accelerate the convolution. Therefore, approximate determinant of Hessian matrix [8] is:

$$\det(H_{approx}) = D_{xx} \times D_{yy} - (0.9 \times D_{xy})^2 \tag{2}$$

where D_{xx}, D_{yy} and D_{xy} are the approximate of L_{xx}, L_{yy} and L_{xy}.

Through different sizes of box filter, the pyramid of multiscale image can be constructed. On the pyramid of multiscale image, SURF performs the non-maximal suppression on each point by comparing its 26 neighborhoods at the current scale and adjacent scales. After non-maximal suppression, a set of candidate feature points are found. Then, in the scale space and the image space, the interpolation operation is carried out to obtain the final feature points. Finally, in order to ensure the invariance of rotation, SURF computes the Haar wavelet characteristics in the domain of feature point to assign dominant orientation for each feature point.

2.2   Feature Descriptor of BRISK

BRISK utilizes a uniform sampling pattern used for sampling the neighborhood of the feature point. Based on the feature point as the center, BRISK constructs concentric circles of different radii, and N sampling points are obtained by equally spaced sampling on each circle. For avoiding aliasing effects, BRISK performs Gaussian filtering on the sampling points of the concentric circles. Since there are N sampling points, these sampling points are combined to form $N(N-1)/2$ point pairs, which are represented by a set $\mathcal{A}$ as Eq. (3). After Gaussian filtering, the smoothed intensity values of the two sampling points are $I(p_i, \sigma_i)$ and $I(p_j, \sigma_j)$ respectively, which are made use of calculating the local gradient $g(p_i, p_j)$ as Eq. (4).

$$\mathcal{A} = \{(p_i, p_j) \in \mathbb{R}^2 \times \mathbb{R}^2 | i < N, j < i, \, i,j \in \mathbb{N}\}. \tag{3}$$

$$g(p_i, p_j) = (p_i - p_j) \cdot \frac{I(p_j, \sigma_j) - I(p_i, \sigma_i)}{\|p_i - p_j\|^2}, \tag{4}$$

A subset of short-distance pairings $\mathcal{S}$ and another subset of long-distance pairings $\mathcal{L}$ [9] are proposed respectively as follows:

$$\mathcal{S} = \{(p_i, p_j) \in \mathcal{A} | \|p_j - p_i\| < \delta_{\max}\} \subseteq \mathcal{A} \tag{5}$$

$$\mathcal{L} = \{(\mathbf{p_i}, \mathbf{p_j}) \in \mathcal{A} | \|\mathbf{p_j} - \mathbf{p_i}\| < \delta_{\min}\} \subseteq \mathcal{A} \tag{6}$$

Using the above formula, the main direction of the feature point is estimated to be:

$$g = \begin{pmatrix} g_x \\ g_y \end{pmatrix} = \frac{1}{L} \cdot \sum_{(p_i, p_j) \in \mathcal{L}} g(p_i, p_j). \tag{7}$$

where g_x is the gradient in the x-direction; g_y is the gradient in the y-direction; L refers to the number of $\mathcal{L}$.

In order to keep the rotation invariance, BRISK rotated the sampling area at $\alpha = arctant2(g_x, g_y)$. After rotation, in the new subset of short-distance pairings S, intensity values of point pair $(p_i^\alpha, p_j^\alpha,)$ is compared to construct the bit-vector descriptor [10] of length 512. Each bit b corresponds to:

$$b = \begin{cases} 1, & I(p_j^\alpha, \sigma_j) > I(p_i^\alpha, \sigma_i) \\ 0, & \text{otherwise} \end{cases} \quad \forall (p_i^\alpha, p_j^\alpha) \in \mathcal{S} \tag{8}$$

2.3   Spark

Apache Spark developed by UC Berkeley AMPLab is a large-scale data processing system [11]. Furthermore, Spark is extending a variety of specific domain, such as

GraphX for graph-parallel computation, MLlib for machine learning, and Spark SQL for working with structured data. Diverse data sources can be access on Spark, including distributed file system (like Hadoop HDFS [12]), distributed database (like Apache HBase), and object storage (like Amazon S3). The intermediate result of the task can be saved to memory on Spark, so that Spark is better suited for iterative and interactive computation. The all distributed operation of the Spark are based on RDD (Resilient Distributed Dataset) [13, 14], which is a parallelizable data collection. Spark provides two types of operations: Transformation and Action. Transformation produces a new RDD by processing from input RDD, and Action computes a desired result and returns the result to the driver. The operations on the RDD are done in memory, so that it reduces a lot of disk operations and improves performance during the distributed computing. Therefore, Spark framework is more suitable for requiring real-time processing, data mining, machine learning and so on.

3   Proposed CBIR System

3.1   Extraction of Feature Vector

In this step, our system adapts SURF-BRISK algorithm to extract feature vectors. The SURF is almost three times faster than SIFT due to the Determinant of Hessian (DOH) operator and the integral image [15] in addition to better robustness and reliability. Although the SURF improved a lot on the basis of SIFT, SURF still has some flaws such as wasting plenty of time to build local feature descriptors so that it is difficult to meet the real-time requirements and limited memory space in some specified occasions. Based on this situation, our system uses BRISK to create feature descriptors which optimize the original SURF. Unlike the descriptor of floating number which is generated by SURF for each keypoint, BRISK builds binary descriptor to store the feature. During the feature matching phase, calculating the Hamming distance of the binary descriptor is much faster than the usual calculation of the Euclidean distance of floating number. In terms of data storage, lower memory [16] is required by the BRISK descriptor.

Specific steps of SURF-BRISK algorithm are as follows:

1. SURF uses different sizes of box filter to generate the pyramid of multiscale image.
2. SURF obtains the feature point based on the local maximum of the Hessian matrix determinant and executes the non-maximal suppression on each point to find a set of candidate feature points on the pyramid of multiscale image.
3. Execution of the interpolation operation gets the final feature points in the scale space and the image space.
4. BRISK constructs concentric circles with different radii, the center of which is the feature points detected by SURF and utilizes uniform sampling pattern to obtain N sampling points which are combined to form $N(N - 1)/2$ point pairs.
5. Local gradient, a subset of short-distance pairings and another subset of long-distance pairings are obtained based on $N(N - 1)/2$ point pairs,. Finally, the main direction α of the feature point is determined.

6. By rotating the sampling area at the main direction α of the feature point, the new subset of short-distance pairings is obtained. The bit-vector descriptor is generated by comparing intensity values of point pair in the new subset of short-distance pairings.

3.2 Construction of Image Database

In preprocessing phase of image files, the original images are uploaded to HDFS, each of which is renamed with a unique identifier ID. Then, the SURF-BRISK algorithm starts to extract image features and descriptor of each keypoint will be obtained. In order to represent each image, we define an image class which contains collection of feature descriptor and the unique identifier ID of image. Finally, through the object serialization technology, all the image objects are saved to HDFS, which completes construction of image database.

3.3 Image Match

The image uploaded to HDFS contains abundant descriptors. Given a SURF-BRISK descriptor $S = [x_1, x_2, x_3 \cdots x_{512}]$, where x_j is the j-th value $1 \leq j \leq 512$ in this set and $x_j \in \{0, 1\}$. Based on the specific structure of SURF-BRISK descriptor, we could achieve a quick matching by using the Hamming distance [17]. Assuming two SURF-BRISK descriptors S_1, S_2, the Hamming Distance D can be defined as follows:

$$D(S_1, S_2) = \sum_{i=1}^{512} (x_i \oplus y_i) \tag{9}$$

The smaller the value of D is, the higher the similarity between S_1 and S_2 becomes. If the similarity beyond threshold, this descriptor is matched. However, these descriptors may contain some mismatched point pairs. To solve this problem, the RANSAC algorithm [18] is used to eliminate mismatched point pairs. The RANSAC algorithm is robust to noise points and mismatched points, moreover, its calculation process is stable and reliable. If the number of matched points beyond threshold, we define two images are match and the identifier ID of matched image is outputted.

3.4 Workflow of System

The implementation of improved CBIR system is shown in Fig. 1. At first, we detect the keypoints of image via SURF and use BRISK to create binary descriptors. By comparing the feature descriptors of the query image with the feature descriptors of the images in the database, candidate matching point pairs are generated. Then, the RANSAC is used to eliminate the mismatched point pairs, so that it would get more accurate matching point. These operations are performed under the Spark platform. Finally, if the number of matched points beyond threshold, we define two images are similar and the identifier ID of matched image is output.

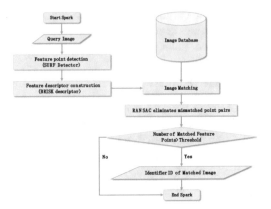

Fig. 1. Workflow summary of our improved CBIR system

4   Experiments and Evaluation

In order to verify the reliability and effectiveness of our system which is evaluated by a large number of simulation experiments in this section. Our system is deployed on Spark cluster consists of 4 nodes with Master Node and three Slave Nodes. Configuration of experiments and environment is shown in Table 1.

Table 1. Configuration of Experiments environment

Nodes	Memory(G)	CPU(G)	Disk(T)
Master node	Xeon E5-2609 v3	32	3
Slave node	Xeon E5-2609 v3	16	3
Slave node	Xeon E5-2609 v3	16	3
Slave node	Xeon E5-2609 v3	16	3

In this section, we evaluate our system on four image datasets with different capacity: the capacity of image database is 100, 1 K, 10 K, 100 K, 1000 K respectively. In order to evaluate, we test three kinds of CBIR systems: improved algorithm on single machine, original algorithm on Spark and improved algorithm on Spark. The results of the test are presented in Table 2.

Table 2. Execution Time on different CBIR system

Image database size	Improved algorithm on single machine (seconds)	Original algorithm on spark (seconds)	Improved algorithm on spark (seconds)
100	1.053	0.156	0.131
1 K	8.509	1.043	0.805
10 K	70.568	7.196	4.604
100 K	582.351	47.261	28.199
1000 K	4795.661	304.057	168.827

Speedup ratio can be used more generally to show the effect on performance after any resource enhancement, by which, comparison analysis is presented in Fig. 2.

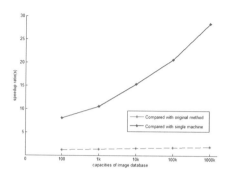

Fig. 2. Comparison of speedup ratios in different occasions

As is shown in Table 2 and Fig. 2, Spark cluster has a distinct advantage over single machine. And our improved algorithm further accelerates the matching speed significantly on Spark platform. By comparing the improved algorithm and the original algorithm on Spark platform, the result shows that when the database capacity is small, our improved algorithm has no obvious advantages. However, with the increase of image data capacity, the image retrieval speed is remarkably improved, in case of 1000 k, our algorithm optimized up to 44.7%. From the experimental results, we draw a conclusion that the performance of our CBIR system outperforms the traditional CBIR system.

5 Conclusion

In this paper, an improved CBIR system optimized by SURF-BRISK algorithm, which extract image features and construct feature descriptors, is proposed on Spark. We utilize binary descriptors, which was proposed in the BRISK algorithm, to accelerate retrieval speed and reduce memory consumption instead of floating number descriptors in SURF. On that basis, the RANSAC is introduced to filter out some false detection of matching points. The CBIR system is deployed on Spark distributed computing environment, stores image feature descriptors in HDFS, and utilizes RDD to speed up image retrieval among massive image data. Though a large number of experiments, the results demonstrate that the proposed CBIR system has better retrieval performance for large-scale image data.

For future work, we will be committed to studying how to further improve the performance of feature matching and select a more efficient matching strategy on Spark.

Acknowledgement. The research was partly supported by the program for Professor of Special Appointment (Eastern Scholar) at Shanghai Institutions of Higher Learning, USST incubation project (15HJPY-MS02), National Natural Science Foundation of China (No. U1304616, No. 61502220).

References

1. Deb, S., Zhang, Y.: An overview of content-based image retrieval techniques. International Conference on Advanced Information NETWORKING and Applications. IEEE Xplore, vol. 1, pp. 59–64 (2004)
2. Lowe, D.G.: Object recognition from local scale-invariant features. In: International Conference on Computer Vision. IEEE Computer Society, vol. 2, pp. 1150 (2001)
3. Bay, H., Tuytelaars, T., Gool, L.V.: SURF: speeded UP robust features. Comput. Vis. Image Underst. 110(3), 404–417 (2006)
4. Rublee, E., Rabaud, V., Konolige, K., et al.: ORB: an efficient alternative to SIFT or SURF, vol. 58(11), pp. 2564–2571 (2011)
5. Leutenegger, S., Chli, M., Siegwart, R.Y.: BRISK: binary robust invariant scalable keypoints. In: IEEE International Conference on Computer Vision. IEEE, pp. 2548–2555 (2011)
6. Tungkasthan, A., Premchaiswadi, W.A.: Parallel processing framework using MapReduce for content-based image retrieval. In: International Conference on ICT and Knowledge Engineering, pp. 1–6 (2013)
7. Alhammadi, M.M., Emmanuel, S.: Improving SURF based copy-move forgery detection using super resolution. In: IEEE International Symposium on Multimedia. IEEE Computer Society, pp. 341–344 (2016)
8. Cheon, S.H., Eom, I.K., Ha, S.W., et al.: An enhanced SURF algorithm based on new interest point detection procedure and fast computation technique. J. Real-Time Image Process. 1, 1–11 (2016)
9. Baroffio, L., Canclini, A., Cesana, M., et al.: Briskola: BRISK optimized for low-power ARM architectures. In: IEEE International Conference on Image Processing. IEEE, pp. 5691–5695 (2015)
10. Choi, S.G., Han, S.W.: New binary descriptors based on BRISK sampling pattern for image retrieval. In: International Conference on Information and Communication Technology Convergence. IEEE, pp. 575–576 (2014)
11. Shi, W., Zhu, Y., Huang, T., et al.: An integrated data preprocessing framework based on Apache spark for fault diagnosis of power grid equipment. J. Signal Process. Syst. 86, 1–16 (2016)
12. Tiwary, M., Sahoo, A.K., Misra, R.: Efficient implementation of apriori algorithm on HDFS using GPU. In: International Conference on High Performance Computing and Applications. IEEE, pp. 1–7 (2015)
13. Gu, R., Wang, S., Wang, F., et al.: Cichlid: efficient Large scale RDFS/OWL reasoning with spark. In: IEEE International Parallel and Distributed Processing Symposium. IEEE, pp. 700–709 (2015)
14. Zaharia, M., Chowdhury, M., Franklin, M.J., et al.: Spark: cluster computing with working sets. In: Usenix Conference on Hot Topics in Cloud Computing. USENIX Association, vol. 10(10), p. 95 (2010)

15. Grycuk, R., Knop, M., Mandal, S.: Video key frame detection based on SURF algorithm. In: International Conference on Artificial Intelligence and Soft Computing. Springer, vol. 9119, pp. 566–576 (2015)
16. Bong, K., Kim, G., Hong, I., et al.: A 1.61mW mixed-signal column processor for BRISK feature extraction in CMOS image sensor. In: IEEE International Symposium on Circuits and Systems. IEEE, pp. 57–60 (2014)
17. Griffith, E.J., Chi, Y., Jump, M., et al.: Equivalence of BRISK descriptors for the registration of variable bit-depth aerial imagery. In: IEEE International Conference on Systems, Man, and Cybernetics. IEEE, pp. 2587–2592 (2013)
18. Zhu, W., Sun, W., Wang, Y., et al.: An improved RANSAC algorithm based on similar structure constraints. In: International Conference on Robots & Intelligent System. IEEE, pp. 94–98 (2016)

Fish Species Recognition Based on CNN Using Annotated Image

Tsubasa Miyazono and Takeshi Saitoh[(✉)]

Kyushu Institute of Technology, 680-4 Kawazu, Iizuka, Fukuoka 820-8502, Japan
saitoh@ces.kyutech.ac.jp

Abstract. The objective of our research project is identify the fish species by using image processing technique. This paper proposes a novel feature-points representation method named annotated image. Furthermore, this paper proposes a fish species recognition method based on CNN using proposed annotated image. We collected 50 species of fish images, and applied to the proposed method. As the results, it was confirmed that the annotated image in which all four feature points are plotted into a channel, was obtained the highest recognition accuracy.

Keywords: Fish image · Convolutional neural network · Annotated image

1 Introduction

The objective of our research project is the development an image-based fish species identification system. Since, some poisonous fishes can appear quite similar to non-poisonous fishes, and it is difficult for not only amateurs but specialists to differentiate fish species. By using the image-based identification system, the user can check the name and characteristics of the fishing fish easily before eating the poisonous fish by mistake. Moreover, the system can be used for education by linking with fish picture book.

There are many image based recognition methods for fish species identification [1–10]. Traditional researches used manually designed features, called hand-craft features, such as geometric features [1, 3, 10], color features [11], and texture features [2, 3, 10]. Furthermore, most traditional approaches used a fish image in which it is easy to extract a fish region given a white or uniform background. These photography condition gives a burden on the user. In our research, we adopt an approach that presents several feature points based on manual operations by the user as shown in Fig. 1, however, our approach is able to accept fish images with complicated backgrounds, including against rocky backgrounds. There is a method to automatically detect the feature points. However, the patterns of fish and background are various, and it is difficult to obtain sufficient detection accuracy since the number of data is insufficient. We plan to develop the fish identification system as a smartphone application. Thus, the manual operations for inputting feature points can be easily done by the user.

© Springer Nature Singapore Pte Ltd. 2018
K.J. Kim et al. (eds.), *IT Convergence and Security 2017*,
Lecture Notes in Electrical Engineering 449,
DOI 10.1007/978-981-10-6451-7_19

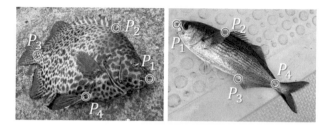

Fig. 1. Target fish images with four feature points.

Recently, deep learning has attracted attention in the fields of artificial intelligence and machine learning. In particular, the Convolutional Neural Network (CNN) has attracted much attention due to its high performance in many tasks in the field of image recognition [12]. In this paper, we propose a novel feature-points representation method named annotated image, and propose a fish species recognition method based on CNN using proposed annotated image. Evaluation experiments are carried out using 50 fish species, and the effectiveness of the proposed method is shown.

2   Proposed Method

2.1   Image Size Normalization

The image sizes of fish images collected in this research are not uniform, however, it is desirable that the image sizes are unified to use a fish image as input data of CNN. Therefore, an image is resized to 256×256 pixels. At this time, the original image is not necessarily a square shape. In this normalization process, the image is resized that the long side becomes 256 pixels, and give the pixel value of $(R, G, B) = (127, 127, 127)$ for all the margin pixels. Sample images of Fig. 2 are normalized images of Fig. 1.

Fig. 2. Resized fish images.

2.2 Annotated Image

The fish image used in this research is contained four feature points; mouth P_1, dorsal fin P_2, caudal fin P_3, and anal fin P_4 as shown in Fig. 1. In order to add these feature points to the image, one of the approaches is to plot the feature points on the original image as shown in Fig. 3. However, in this case, the original image information is lost by plotting points. Therefore, instead of giving information to the original image, one or more feature-point channels having the same size as the original image is prepared, and feature points are plotted on this channel. That is, the original image is a color (RGB) image composed of three channels of red, green, and blue, and a proposed annotated image is composed $(3+n)$ channels as shown in Fig. 4 including generated n feature-points channel.

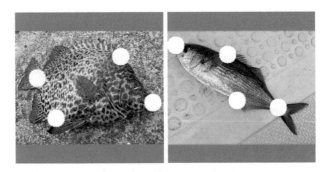

Fig. 3. Target fish images with four feature points.

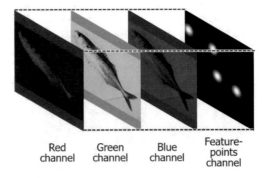

Red channel   Green channel   Blue channel   Feature-points channel

Fig. 4. Four-channels image.

As mentioned above, our fish images have four feature points information. We can consider various patterns of feature-points channel, e.g., a channel image P_{1234} which all four feature points are plotted on one channel, a channel image P_i which i-the feature point is plotted on the channel. Regarding the plotting method of feature points, it is conceivable to plot only one pixel for one feature point. However, in this plotting method, it is considered that the plot point is too small compared with the whole image size, and information on the feature point is useless. Therefore, in this research, we prepare a background image in which all pixel values set to zero, give 255 to the pixel value of the feature point, and apply Gaussian filter with σ. We call this image as the feature-point channel image.

Figure 5 shows some channel images of the right side of Fig. 2. Figure 5(a), (b) and (c) are three channel images of R, G, and B. Figure 5(d) is a channel image P_{1234} which plotted four feature points, Fig. 5(e), (g) and (h) are four channel images in which one feature point is plotted, respectively. Figure 5(i) is a channel image P_{24} which plotted two feature points P_2 and P_4, and Fig. 5(j) is a channel image P_{234} which plotted three feature points P_2–P_4.

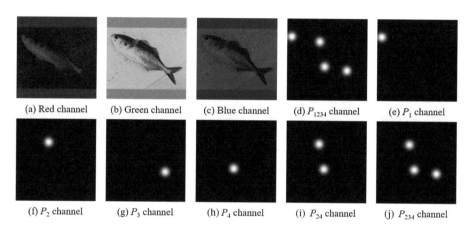

(a) Red channel	(b) Green channel	(c) Blue channel	(d) P_{1234} channel	(e) P_1 channel
(f) P_2 channel	(g) P_3 channel	(h) P_4 channel	(i) P_{24} channel	(j) P_{234} channel

Fig. 5. Channel images.

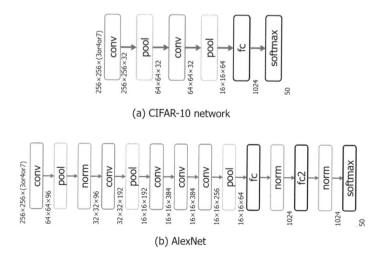

(a) CIFAR-10 network

(b) AlexNet

Fig. 6. CNN architectures.

2.3   CNN Architectures

This paper uses two well-known CNN architectures of CIFAR-10 network and AlexNet. CIFAR-10 network consists of two convolutional layers, a pooling layer, and a fully-connected layer. On the other hand, AlexNet consists of five convolutional layers, three pooling layers, and a fully-connected layer. The structure of both is shown in Fig. 6.

In CNN, there is a learning approach by using parameters of pre-training network. However, our approach has feature points, and it is difficult to prepare data for pre-training. Therefore, this paper adopts an approach that gives random initial values without applying pre-training.

3   Experiment

3.1   Dataset

We selected 50 fish species in the northern area of Kitakyushu in Japan as the recognition targets. Sample fish images are shown in Fig. 7. We uniquely collected 20 distinct samples for each of these species through the Web, and totally collected 1000 fish images. The photographer is an unspecified number, the shooting environment is different, and the image size is also different. Note that, all fish images were checked and confirmed its name by experts. We prepared four feature points: mouth, dorsal fin, caudal fin, and anal fin, by the manual for all fish images.

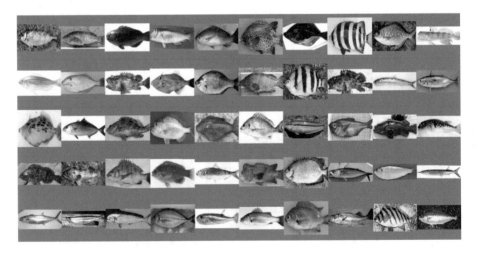

Fig. 7.   50 species of fish images.

3.2   Recognition Results

In our dataset, the sample number of each species was 20. We divided these samples into four groups, and applied 4-fold cross validation, i.e., of the 20 samples for each species, 15 samples represented the training set, while the remaining sample was a test set. By varying one sample, the total number of recognition trials was four for each species. The average recognition rate of four results was used to calculate the resulting accuracy measures.

In order to verify the effectiveness of image with feature-points, seventeen patterns images were generated and recognition experiments were done. Average recognition accuracy of all patterns are shown in Table 1. In the table, N_p is the number of plotted feature points, N_c is the number of channels of input image, R_C is the recognition rate when using the CIFAR-10 network model, and R_A is the recognition rate using the AlexNet model.

Table 1. Recognition results with various conditions.

#	Input pattern	N_p	N_c	R_C [%]	R_A [%]
(1)	RGB	0	3	41.3	52.1
(2)	RGB + P_1	1	4	41.4	53.1
(3)	RGB + P_2	1	4	49.8	59.7
(4)	RGB + P_3	1	4	43.2	56.0
(5)	RGB + P_4	1	4	49.1	60.9
(6)	RGB + P_{12}	2	4	50.2	57.7
(7)	RGB + P_{13}	2	4	44.3	57.9
(8)	RGB + P_{14}	2	4	46.5	61.2
(9)	RGB + P_{23}	2	4	52.5	64.4
(10)	RGB + P_{24}	2	4	58.5	67.4
(11)	RGB + P_{34}	2	4	50.2	65.4
(12)	RGB + P_{124}	3	4	58.1	67.9
(13)	RGB + P_{234}	3	4	58.3	69.9
(14)	RGB$P_1P_2P_3P_4$	4	3		65.4
(15)	RGB + P_{1234}	4	4	**60.9**	**71.1**
(16)	RGB + $P_1P_2P_3P_4$	4	7	57.1	70.9
(17)	$P_1P_2P_3P_4$	4	1		57.8

In case of pattern (1), that is, when the inputting original color image RGB, recognition accuracy is low, and recognition accuracy is improved by adding one or more feature-points channels. Pattern (14) is used the annotated image which all feature points are plotted on the original image as shown in Fig. 3. The annotated image of pattern (16) has the largest N_c among all patterns. On the other hand, the annotated image of pattern (17) has the smallest N_c, and this image is not included the original image, but only feature-points channel. The highest recognition rate was obtained with (15) RGB + P_{1234} using all the feature points. Also a trend of $R_C < R_A$ is observed. This is guessed that the architectural complex AlexNet model is better than the CIFAR-10 network model.

Since our objective is a fish identification system, we note that it is better to consider not only the first candidate, but also several candidates. Figure 8 shows performance curves of four experimental conditions that indicate how often the correct classes for a query were placed among the top k matches, with k varying from one to 10. The recognition accuracy of 91.4% was obtained by considering five candidates, when condition (14) RGB + P_{1234} was inputted and AlexNet was used.

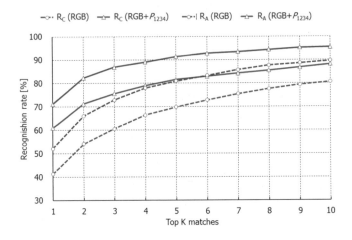

Fig. 8. Recognition results.

4   Conclusion

The key contributions of this paper are as follows: this paper proposes a novel feature-points representation method named annotated image which obtained higher recognition accuracy than the original RGB color image. This paper proposed a fish species recognition method based on CNN using the proposed annotated image. We collected 50 species of fish images through the Internet. We carried out recognition experiments with some patterns of annotated image, and obtained the recognition accuracies of 71.1 and 91.4% by considering one and five candidates, respectively.

In the experiment, the target number of species was only 50. The future task is to demonstrate the effectiveness of the proposed method by carrying out the recognition experiment with an increased number of species.

References

1. Storbecka, F., Daan, B.: Fish species recognition using computer vision and a neural network. Fish. Res. **51**, 11–15 (2001)
2. Alsmadi, M.K., Omar, K.B., Noah, S.A., Almarashdeh, I.: Fish classification based on robust features extraction from color signature using back-propagation classifier. J. Comput. Sci. **7**, 52–58 (2011)
3. Huang, P.X., Boom, B.J., Fisher, R.B.: Underwater live fish recognition using a balance-guaranteed optimized tree. In: ACCV, vol. 7724, pp. 422–433 (2012)

4. Mushfieldt, D., Ghaziasgar, M., Connan, J.: Fish identification system. In: Proceedings of South African Telecommunication Networks and Applications Conference (SATNAC 2012), pp. 231–236 (2012)
5. Pornpanomchai, C., Lurstwut, B., Leerasakultham, P., Kitiyanan, W.: Shape- and texture-based fish image recognition system. Kasetsart Journal (Natural Science) **47**, 624–634 (2013)
6. Chuang, M.C., Hwang, J.N., Williams, K.: Supervised and unsupervised feature extraction methods for underwater fish species recognition. In: 2014 ICPR Workshop on Computer Vision for Analysis of Underwater Imagery, pp. 33–40 (2014)
7. Nasreddine, K., Benzinou, A.: Shape-based fish recognition via shape space. In: Proceedings of 23rd European Signal Processing Conference (EUSIPCO 2015), pp. 145–149 (2015)
8. Spampinato, C., Palazzo, S., Joalland, P.H., Paris, S., Glotin, H., Blanc, K., Lingrand, D., Precioso, F.: Fine-grained object recognition in underwater visual data. In: Multimedia Tools and Applications, pp. 1–20 (2015)
9. Wang, G., Hwang, J.N., Williams, K., Wallace, F., Rose, C.S.: Shrinking encoding with two-level codebook learning for fine-grained fish recognition. In: 2nd Workshop on Computer Vision for Analysis of Underwater Imagery, pp. 31–36 (2016)
10. Saitoh, T., Shibata, T., Miyazono, T.: Feature points based fish image recognition. Int. J. Comput. Inf. Syst. Ind. Manag. Appl. **8**, 12–22 (2016)
11. Chambah, M., Semani, D., Renouf, A., Courtellemont, P., Rizzi, A.: Underwater color constancy: Enhancement of automatic live fish recognition. In: Proceedings of SPIE. vol. 5293, pp. 157–168 (2003)
12. Krizhevsky, A., Sutskever, I., Hinton, G.E.: Imagenet classification with deep convolutional neural networks. In: Advances in neural information processing systems, pp. 1097–1105 (2012)

Head Pose Estimation Using Convolutional Neural Network

Seungsu Lee and Takeshi Saitoh[✉]

Kyushu Institute of Technology,
680-4 Kawazu, Iizuka, Fukuoka 820-8502, Japan
saitoh@ces.kyutech.ac.jp

Abstract. Estimating the head pose is an important capability of a robot when interacting with humans. But there are many difficulties of human head pose estimation, such as extreme pose, lighting, and occlusion, has historically hampered. This paper addresses the problem of head pose estimation with two degrees of freedom (pitch and yaw) using a low-resolution image. We propose a method that uses convolutional neural networks for training and classifying various head poses over a wide range of angles from a single image. Evaluations on public head pose database, Prima database, demonstrate that our method achieves better results than the state-of-the-art.

Keywords: Head pose estimation · Convolutional neural network · Low-resolution image

1 Introduction

Estimation of head pose can be used for human-robot interaction (HRI), driver assistance systems (DAS). But, it still has the problem with large angle estimation using a low-resolution image. It is difficult to estimate facial feature points when occlusion is occurred in the face. This paper focuses on image-based head pose classification to solve that kinds of problems.

In recent years, deep learning is very effective in artificial intelligence and machine learning. Especially, convolutional neural network (CNN) has been successfully applied to computer vision tasks such as image classification [1]. Unlike the ordinary neural network, CNN is composed of convolution layers, pooling layers and fully-connected layers. In traditional machine learning, feature selection is a time-consuming manual process. However, CNNs can learn to extract generic features that can be used to train a new classifier to solve the classification problem automatically. Benefit from this advantage, this paper proposes head pose estimation based on CNN. Our CNN structures are consisted of three convolutional layers, two max-pooling layers, and one or no fully-connected layers. Experimental evaluations show that our head pose estimation method outperforms state-of-the-art methods.

The remainder of the paper is organized as follows: Sect. 2 describes related head pose estimation methods. Section 3 outlines the CNN architecture and data augmentation. Dataset, experimental conditions and results are explained in Sect. 4 and conclusions are drawn in Sect. 5.

© Springer Nature Singapore Pte Ltd. 2018
K.J. Kim et al. (eds.), *IT Convergence and Security 2017*,
Lecture Notes in Electrical Engineering 449,
DOI 10.1007/978-981-10-6451-7_20

2   Related Works

In this section, we briefly discuss approach of the head pose estimation. There is head pose estimation literature that present many approaches of head pose estimation [2]. Local or global approaches exist for head pose estimation. The former approaches usually estimate head pose using facial landmarks such as eyes, lip corners and nose tip. However, the detection of facial features tends to be sensitive to partial changes of illumination, person and pose variations. Robust techniques have been proposed to handle such variations [3, 4]. But, these require high resolution images of the face and tracking can fail when certain facial features are occluded. On the other hand, the latter approaches use the entire image of the face to estimate head pose. There are some advantages, no facial landmark, no face model, are required. These approaches can accommodate very low resolution images of the face. A global approach using linear auto-associative neural networks is proposed by Gourier et al. [5]. Local approach using probabilistic high-dimensional regression method is proposed by Drouard et al. [6]. They proposed method based on learning a mixture of linear regression model that maps high-dimensional HOG-based descriptors onto the low-dimensional space of head poses. Both of researches make the results of head pose estimation using Prima database.

3   Proposed Methods

3.1   CNN Architecture

Prima database has only low resolution images and no many samples. It is difficult to use high resolution input of image and very deep CNN structure because of vanishing/exploding gradients problem. Then, our CNN architectures are based on LeNet-5 [7] which is proposed by LeCun et al. and Malagavi et al. [8]. Each of our model is shown in Fig. 1. Two kinds of our architectures have same number of convolution layers and pooling layers and also designed for same input image size of 32×32. We changed the output of feature map size in every convolution layers and number of fully connected layers. Two models have difference with number of fully-connection and outputs of convolution, fully-connection. After the image convolution, rectified linear unit (ReLU) is used for the activation function and max pooling is used in pooling layers. Finally, output layer shows thirteen outputs of horizontal direction and nine outputs of vertical direction.

In our experiments, we used adaptive moment estimation method for gradient descent optimization. Batch normalization also used in conv1 and conv2 layers to reduce internal covariate shift. Batch normalization allows us to use much higher learning rates and be less careful about initialization [9].

3.2   Data Augmentation

Because of lack of training and test dataset, overfitting problem occur easily, and getting many samples of dataset consume bags of times. As known as data augmentation can solve lack of dataset problem to make parallel shift images, mirror images,

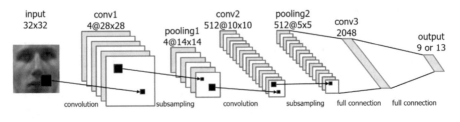

(a) Model1

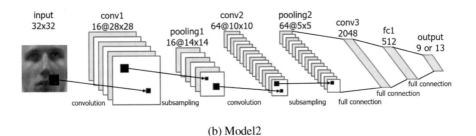

(b) Model2

Fig. 1. CNN architectures.

rotated images, gamma corrected images, etc. To get the good performance, we generate extra images of dataset using seven kinds of data augmentation.

As shown in Fig. 2, we generated seven extra images from every single original image. (1) Original image is shown in Fig. 2(a) and after the 10×10 moving average filtered image is shown in Fig. 2(b). (2) Histogram equalization image is shown in Fig. 2(c). (3) Two kinds of linear density transformation shown in the following formulas.

$$z' = \frac{z - a}{b - a} z_m, \tag{1}$$

$$z' = \frac{b - a}{z_m} z + a, \tag{2}$$

where two parameters z', z are meaning of density value of output and input images, $z_m = 255$ means maximum density value. If given $a = 50$ and $b = 205$, we can get two image as shown in Figs. 2(d) and (e) from Eqs. 1 and 2. (4) To apply gamma correction using the equation below.

$$z' = z_m \left(\frac{z}{z_m}\right)^{1/\gamma} \tag{3}$$

We set the $\gamma = 0.75$, $\gamma = 1.5$. Generated image is shown in Figs. 2(f) and (g). (5) Make a copy for original image.

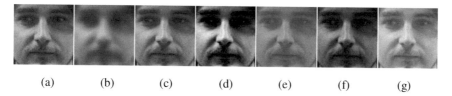

   (a)   (b)   (c)   (d)   (e)   (f)   (g)

Fig. 2. Example of data augmentation.

4 Experiments

4.1 Dataset

Collecting data for head pose estimation and annotating collected images with an exact head orientation are difficult tasks. To evaluate and compare head pose estimation systems, we try to test of performance using the Prima database [3].

    This public benchmarking database is to ask the participants to look at a set of markers that are located in predefined direction in the measurement room. There are 2790 monocular face images of fifteen persons with variation of horizontal and vertical angles from −90 to +90 degrees. For every person, two series of 93 images are available. A sample of Prima database images is shown in Fig. 3. The subjects range in age from 20 to 40 years old, five possessing facial hair and seven searing glasses. In addition, face positions on each image are labeled also. We crap the image from the label to use our experiments. 93 different poses of subject are shown in Fig. 4. Actually, face label size of Prima database is not constant. We generated the 32×32 resized images and gray scaled too.

Fig. 3. Example of original image from Prima Database

4.2 Experimental Conditions

The purpose of having two series per person is to be able to training and test algorithms on known and unknown faces. In the known faces experiment, estimating the head pose of known faces is done by splitting the database into two groups. Each group gathers sets of the same series of all persons. The test is done by doing a 2-fold cross

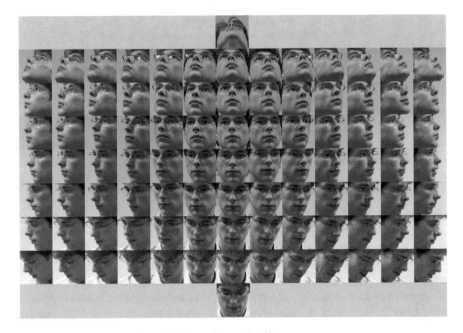

Fig. 4. Example of 93 different poses.

validation on these two groups. In the unknown faces experiment, estimating the head pose of unknown faces is done by doing a leave-one-out algorithm on the persons of the database. All images are used for training, except images of one subject, which will be used for testing. In our research, consider each head poses as a class. Also horizontal and vertical angles are separated with two kinds of experiments. Therefore, we need to classify thirteen classes for horizontal angles and nine classes for vertical angles.

4.3   Results

This section will compare the performance between proposed method and other methods. We got the head pose classification results about model 1 and model 2, are shown in Table 1. This table shows the competitive result from Gourier et al. [5] and Drouard et al. [6]. If the mean absolute error of angle shows smaller value, it means high accuracy of classification. Both of our models shows high accuracy and model 2 shows more high accuracy than model 1 which have no fully-connected layer. There are confusion matrices from the known face experiment result using model 2, are shown in Table 2. Confusion matrices from the unknown face experiment result using model 2, are shown in Table 3.

We calculated the average classification rate in the horizontal and vertical angles under each condition when allowing an error of ±15 degrees. As for the known face, the horizontal and vertical angles of 97.61 and 89.29% were obtained, respectively. As for the unknown face, the horizontal and vertical angles of 96.01 and 84.25% were obtained, respectively.

Table 1. Head pose classification results.

(a) Known face.

Method	Error [deg]		Classification [%]	
	Hor.	Ver.	Hor.	Ver.
Gourier et al. [5]	7.3	12.1	61.3	53.8
Drouard et al. [6]	6.7	7.2	–	–
Model1	5.53	6.15	67.58	74.02
Model2	5.17	5.36	69.88	77.87

(b) Unknown face.

Method	Error [deg]		Classification [%]	
	Hor.	Ver.	Hor.	Ver.
Gourier et al. [5]	10.3	15.9	50.4	43.9
Drouard et al. [6]	7.5	7.3	–	–
Model1	7.19	9.07	60.34	62.83
Model2	6.50	7.73	62.45	66.37

Table 2. Confusion matrices from the known face results using model2.

(a) Horizontal

	-90	-75	-60	-45	-30	-15	0	15	30	45	60	75	90
-90	76	21	2	0	0	0	0	0	0	0	0	0	1
-75	21	64	14	1	0	0	0	0	0	0	0	0	0
-60	1	12	72	14	0	0	0	0	0	0	0	0	0
-45	0	1	15	73	11	0	0	0	0	0	0	0	0
-30	0	0	0	12	73	15	0	0	0	0	0	0	0
-15	0	0	0	0	11	77	11	0	0	0	0	0	0
0	0	0	0	0	0	6	81	11	0	0	0	0	1
15	0	0	0	0	0	0	7	80	13	0	0	0	0
30	0	0	0	0	0	0	1	14	72	13	0	0	0
45	0	0	0	0	0	0	0	1	11	64	20	30	1
60	0	0	0	0	0	0	0	0	1	17	56	21	4
75	0	0	0	0	0	0	0	0	0	3	21	49	27
90	0	0	0	0	0	0	0	0	0	1	2	24	72

(b) Vertical

	-90	-60	-30	-15	0	15	30	60	90
-90	59	39	2	0	0	0	0	0	0
-60	1	84	13	1	0	0	0	0	0
-30	0	11	74	14	1	1	0	0	0
-15	0	0	15	69	14	1	0	0	0
0	0	0	1	12	77	10	0	0	0
15	0	0	0	0	13	71	16	0	0
30	0	0	0	0	0	9	86	4	0
60	1	1	0	0	0	0	3	94	2
90	0	0	0	0	0	0	3	10	87

Table 3. Confusion matrices from the unknown face results using model2.

(a) Horizontal

	-90	-75	-60	-45	-30	-15	0	15	30	45	60	75	90
-90	71	24	3	0	0	0	0	0	0	0	0	0	1
-75	23	55	20	1	1	0	0	0	0	0	0	0	0
-60	1	16	66	17	1	0	0	0	0	0	0	0	0
-45	0	1	13	74	11	0	0	0	0	0	0	0	0
-30	0	0	0	13	70	16	0	0	0	0	0	0	0
-15	0	0	0	0	15	72	12	0	0	0	0	0	0
0	0	0	0	0	0	13	74	12	0	0	0	0	1
15	0	0	0	0	0	0	16	65	18	1	0	0	0
30	0	0	0	0	0	0	2	24	57	15	1	0	0
45	0	0	0	0	0	0	0	3	20	54	19	3	2
60	0	0	0	0	0	0	0	0	6	21	44	21	7
75	0	0	0	0	0	0	0	0	1	7	23	40	29
90	0	0	0	0	0	0	0	0	0	1	7	23	69

(b) Vertical

	-90	-60	-30	-15	0	15	30	60	90
-90	60	39	1	0	0	0	0	0	0
-60	2	78	18	1	0	0	0	0	0
-30	0	21	57	17	3	2	0	0	0
-15	0	3	23	53	18	3	0	0	0
0	0	0	4	23	53	19	1	0	0
15	0	0	0	3	21	52	23	1	0
30	0	0	0	0	3	18	68	11	0
60	0	0	0	0	0	0	9	89	2
90	0	0	0	0	0	0	1	12	87

5   Conclusion

In this research, we proposed head pose classification method using CNN. It was highly accurate using public dataset, and confirmed that higher classification rates can be obtained compared with the conventional methods.

The horizontal angles and vertical angles are classified separately in our method. For the separated classification, we consume additional computation times and cost now. We need to consider the CNN architecture that can train and classify at once.

References

1. Krizhevsky, A., Sutskever, I., Hinton, G.E.: Imagenet classification with deep convolutional neural networks. In: Advances in Neural Information Processing Systems, pp. 1097–1105 (2012)
2. Murphy-Chutorian, E., Trivedi, M.M.: Head pose estimation in computer vision: a survey. IEEE Trans. Pattern Anal. Mach. Intell. **31**, 607–626 (2009)

3. Gourier, N., Hall, D., Crowley, J.L.: Estimating face orientation from robust detection of salient facial structures. In: FG Net Workshop on Visual Observation of Deictic Gestures (2004)
4. Wu, J., Pedersen, J.M., Putthividhya, D., Norgaard, D., Trivedi, M.M.: A two-level pose estimation framework using majority voting of gabor wavelets and bunch graph analysis. In: ICPR Workshop Visual Observation of Deictic Gestures, Citeseer (2004)
5. Gourier, N., Maisonnasse, J., Hall, D.: Head pose estimation on low resolution images. In: First International Evaluation Workshop on Classification of Events, Activities and Relationships, pp. 270–280 (2006)
6. Drouard, V., Ba, S., Evangelidis, G., Deleforge, A., Horaud, R.: Head pose estimation via probabilistic high-dimensional regression. In: IEEE International Conference on Image Processing (ICIP), pp. 4624–4628 (2015)
7. LeCun, Y., Bottou, L., Bengio, Y., Haffner, P.: Gradient-based learning applied to document recognition. IEEE **86**, 2278–2324 (1998)
8. Malagavi, N., Hemadri, V., Kulkarni, U.P.: Head pose estimation using convolutional neural networks. J. Innov. Sci. Eng. Technol. **1**, 470–475 (2014)
9. Ioffe, S., Szegedy, C.: Batch normalization: accelerating deep network training by reducing internal covariate shift. arXiv preprint arXiv:1502.03167 (2015)

Towards Robust Face Sketch Synthesis with Style Transfer Algorithms

Philip Chikontwe[1(✉)] and Hyo Jong Lee[2]

[1] Division of Computer Science and Engineering, Chonbuk National University,
Jeonju, South Korea
chamaphilip@gmail.com
[2] Center for Advanced Image and Information Technology, Jeonju 561-756, South Korea
hlee@chonbuk.ac.kr

Abstract. We propose an approach for face sketch synthesis by employing deep image transformations using an artistic style transfer algorithm. Face sketch synthesis remains an area of great interest in the research community as well as its applications in law enforcement towards face recognition. Recent methods for this problem typically employ traditional approaches to synthesize face sketches to digital images. However, most approaches are gradually shifting towards convolutional neural networks for robust feature learning and image transformations. In this paper, we propose an approach that uses recent artistic style transfer algorithms for face sketch synthesis. Additionally, we show that poorly synthesized images can be improved with a denoising autoencoder for better facial feature reconstruction. Further, the approach is extended to perform face verification of heterogeneous image samples to assess the effectiveness of the proposed approach and gives a better view into the potential applications for styling algorithms for face image synthesis and transformation problems alike.

Keywords: Convolutional neural networks · Deep learning · Face recognition · Face synthesis · Style transfer

1 Introduction

In the domain of heterogenous face recognition [1, 2], cross-modal face matching problems exist mainly due to the modality differences between images. A domain that encompasses a vast array of image types such as infrared, 2D as well as 3D images for matching for classification tasks and recognition related challenges has been considered an ill posed problem. To that end, in the case of face sketch-photo matching, face image style transformations have played a great role in addressing this problem. More specifically, face synthesis [3–5] has enabled the effective use of modern face recognition algorithms with seamless ease. In addition, this research domain's possible application areas remain a matter of interest in law enforcement agencies and industry alike.

To achieve robust face synthesis with visually appealing results we consider image synthesis to be largely a problem of texture transfer. The difference in modality between

© Springer Nature Singapore Pte Ltd. 2018
K.J. Kim et al. (eds.), *IT Convergence and Security 2017*,
Lecture Notes in Electrical Engineering 449,
DOI 10.1007/978-981-10-6451-7_21

sketch images and images can be largely attributed to the lack of texture and minute features in sketch images [2]. The problem of face recognition is considered more challenging as compared to what is considered homogeneous recognition, by leveraging a texture transfer approach [6] it is possible to preserve the semantic content of the face image as well as retain a realistic structure of the sketch image like the photo. To that end, we assess the effectiveness of texture synthesis incorporated in the artistic style algorithm for our task. The recent algorithms [6–9] have shown to produce visually outstanding synthetic images that seamlessly retain texture and structure of a given content and style image.

In this work, we investigate the proposed approach on viewed face sketches. Viewed sketches are face sketch images drawn by observation of a corresponding photo image for a duration of time and are generally considered to have visually similar structures and features. Thus, it may be considered rather feasible to perform face matching between samples due to this, however may produce undesired results and thus remains challenging. Therefore, we employ the Chinese University of Hong Kong CUFS [3, 10–12] database to evaluate the proposed approach. In this work, our goal is to investigate the effectiveness of recent neural style algorithms applied to the problem of face sketch synthesis and consequently extend the results to perform face verification.

The rest of the paper is organized as follows: Sect. 2 gives insight into related works for this problem, Sect. 3 presents the neural style transfer algorithm used. Section 4 gives an overview of denoising convolutional autoencoder employed for image denoising and restoration. Section 5 presents the experimental procedures used and based on the results observed in Sect. 6 conclusions are drawn in Sect. 7.

2 Related Work

The authors in [5] proposed an automatic sketch-photo synthesis and retrieval algorithm by leveraging sparse representations at patch level to improve the quality of synthesized images. By learning mappings of sketch patches and photo patches using the sparse representations the approach was extended to retrieval with extensive experimentation. In contrast, authors in [13] used 2D discrete Haar wavelet transform to bring sketches and photos to a new dimension followed by a negative approach with Principal Component Analysis (PCA) for feature extraction eventually using KNN and SVM for more robust classification.

In recent works, the impact of design choices through lengthy benchmarking of results on different databases is evaluated on a CNN pre-trained on visible spectrum images for heterogenous recognition [14]. As deep learning approaches become more prevalent, most researchers continue to transition from traditional approaches towards deep representation driven approaches as in [15] where face sketches are inverted using deep neural networks to produce realistic synthetic images in both controlled and uncontrolled settings. A large part of the research makes use of batch normalization and stochastic optimization to produce appealing images. In addition, the work in [16] addressed the challenge of image restoration using a denoising autoencoder network

[17] as prior and achieved good results on non-blinded deconvolutions and super-resolution [18].

Inspired by recent approaches, we employ a denoising autoencoder to restore poorly synthesized images in contrast to approach used in [5]. Additionally, we employ an algorithm for artistic styling [6, 8] of images for synthesis that is based on convolutional neural networks. Thus, we further show face verification can be performed on the synthesized images for matching using recent deep representation models trained on large image datasets similar to work in [19].

3   Artistic Neural Style Algorithm

The algorithm [6, 8] originally designed for artistic styling of images can be leveraged to perform robust texture synthesis and enables us to render the semantic content of the face image with the style of the face sketch image. The challenge of separating content and style of given input images has been mitigated by recent advances in computer vision with deep learning systems able to extract high level semantic information. The algorithm further shows that feature representations from large convolutional networks such as the VGG-16 can be leveraged and reduced to an optimization problem within one deep network. VGG-16 is ideal for this problem considering it is a large deep network trained on large datasets learning high level features for image recognition.

Formally, a style transfer method captures the content of an input image and style of a given artistic image. To that end, with these inputs a new image that captures the style and content of both can be generated. A common loss function defines the relationship between two feature representations and minimizes the difference between the entries given by feature correlations from the Gram matrix. The feature correlations are given by $G^l \in R^{N_l \times N_l}$ with G^l_{ij} as the inner product between the vectorised features maps in different layers. The formal definition is illustrated in Eq. 1 as:

$$G^l_{ij} = \sum_k F^l_{ik} F^l_{jk}.$$

(1)

where the total loss of style and content is defined as:

$$L_{total}(\vec{p}, \vec{a}, \vec{x}) = \alpha L_{content}(\vec{p}, \vec{x}) + \beta L_{style}(\vec{a}, \vec{x})$$

(2)

4   Denoising Autoencoder

The difficulties encountered in learning robust generative models with deep learning can be addressed the use of unsupervised learning to map inputs to outputs [20]. In this work, convolutional autoencoders are employed to address the issue of robust image reconstruction and denoising. Denoising autoencoders are a branch of classical autoencoder inertly designed to map noisy images to a clear noise free image. To recall, a classical autoencoder is considered to take an input vector $x \in [0, 1]^d$ and maps it to a hidden representation $y \in [0, 1]^{d'}$ through a deterministic mapping function $y = f_\theta(x) = s(Wx + b)$ with

$\theta = \{W, b\}$ serving as parameters. Where W is defined as a weighting matrix and b is a bias vector. Further, each training sample x is mapped to corresponding y, the model's parameters are optimized to minimize the average reconstruction error based on the equation:

$$\theta^*, \theta'^* = \arg\min \frac{1}{n} \sum_{i=1}^{n} L\left(x^{(i)}, z^{(i)}\right)$$

$$= argmin_{\theta,\theta'} \frac{1}{n} \sum_{i=1}^{n} L\left(x^{(i)}, g_{\theta'}\left(f_\theta\left(x^{(i)}\right)\right)\right)$$

(3)

where the traditional squared error is defined as $L(x, z) = ||x - z||^2$.

Poorly reconstructed images are considered as noisy images to be feed to the denoising model that consists of two convolutional layers, where we use 7×7 filters and 5×5 filters followed by rectified linear units, batch normalization and pooling layers respectively. This section of the network is considered an encoder and maps grayscale inputs of size 64×64 dimensions to a 4096-feature vector. The decoder mirrors the configuration of encoder as it aims to reconstruct the feature vector to its original form without noise.

5   Experiments

The Chinese University Face Sketch Database (CUFS) sample face sketch images are used in our experimental procedures. The CUFS consists of 188 face sketch pairs from the Chinese University of Hong Kong as well as 1194 face sketch pairs from the FERET [21, 22] database which can be obtained freely for research purposes. For training the autoencoder and evaluation of the style network, an initial preprocessing step is performed on the samples where each face image is aligned and normalized to contain only the frontal face region and resized to 64×64 dimensions. The autoencoder was trained for 5000 iterations on a Nvidia Titan X GPU system and train test splits are performed on the samples by preserving 80% samples for training and the remainder for testing the model. Figure 1 shows an overview of the proposed framework that goes beyond synthesis towards face verification.

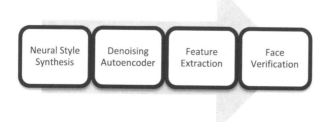

Fig. 1. Proposed pipeline for the framework starting from synthesis to face verification.

To evaluate the effectiveness of the style transfer algorithm, we consider a given face photo input image as style image and a corresponding face sketch image as the content image respectively. By employing the deep network pretrained weights of the VGG-16 model we can leverage high level features to produce a new generated image that captures both the texture and style of the given inputs to produce a photo realistic image. Arbitrary values are chosen as an initial step for two given inputs, considering the approach is a slow styling procedure i.e. for given inputs 500 to 1000 epochs are run to produce realistic synthetic images. A high value for the content weight parameter is chosen as initial and from observed styling results after a given number of epochs the total variation loss is adjusted to attain better results. The experimental procedure is maintained for different samples and the styling results per image are analyzed to observe the effect of using high or relatively low weighting parameters.

6   Results

The results of the procedures described in the experiments section are presented in this section. A reconstruction loss is reported based on the mean squared error as shown in Fig. 2. To that end, given samples from the datasets used in our approach it was noted the styling algorithm did not successfully reconstruct certain images as can be observed in Fig. 3. To resolve the poor synthesis, the proposed denoising autoencoder was employed to produce visibly recognizable images. To this end, we perform face verification by employing the VGG-16 for feature extraction and Support Vector Machine algorithm SVM [23, 24] is used for matching.

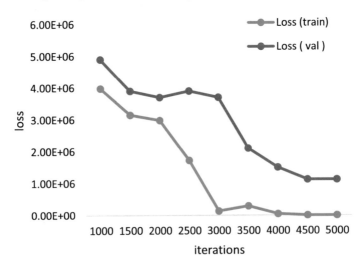

Fig. 2. An overview of the train loss and validation loss of the denoising autoencoder on CUFS face image samples.

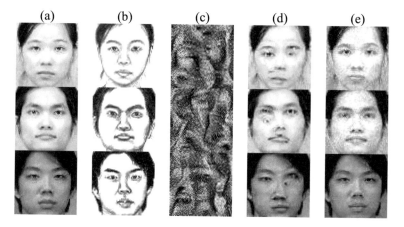

Fig. 3. Results of applying the style transfer algorithm applied to samples from the CUFS dataset. (a) style image (b) content image (c) random generated image after 1 iteration (d) final reconstructed image (e) denoised image using autoencoder.

The resulting images obtained are further used for matching based on features. By employing the VGG-16 deep model for feature extraction, matching is performed with SVM classifier achieving an encouraging similarity score of 90% on average for well reconstructed images. In addition, the style transfer algorithm otherwise considered slow style employed involves a lengthy processing time for a single image taking 3 to 7 min for reconstruction with 1000 epochs. Consequently, only a few samples are presented in this paper and a recognition approach is avoided as would require more time to synthesize larger datasets using the approach proposed. However, the results show the potential of the approach to be applied to larger datasets without using pretrained models as in this work as well as give a better view of style algorithms weighting configurations applicable in fast styling approaches.

7 Conclusion

In this work, the use of a style transfer algorithm was investigated on its effectiveness in the problem of face sketch synthesis. Further, accounted for poorly synthesized images by application of a denoising autoencoder and explored face matching based on features extracted from a deeply learned model. This work presented the advantages of using deep generative models for robust synthesis. In addition, highlighted a possible framework for real applications from synthesis towards face verification. We showed the approach is effective for producing photorealistic images with visually appealing structure and texture.

Acknowledgements. This research was supported by Basic Science Research Program through the National Research Foundation of Korea (NRF) funded by the Ministry of Education (GR 2016R1D1A3B03931911). This research was supported by the MSIT(Ministry of Science and ICT), Korea, under the ITRC(Information Technology Research Center) support program

(IITP-2017-2015-0-00378) supervised by the IITP(Institute for Information & Communications Technology Promotion).

References

1. Ashwini, B., Prajakta, K.: A survey of face recognition from sketches. Int. J. Latest Trends Eng. Technol. **6**(3), 150–158 (2016)
2. Ouyang, S., et al.: A survey on heterogeneous face recognition: Sketch, infra-red, 3D and low-resolution. arXiv preprint arXiv:1409.5114 (2014)
3. Wang, X., Tang, X.: Face photo-sketch synthesis and recognition. IEEE Trans. Pattern Anal. Mach. Intell. **31**(11), 1955–1967 (2009)
4. Gao, X., et al.: Face sketch synthesis algorithm based on E-HMM and selective ensemble. IEEE Trans. Circuits Syst. Video Technol. **18**(4), 487–496 (2008)
5. Gao, X., et al.: Face sketch–photo synthesis and retrieval using sparse representation. IEEE Trans. Circuits Syst. Video Technol. **22**(8), 1213–1226 (2012)
6. Gatys, L.A., Ecker, A.S., Bethge, M.: Image style transfer using convolutional neural networks. In: Proceedings of the IEEE Conference on Computer Vision and Pattern Recognition (2016)
7. Johnson, J., Alahi, A., Fei-Fei, L.: Perceptual losses for real-time style transfer and super-resolution. In: European Conference on Computer Vision. Springer (2016)
8. Gatys, L.A., Ecker, A.S., Bethge, M.: A neural algorithm of artistic style. arXiv preprint arXiv: 1508.06576 (2015)
9. Ulyanov, D., Vedaldi, A., Lempitsky, V.: Instance normalization: the missing ingredient for fast stylization. arXiv preprint arXiv:1607.08022 (2016)
10. Jain, A.K., Klare, B., Park, U.: Face matching and retrieval in forensics applications. IEEE Multimed. **19**(1), 20 (2012)
11. Tang, X., Wang, X.: Face photo recognition using sketch. In: Image Proceedings 2002 International Conference on IEEE (2002)
12. Liu, Q., et al.: A nonlinear approach for face sketch synthesis and recognition. In: 2005 IEEE Computer Society Conference on Computer Vision and Pattern Recognition (CVPR 2005) (2005)
13. Pramanik, S., Bhattacharjee, D.: An approach: modality reduction and face-sketch recognition (2013)
14. Saxena, S., Verbeek, J.: Heterogeneous face recognition with CNNs. In: Computer Vision–ECCV 2016 Workshops. Springer (2016)
15. Güçlütürk, Y., et al.: Convolutional sketch inversion. arXiv preprint arXiv:1606.03073 (2016)
16. Bigdeli, S.A., Zwicker, M.: Image restoration using autoencoding priors. arXiv preprint arXiv: 1703.09964 (2017)
17. Burger, H.C., Schuler, C.J., Harmeling, S.: Image denoising: can plain neural networks compete with BM3D? In: 2012 IEEE Conference on IEEE Computer Vision and Pattern Recognition (CVPR) (2012)
18. Dong, C., et al.: Image super-resolution using deep convolutional networks. IEEE Trans. Pattern Anal. Mach. Intell. **38**(2), 295–307 (2016)
19. Zhang, W., Wang, X., Tang, X.: Coupled information-theoretic encoding for face photo-sketch recognition. In: 2011 IEEE Conference on IEEE Computer Vision and Pattern Recognition (CVPR) (2011)
20. Vincent, P., et al.: Extracting and composing robust features with denoising autoencoders. In: Proceedings of the 25th International Conference on Machine Learning. ACM (2008)

21. Phillips, P.J., et al.: The feret evaluation, in face recognition, pp. 244–261. Springer (1998)
22. Phillips, P.J., et al.: The FERET database and evaluation procedure for face-recognition algorithms. Image Vis. Comput. **16**(5), 295–306 (1998)
23. Weston, J., Watkins, C.: Support vector machines for multi-class pattern recognition. In: ESANN (1999)
24. Hearst, M.A., et al.: Support vector machines. IEEE Intell. Syst. Appl. **13**(4), 18–28 (1998)

Object Segmentation with Neural Network Combined GrabCut

Yong-Gyun Choi[1] and Sukho Lee[2($\boxtimes$)]

[1] Department of Ubiquitous IT, Dongseo University,
Jurye-Ro 47, Sasang-Gu, Busan, Korea
[2] Department of Software Engineering, Dongseo University,
Jurye-Ro 47, Sasang-Gu, Busan, Korea
petrasuk@gmail.com

Abstract. Style transfer refers to the technique which applies the style of an artistic image to a real image which contains different contents than the artistic image. Nowadays, state-of-the-art results are obtained by using deep neural networks based style transfer methods. Recently, researches have been performed that apply the style of a single image to a whole video sequence. To give a feeling of a mixture of a real world and an animated world, we proposed a method that can apply the style of a single still image only on a selected object in the video sequence. In this paper, we propose an improved version of this method to obtain a more correct region of the object. The method combines the level set based segmentation and the GrabCut method together. The level set based segmentation suggests the foreground and the background colors to the Gaussian mixture model in the GrabCut method, and using this color suggestions, the GrabCut method cuts out the object region. Experimental results show that the proposed method can accurately select the object on which the object selective style transfer is applied.

Keywords: Style transfer · GrabCut · Level set · Video processing

1 Introduction

Since the work of [1] which successfully applied the style of an artistic image to the real image with different contents, many researches have been proposed to apply it to wider applications. One of these research is to apply the style of a single image on a whole video sequence [2, 3]. In [5], we proposed a method that can stylize a selected object in the video based on the use of a gradient image. The main advantage of this method is that it uses no extra network to extract the object region. However, since the output node depends not exactly only on the object region, the gradient image does not provide for an accurate object region. Therefore, in this paper, we propose a method in which the gradient image is used as a suggestion clue for the background and foreground colors that are used in the Gaussian mixture model in the GrabCut method. The gradient image is processed by two level set based segmentation steps, where the first one tries to include only the foreground colors while the second one only the background colors. Both clues are provided as inputs to the Gaussian mixture model in the

© Springer Nature Singapore Pte Ltd. 2018
K.J. Kim et al. (eds.), *IT Convergence and Security 2017*,
Lecture Notes in Electrical Engineering 449,
DOI 10.1007/978-981-10-6451-7_22

GrabCut model. The GrabCut model cuts out the object region with the provided clues such that a clean region is obtained. The proposed model can work with partial occlusion as the output node of the neural network is only affected by the object region.

2 Related Works

2.1 GrabCut Method

The GrabCut method [4] uses a Gaussian mixture model (GMM) with k components, which parameters are obtained by minimizing the following energy functional:

$$E(\alpha, \mathbf{k}, \boldsymbol{\theta}, \mathbf{z}) = U(\alpha, \mathbf{k}, \boldsymbol{\theta}, \mathbf{z}) + V(\alpha, \mathbf{z})$$

The data term U takes the color GMM models into account, as

$$U(\alpha, \mathbf{k}, \boldsymbol{\theta}, \mathbf{z}) = \sum_n D(\alpha_n, k_n, \boldsymbol{\theta}, z_n)$$

where $D(\alpha_n, k_n, \boldsymbol{\theta}, z_n) = -\log p(z_n | \alpha_n, k_n, \boldsymbol{\theta}) - \log \pi(\alpha_n, k_n)$, and $p(\cdot)$ is a Gaussian probability density function, and $\pi(\cdot)$ are mixture weighting coefficients. For the GMM model to work well the means, the weight coefficients, and the covariances of the Gaussian components have to be found properly.

2.2 Gradient Image Based Object Segmentation

In [5], we proposed the use of the gradient of the output node with respect to the input image as the clue for the object region in the image. Let $\mathbf{x}$ be the input image and y_{Object} the output node of the object under interest. Then, the image $\left| \nabla_{\mathbf{x}} y_{Object} \right|$ shows large gradient values at pixels that lie inside the object and small that lie outside. Using this information, we get a rough clue for the object region. However, this region is not exact, and therefore, resulting in a noisy object selective stylization. Therefore, here we propose a better method how the gradient image can be postprocessed to result in a more correct object region.

3 Proposed Method

Figure 1 shows the overall system diagram of the proposed method. For the GrabCut to work well, it is important to have a good foreground/background color proposal method which goes as an input to the Gaussian mixture model. Therefore, the initially segmented background region should not include the foreground region, such that the background color data model does not include foreground colors. To this aim, we propose the use of the bimodal segmentation on the gradient image with two different parameter settings: one for the background color extraction and the other for the foreground color extraction. This is different from the conventional bimodal segmentation which decides the background and foreground regions by a single segmentation.

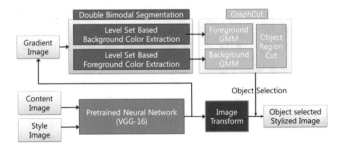

Fig. 1. Overall system diagram.

In fact, we are not interested in the exact background and foreground regions, but are interested in the regions which 'surely' belong to the background or foreground. This is done by the minimization of the following functional with different parameter settings:

$$E(\phi) = \lambda_1 \int_\Omega \left[\left| \left| \nabla_{\mathbf{x}} y_{Object} \right| (\mathbf{r}) - ave_{\{\phi \geq 0\}} \right|^2 + \alpha \right] \phi(\mathbf{r}) d\mathbf{r}$$

$$- \lambda_2 \int_\Omega \left| \left| \nabla_{\mathbf{x}} y_{Object} \right| (\mathbf{r}) - ave_{\{\phi < 0\}} \right|^2 \phi(\mathbf{r}) d\mathbf{r} + \lambda_3 \int_\Omega \left| \nabla \phi(\mathbf{r}) \right|^2 d\mathbf{r} \tag{6}$$

This is the level set bimodal segmentation model with regularization model [6], but instead of using the image as the input, we use the gradient image as the input. Furthermore, the purpose of using (6) is 'not' to exactly obtain an region of interest but to obtain a region which includes the region of interest, and also to obtain a region which surely lies within the region of interest. Thus, the minimization of the energy in (6) is used twice, with different parameter settings. Here, $ave_{\{\phi \geq 0\}}$ and $ave_{\{\phi < 0\}}$ are the average intensity values in the regions $\{\mathbf{r} \mid \phi(\mathbf{r}) \geq 0\}$ and $\{\mathbf{r} \mid \phi(\mathbf{r}) < 0\}$, respectively, and ϕ is the level set function. We use an extra parameter α which adjusts the adaptive threshold and use different λ_1, λ_2 parameters for the background and foreground color extraction. Here, we use $\alpha = 1$ for the background extraction, and $\lambda_1 = 0.8, \lambda_2 = 0.2, \lambda_3 = 0.01$. The difference in the parameters λ_1 and λ_2 gives an unbalanced competition between the gradient values of the gradient image, which results in a region larger than the object, i.e., a region which contains the object as can be seen in Fig. 2(a). For the foreground extraction, we use $\lambda_1 = \lambda_2 = 1$ and a regularization parameter of $\lambda_3 = 0.05$ which is larger than in the background region extraction. The large regularization parameter will make the extracted foreground region small enough such that the extracted region lies surely inside the object region (Fig. 2(c)). The extracted background colors is inserted in the Gaussian mixture model as 'sure' background colors, while the extracted foreground colors are inserted as 'probable' foreground colors. Figure 2(b) shows the extracted background overlapped on the original image to show that there is no overlap of the extracted sure background region with object. Figure 2(d) shows the fact with the possible foreground region. Figure 2(f) shows the object region obtained by the GrabCut model with Fig. 2(a) and (c) as the input. Figure 2(g) shows the style transferred image on the whole domain while Fig. 2(h) shows the style transferred result on the object region only.

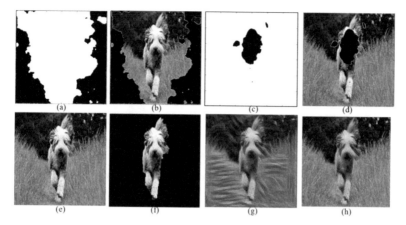

Fig. 2. Experimental results: (a)(b) sure background region (black) (c)(d) possible foreground region (black) (e) original (f) segmented region (g) fully stylized (h) object selective stylized.

Acknowledgements. This work was funded by the "Dongseo Frontier Project" Research Fund of 2016.

References

1. Gatys, L.A., Ecker, A.S., Bethge, M.: Image style transfer using convolutional neural networks. In: Computer Vision and Pattern Recognition (CVPR) (2015)
2. Ruder, M., Dosovitskiy, A., Brox, T.: Artistic style transfer for videos. Lecture Notes in Computer Science (LNCS), vol. 9796, pp. 26–36. Springer, Heidelberg (2016)
3. Johnson, J., Alahi, A., Fei-Fei, L.: Perceptual losses for real-time style transfer and super-resolution. Lecture Notes in Computer Science book (LNCS), vol. 9906, pp. 694–711. Springer, Heidelberg (2017)
4. Rother, C., Kolmogorov, V., Blake, A.: GrabCut - interactive foreground extraction using iterated graph cuts. ACM Trans. Graph. **23**, 309–314 (2004)
5. Perez, M., Lee, B.K., Lee, S.: Gradient-based object selective style transfer for videos. In: Proceedings of the Korean Multimedia Conference, vol. 20(1), pp. 92–95 (2017)
6. Lee, S., Seo, J.K.: Level set-based bimodal segmentation with stationary global minimum. IEEE Trans. Image Process. **15**(9), 2843–2852 (2006)

From Voxels to Ellipsoids: Application to Pore Space Geometrical Modelling

Alain Tresor Kemgue[1,2,3](✉) and Olivier Monga[2,3,4](✉)

[1] University of Yaoundé I, PO Box 812, Yaoundé, Cameroon
kemguealain@yahoo.fr
[2] University of Paris 6 (UPMC), 4 Place Jussieu, 75005 Paris, France
olivier.monga@ird.fr
[3] UMI 209, UMMISCO, 93143 Bondy Cedex, France
[4] IRD, 93140 Bondy, France

Abstract. This paper deals with the representation of complex volume shapes using a piece wise approximation by ellipsoids. The principle consists in optimizing a functional including an error term and a scale term. The result is a set of tangent or disjoint ellipsoids representing the shape in an intrinsic way. We propose a general scheme that we apply to 3D soil pore space modelling from volume Computed Tomography images. Within this specific context, we validate our geometrical modelling by using it for water draining simulation in porous media.

Keywords: Complex volume · Tomography · Optimization · Ellipsoids · Porous media

1 Introduction

In most real cases, voxel-based shape descriptions are difficult to use for computational and modeling issues. The reason is twofold: first, the number of voxels is generally very high, typically hundreds millions, and second, raw voxel representation does not give any explicit shape descriptors. Therefore, for most real applications, it is important to compute more compact and relevant geometrical structures representations. This issue is related to 3D image segmentation whose goal is to provide intrinsic shape representation from rough sensors data (Monga 2007). The basic principle of most shape modeling methods consists in looking for piecewise approximations by analytic surface or volume primitives. The representation by piecewise primitives is computed thanks to the explicit or sometimes implicit optimization of a non-linear functional integrating an approximation error term and an antagonist scale term (Monga 2007; Ngom et al. 2007; Ngom et al. 2012, 2011). In the case of natural shapes, 3D image segmentation remains a difficult and open problem due to the difficulty to find out suitable functional and also tractable minimization schemes. The basic reason is that natural shapes have not been designed as manufactured items using analytic surfaces and volumes. Then, in a certain sense, the algorithm has to invent a meaningful analytic representation of the shape initially described, typically, by hundreds of millions of voxels. By meaningful, we mean that the shape representation should approximate well the shape, be robust to small shape changes, have a kind of "continuity" with respect to initial description, be compact and also fit to its forthcoming use forthcoming use (Monga et al. 2007; Monga et al. 2008, 2009; Ngom et al. 2012, 2011). Indeed, current state of the art shows very few works about 3D shape modeling for complex natural volume shapes. It is due first to the mathematical and algorithmic difficulties that have to be faced, and second to the lack of motivation linked to specific application contexts. The first point will remain true maybe during a few decades, but the second one will invert due to the growing development of natural complex systems modeling area. For instance, references (Monga 2007; Ngom et al. 2007; Ngom et al. 2012, 2011) proposed to use balls, tori and generalized cylinders to represent pore space from computed tomography images in order to simulate microbial decomposition in soil (Monga et al. 2008, 2009, 2014). In this work, we investigate more sophisticated algorithms to represent complex volume shape using ellipsoids (Alsallakh 2014; Banegas et al. 2001). We apply our method to porous media where ellipsoids fit well to various cavities. The input of our algorithm is voxel-based shape descriptions from soil samples volume Computed Tomography. The output consists in piece-wise shape approximation using ellipsoids. Typically, we represent hundreds millions of voxels by means of a few thousands primitives. We implement an original scheme consisting in defining hierarchically primitive based representations using balls and then ellipsoids. Afterward, these new shape representations are used for simulation and modeling purposes. In this paper, we present an application of ellipsoid based pore space representation for draining. We show results on real data using same data sets than the one described in (Pot et al. 2015).

2   Method Global Scheme

In this work, we continue and upgrade the work presented in references (Monga 2007; Ngom et al. 2007; Ngom et al. 2012, 2011). First we apply the theoretical framework of reference to the computation of piece-wise shape approximation using ellipsoids. We proceed by defining hierarchical geometrical representations to be calculated from the primary shape voxel-based description. The levels of our hierarchical representation are: the initial set of voxels, the ball based representation (Ngom et al. 2007), and the ellipsoid based representation. We define a functional to be minimized, which characterizes the required ellipsoid based shape approximation. As in (Monga 2007), this functional is the sum of a first term defining approximation error, and a second term attached to the compactness of the representation (Monga 2007). The second term will include a scale criterion and its minimization is antagonist to the one of the first term. Thus, for a given scale value, the minimization of this functional defines a trade-off between the precision of the shape approximation and the compactness representation. We first describe the initial set of voxels by means of balls as in references (Monga 2007; Ngom et al. 2012). Indeed, we compute the minimal set of balls included within the shape (set of voxels) recovering the skeleton (or median axis). Especially for complex porous media shapes at microscopic scale, this first stage will yield much more compact and relevant shape representation than the voxel based one. In a second stage we tackle the optimal approximation of the set of balls by ellipsoids.

In a first step, we use statistical methods to find out connected and compact set of balls. References Alsallakh (2014) and Banegas et al. (2001), have already investigated these representations in other application fields. Statistical methods are variations of the dynamic clustering method, also called k-means: basically, one searches for the best partition of a given cloud of points into k classes. The best partition maximizes the inter-class variance (the variance between distinct classes) and minimizes the sum of the intra-class variances (the variance inside each class). This approach is used for shape recognition and shape matching in medical applications (Banégas's PhD or reference Alsallakh (2014), software CORPUS 2000 by CIRAD). We apply clustering methods to the set of balls representing the porous medium shapes (pore space, aggregates) instead of using the set of voxels. We attach to each ball a weight defined by its volume. Thus, statistical classification provides seeds for ellipsoids based approximation. We have tested different recursive clustering strategies including connectivity and topological criteria, in order to provide initial balls partition where each subset can be approximated by an ellipsoid. Clustering schemes will be linked to optimal functional optimization defined within the theoretical framework. This first step will produce initial set of ellipsoids describing the shape. In a second step we will apply merging strategies in order to decrease the number of ellipsoids. Same as in previous step, merging strategy will take into account minimization criterion from theoretical framework. We have tested various region growing strategies in order to obtain a final limited set of ellipsoids representing in a robust and intrinsic way the shape (Monga 2007). We have validated the whole approach by means of sensitivity studies on synthetic and real porous media data. We have also validated it by simulating draining using same experiments than in reference Pot et al. (2015). Indeed, for draining simulation we have tested several functions of ellipsoids descriptors (axis lengths and directions…) to be used as pore diameter in Young Laplace law.

3   Problem Formal Statement

The goal consists in finding out piece wise approximations of shape S using typical primitives having compactness, robustness and invariance properties. Of course, as in every approximation scheme, the scale notion should be explicitly expressed within this representation. Given that this representation will be used to simulate complex spatialized phenomena taking place inside the shape, deep learning based descriptors cannot be at least directly applied to solve this problem (Fig. 1):

Let S be a volume shape defined by a set of voxels as follows: 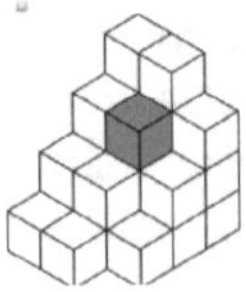
Let $I(i,j,k)$ be a 3D discrete binary image
Let S be the volume defined by all voxels $M(i,j,k)$ of I set to one :
$M(i,j,k) \in S \Leftrightarrow I(i,j,k) = 1$
In the continuous space, each voxel could be considered as a cube:

Fig. 1. Voxels

Let V be the set of voxels defining the volume shape:

$$V = \{v_1, v_2, \ldots v_n\}$$

Let E be a set of ellipsoids such that each pair is either disjoint or tangent:

$$E = \{e_1, e_2, \ldots .e_m\}; \forall (i,j), (e_i \cap e_j = \emptyset) \vee (e_i \cap e_j \in S(E))$$

where S(E) is the set of regular surfaces of affine space E.

Following the basic principle of (Monga 2007) we define the following functional:

$$C(V,E) = (\Phi(V \cup E) + \Phi(E \cup V)) + \lambda \sum_{j=1}^{j=m} \rho(e_j)$$

where Φ is the function which associates to a volume shape the numerical value of its volume, and ρ is the function which associates to a surface shape the numerical value of its area.

We look for sets of ellipsoids E minimizing $C(V, E)$ where λ defines the scale of the representation (Monga 2007). In order to address this optimization problem, we propose using a split and merge strategy thanks to clustering and region growing algorithms.

4   Initial Set of Ellipsoids by Clustering and Merging Scheme

This task cope with the partitioning of ball set into compact and connected set of balls to be approximated by ellipsoids. In a first stage, we define a primary set of ellipsoids by segmenting ball sets using k-means scheme (Alsallakh et al. 2014; Banegas et al. 2001). Second stage deals with the merging of the initial ellipsoids set by means of merging algorithms.

4.1   Clustering Using K-Means Algorithm

In a first step, we apply the k-means algorithm to the set of balls where each ball is attached a weight defining its volume. We introduce heuristics, based on calculation of approximation errors by fitting ellipsoids to balls subsets.

Practically we consider either the set of all maximal balls included within the shape or the minimal set. We set k to an initial value equal to a given percentage of the number of balls (20–30%). We also fix a maximal value for the approximation error of a set of balls by an ellipsoid typically 0.7–0.8. k-means algorithm provides (k) sets of "compact" balls sub-sets. For each ball subset, we determine the connected components, generally one per subset. Afterward we approximate each of these connected components by means of an ellipsoid. To compute the best ellipsoid fitting with the set of balls, we use the following method:

- Sampling spheres corresponding to balls.
- Computation of the convex hull of sampled points; if sampling rate is enough and uniform, this convex hull corresponds to the convex hull of the initial set of balls.
- Determination of the best fitting ellipsoid of the convex hull points thanks to either "Minimum volume enclosing ellipsoid" or classical co-variance matrix based method described in (Bricault and Monga 1997).
- Computation of the fitting error by coming back to the initial voxels; we count the number of voxels of the ellipsoid included within the initial shape and then calculate the ratio between the volume of the shape in the ellipsoid and the ellipsoid volume; this ratio does not strictly characterize the fitting error of the set of balls by the ellipsoid but an estimation of how the ellipsoid is included into the shape.

Then we obtain an ellipsoid and a fitting error equal to the percentage of ellipsoid voxels included within the initial shape. We keep the ellipsoid if the approximation error is greater than a given threshold (0.7–0.8).

We iterate this process for each connected component of the k classes provided by the k-means algorithm. Therefore, we get a set of ellipsoids attached each to a balls subset. We remove from the set of balls to be processed the ones attached to the ellipsoids. Thus we get a first set of ellipsoids and a set of balls that we were not able to approximate. Then we iterate the above scheme using the set of remaining balls, until no more ellipsoids can be obtained.

Thus, we get a set of ellipsoids and a set of maximal balls "not approximated". For the set of maximal balls "not approximated", we compute the connected components. For each connected component, we calculate another k-means where k is set to a percentage of the number of balls of the connected component. We do this step only once and keep the correct ellipsoids and the remaining balls are conserved as an ellipsoid.

Therefore, we get a set of ellipsoids where a small amount are balls. Practically this set of ellipsoids is weakly connected and forms a piece-wise approximation of the shape.

4.2   Merging Using Optimal Region Growing Algorithm

Practically, the first stage can provide a set of ellipsoids not minimal in the sense that adjacent ellipsoids could be merged without increasing the global fitting error. This is the reason why we add a complementary merging stage using optimal region growing (Wrobel and Monga 1987). In a first step we compute the adjacency graph of the ellipsoids provided by the previous stage. We determine the connectivity by means of the corresponding set of balls and not via ellipsoids equations. For each ellipsoid, we keep the corresponding set of balls. Then, for each pair of connected ellipsoids, we sample the points of the union of the set of balls using the method described before. We compute the convex hull and then the best fitting ellipsoid. We calculate the ratio between the ellipsoid volume and the summation of the volumes of the two corresponding ellipsoids. In order to increase the precision of the error evaluation, we can also come back to the initial set of voxels as previously. We take the minimum between the ratio: volume of the ellipsoid divided by summation of the two initial ellipsoids, and the inverse. The merging cost is set to this ratio which is also used to calculate the criterion defining the merging predicate (Wrobel and Monga 1987).

We merge iteratively the best couple of adjacent ellipsoids verifying the merging predicate using the algorithm described in (Monga 2007). Practically, for the merging stage, when using voxel based error computation, we set the approximation error threshold to 80–90%.

5   Application to Pore Space Modelling and to Water Draining

We have validated our approach with practical applications related to soil pore space modelling. In this specific case, the volume shape is the pore space that influences strongly the biological dynamics because containing partly water depending on water pressure. We have used 3D Computed Tomography (CT) images representing the same real soil sample as in (Pot et al. 2015). The soil samples come from Cage site (Versailles, France). It contains 17% Clay, 56% Silt and 27% Sand. The dimensions of the first 3D CT image (P1l) provided by the X-scanner are $132 \times 157 \times 90$ voxels with a 9.2 µm resolution and the dimensions of the second 3D CT image (P4l) provided by the X-scanner are $112 \times 208 \times 90$ voxels with a 9.2 micron meter resolution. Figures 2 and 3 shows some slices of the 3D CT image for two pores p1l and p4l. Figure 4 shows approximation result by ellipsoids

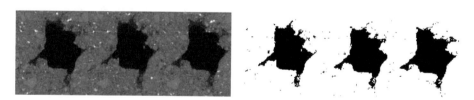

Fig. 2. Left: slices of the 3D CT image from Cage site (slices 48–50), p1l. **Right:** we extract first a voxel based description of pore space by thresholding thanks to an estimated porosity value.

In order to illustrate the interest of our ellipsoid- based pore space representation, we present draining results. We consider the same data than the ones used in reference (Pot et al. 2015). We have compared the draining results provided by the ellipsoids based method to real experimental results same as in (Pot et al. 2015). We have considered two values for the hydric pressure: -2 kPa and -1 kPa, corresponding respectively to threshold diameters of 150 µm and 73.5 µm according to Young Laplace law. At equilibrium point, we consider that there are primitives filled with water and primitives filled with air (void).

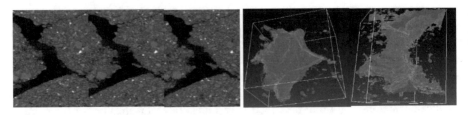

Fig. 3. Left: same as Fig. 2 for p4l. **Right:** 3D visualization p1l and p4l respectively.

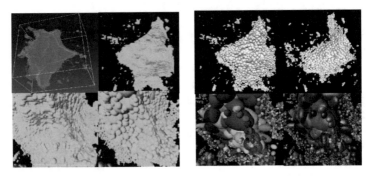

Fig. 4. Left: perspective views of maximal balls. **Right:** the ellipsoids provided from k-means and then region growing for pore p1l.

In Fig. 5, we have compared water distribution images provided by the ellipsoid based draining method to real experiment results, by taking 2D cross sections of the 3D image. In this figure, air is set to black color, water to grey color and solid to white color. Line 1 shows real experiment draining, lines 2, draining processed using minimal ellipsoid radius. Figure 7 presents the same result but using 3D visualization. As in reference (Pot et al. 2015), we have calculated the Mean Absolute Error regarding the spatial agreement of the position of the menisci between simulations and measurements. Same as in (Pot et al. 2015):

$$MAE = \frac{1}{n}\sum_{i=1}^{n}|s_i - m_i|$$

where the summation is over all pore space voxels; s_i corresponds to the simulation and m_i to the real experiment; we set solid voxels to 2, water voxels to 1 and air voxels to 0.

For pore p1l, using the ellipsoid based representation, we found a MAE equal to 0.1758 by selecting using the minimal radius. For the same data, using the ball based representation as described in reference (Pot et al. 2015) we found a MAE equal to 0.2. The computing time was 5 times less using the ellipsoid based representation than using the ball based representation. Indeed, in the general case, the computational complexity of the ellipsoid based draining algorithm is **O(v)** where **v** is the total volumes of the ellipsoids. The algorithmic complexity of the ball based draining method is **O(w)** where **w** is the total volumes of the balls.

Fig. 5. Left: line 1 is real experiment, line 2 is minimal radius.

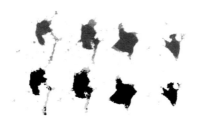

Fig. 6. Right: Same as Fig. 5 for P4l.

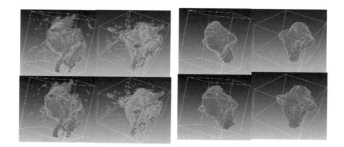

Fig. 7. Left: visualization of draining in 3D at equilibrium point. **Line 1:** real experiment, **Line 2:** draining using minimal radius. **Right:** visualization of air-water interfaces at draining equilibrium line 1: real experiment, line 2: draining using minimal radius.

In practical cases the ratio $\frac{w}{v}$ is between 5 and 100 depending on the data. Indeed, the ellipsoid representation of the pore space defines a piece wise approximation that could be used for other simulation issues including biologic dynamics. In more, the ball based representation used in reference (Pot et al. 2015) is relatively efficient for draining simulation purpose but does not define a piece wise pore space representation. For pore p4l, we found a MAE equal to 0.31 by selecting using the minimal radius. For the same data using the ball based representation as described in reference (Pot et al. 2015) we found a MAE equal to 0.38. The computing time was 10 times less (Figs. 6 and 8).

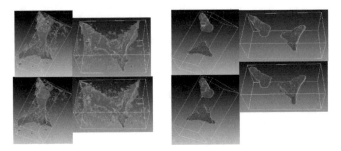

Fig. 8. Same as Fig. 7 for pore P4l.

6 Conclusion

This work deals with innovative algorithms to represent 3D complex shape from voxel-based representation. The basic idea consists in finding an intrinsic piece wise approximation of the shape by means of ellipsoids. The algorithm processes in a hierarchical way by defining a set of successive representations. The scheme is based on a split and merge approach using as initial description the set of maximal balls included in the shape. The splitting stage uses k-means algorithm and the merging stage optimal region growing. We optimize a functional defined by the addition of the approximation error and of a scale term. Thus, we get an intrinsic and scaled ellipsoid based geometrical representation of the shape. The meaning of this representation is ensured by the optimality criteria and also by properties linked to the shape skeleton. We validate our approach in the specific case of soil pore space images. we use this representation for water draining purpose.

Our approach has been validated using porous media data from Computed Tomography images

Acknowledgements of Support. This publication was made possible by NPRP grant #9-390-1-088 from the Qatar National Research Fund. The findings achieved herein are solely the responsibility of the authors.

References

Pot, V., Peth, S., Monga, O., Vogel, L.E., Genty, A., Garnier, P., Vieublé-Gonod, L., Ogurreck, M., Beckmann, F., Baveye, P.C.: Three-dimensional distribution of water and air in soil pores. Adv. Water Res. **84**, 87–102 (2015)

Alsallakh, B.: Visual methods for analyzing probabilistic classification data. IEEE Trans. Vis. Comput. Gr. **20**(12), 1703–1712 (2014)

Banégas, F., Jaeger, M., Michelucci, D., Roelens, M.: The Ellipsoidal Skeleton in Medical, 2001

Monga, O.: Defining and computing stable representations of volume shapes from discrete trace using volume primitives: application to 3D image analysis in soil science. Image Vis. Comput. **25**, 1134–1153 (2007)

Monga, O., Ngom, N.F., Delerue, J.F.: Representing geometric structures in 3D tomography soil images: application to pore space modelling. Comput. Geosci. **33**, 1140–1161 (2007)

Monga, O., Bousso, M., Pot, V., Garnier, P.: 3D geometrical structures and biological activity: application to microbial soil organic matter decomposition in pore space. Ecol. Model. **216**, 291–302 (2008)

Monga, O., Bousso, M., Pot, V., Garnier, P.: Using pore space 3D geometrical modelling to simulate biological activity: impact of soil structure. Comput. Geosci. **35**, 1789–1801 (2009)

Monga, O., Garnier, P., Pot, V., Coucheney, E., Nunan, N., Otten, W., Chenu, C.: Simulating microbial degradation of organic matter in a simple porous system using the 3D diffusion based model MOSAIC. Biogeosciences **11**, 2201–2209 (2014)

Ngom, F., Monga, O., Ould, Mohamed M.: 3D segmentation of soil microstructures using generalized cylinders. Comput. Geosci. **39**, 50–63 (2012)

Ngom, N.F., Garnier, P., Monga, O., Peth, S.: Extraction of three-dimensional soil pore space from microtomography images using a geometrical approach. Geoderma **163**(1–2), 127–134 (2011)

Bricault, I., Monga, O.: From Volume Medical Images to Quadratic Surface Patches, 1997

Wrobel, B., Monga, O.: Image segmentation: towards a methodology, 1987

Investigation of Dimensionality Reduction in a Finger Vein Verification System

Ei Wei Ting[1], M.Z. Ibrahim[1(✉)], and D.J. Mulvaney[2]

[1] Faculty of Electrical and Electronic, University Malaysia Pahang, 26600 Pekan, Pahang, Malaysia
eweiting@rocketmail.com, zamri@ump.edu.my
[2] School of Electronic, Electrical and Systems Engineering, Loughborough University, Loughborough, LE11 3TU, UK
d.j.mulvaney@lboro.ac.uk

Abstract. Popular methods of protecting access such as Personal Identification Numbers and smart cards are subject to security risks that result from accidental loss or being stolen. Risk can be reduced by adopting direct methods that identify the person and these are generally biometric methods, such as iris, face, voice and fingerprint recognition approaches. In this paper, a finger vein recognition method has been implemented in which the effect on performance has of using principal components analysis has been investigated. The data were obtained from the finger-vein database SDMULA-HMT and the images underwent contrast-limited adaptive histogram equalization and noise filtering for contrast improvement. The vein pattern was extracted using repeated line tracking and dimensionality reduction using principal components analysis to generate the feature vector. A 'speeded-up robust features' algorithm was used to determine the key points of interest and the Euclidean Distance was used to estimate similarity between database images. The results show that the use of a suitable number of principal components can improve the accuracy and reduce the computational overhead of the verification system.

Keywords: Finger vein recognition · Repeated line tracking · Principal component analysis · Speeded-up robust features

1 Introduction

It is becoming general recognized that with the increasing need to provide both physical and online security, allowing access according to what an individual is carrying or what an individual knows is less secure than limiting access depending on the characteristics of the individual themselves. To meet these demands, biometric personal identification and verification system such as fingerprint, voice, and iris recognition are gradually replacing smart cards and personal identification numbers. However, those systems have the major disadvantage that the biometric traits they measure can easily be duplicated as they depend on features that are visible to the naked eye. A further drawback is that a number of biometric traits are known to change as a result of child development or

© Springer Nature Singapore Pte Ltd. 2018
K.J. Kim et al. (eds.), *IT Convergence and Security 2017*,
Lecture Notes in Electrical Engineering 449,
DOI 10.1007/978-981-10-6451-7_24

ageing and it is often necessary to update the recorded biometric patterns at regular intervals, so introducing a potential security risk.

Information obtained from veins in the finger has been investigated as a possible biometric only relatively recently, but it has been recognized that as part of a security verification system they may have a number of advantages. Firstly, as the vein is located under the skin, it cannot be viewed directly, making the information harder to copy [1]. Secondly, although during an individual's lifetime the finger will grow and change appearance externally, the underlying vein pattern does not change. Finally, as the finger vein pattern can only be taken from a live body, the individual needs to be present during verification.

The use of finger veins for individual verification is currently being investigated by a number of researchers and a number of challenges remain in order to achieve appropriate accuracy and reliable performance. One main problem that affects the performance of all biometric identification systems is the quality of the data obtained during the acquisition process [2]. A finger vein verification system, shown in Fig. 1, typically consists of a suitable method of acquiring the image, preprocessing to reduce noise content, the use of feature extraction to reduce the dimensionality of the problem and matching against features obtained from previous finger vein images.

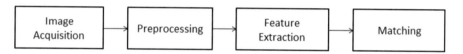

Fig. 1.   The stages in a typical finger vein verification system

In previous studies, finger vein identification and verification has been able to deliver some success. Park et al. [3] implemented finger vein pattern extraction using local binary patterns combined with wavelet transforms and support vector machines. Wu and Liu [4] used principal components analysis (PCA) and linear discriminant analysis (LDA) for feature extraction in a finger vein identification system A substantial hurdle to the commercial realization of finger vein verification systems is that the reliability of the identification is significantly affected by the poor contrast in the images obtained by the available acquisition methods.

2   Proposed Method

This paper proposes pre-processing using noise filtering, region of interest segmentation and image enhancement to improve the finger vein image quality. The finger veins themselves are extracted using repeated line tracking and PCA to reduce the dimensionality of the feature vector. Speeded-up robust features (SURF) matching to extract key points of interest and classification is carried out using the Euclidean distances between points. Results were then obtained which allow the analysis of the effect of PCA on finger vein verification performance.

2.1   Finger Vein Database

The images were obtained from the SDMULA-HMT vein database [5]. This database was found to be the most suitable among those available, with other candidate finger vein databases containing poor quality images, badly aligned images or images that were not useful as they has been obtained in a manner that made them suited to only specialized applications. The SDMULA-HMT vein database contains images obtained from 106 subjects with six images being provided for each subject, one for the fore, middle and ring fingers of each hand. The images are stored in the bitmap format and have a resolution of 320 × 240 pixels and samples are shown in Fig. 2.

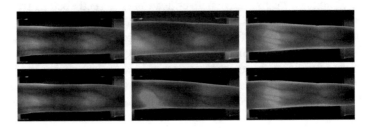

Fig. 2.   Samples of finger vein images obtained from the SDMULA-HMT database

2.2   Image Preprocessing

Image preprocessing stage is used to improve image contrast and quality. In this work, this includes image enhancement, edge detection, region of interest (ROI) detection, and noise filtering.

Firstly, each pixel in the finger vein image was converted from the red, green and blue pixel values in the bitmap RGB format to a luminance value (grayscale), using the formula shown in Eq. 1.

$$Y = 0.299 \times R + 0.587 \times G + 0.114 \times B \tag{1}$$

Since finger vein images tend to be of low quality due to presence of bones and tissues, a good image enhancement method is important to improve image quality. In this work, the luminance image is enhanced using the conventional contrast-limited adaptive histogram equalization (CLAHE) proposed by Lu et al. [6] for contrast enhancement purposes. CLAHE computes several histograms, each of which corresponds to a distinct region of the finger vein image while simultaneously redistributing the brightness of the image.

Finger edge detection and finger region localization are then applied to remove unwanted regions and regions containing no finger information. Edge detection was achieved using a canny edge detection operator. The ROI was then cropped and resized to the required new width and height. Finally, a 3 × 3 median filter was applied in order to provide noise reduction. Figure 3 shows the outcomes of each of the image preprocessing steps.

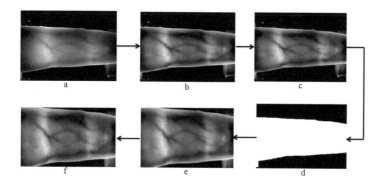

Fig. 3. Image pre-processing outcomes (a) original image, (b) CLAHE enhancement, (c) finger edge detection, (d) finger region detection, (e) cropped ROI of finger vein and (f) noise reduction following median filtering

2.3   Feature Extraction

In this paper, the finger vein pattern was extracted using the repeated line tracking (RLT) method proposed by Miura et al. [7] in their work on using finger veins for subject identification. In the current work, this technique was also found to be able to generate useful data for authentication from finger veins.

This method initially determines a starting point for line tracking and the tracking then follows connected pixels in the finger vein image according to a specified orientation and diameter requirements. The perpendicular to the direction of a vein has a luminance profile that includes a minimum near to the axis of the vein and so the line tracking involves moving to the next unexplored pixel that is the closest (in a straight line sense) local minima within a suitably small local region. By selecting a number of starting points it is possible to produce line segments that correspond to all the veins in the original image. The tracking ends when there remains no further minima to be processed.

2.4   Dimensionality Reduction

PCA is a linear approach to reducing data dimensionality. It is able to identify those variables in the data that are best at distinguishing between categories, while often also having the effect of removing those variables that only contribute to noise information that confuses the categorization process.

PCA and LDA were used by Wu et al. [4] to extract finger vein patterns. The authors' results showed that the combination of PCA and LDA methods we able to reduce the dimensionally of an image and they obtained successful identification in 98% of cases when using seven PCA features and four LDA features.

2.5   Speeded up Robust Feature (SURF) Algorithm

SURF is a robust local feature detector, that is part of the scale-invariant feature transform (SIFT) [8]. SURF includes a method for the rapid detection of sub-Hessians to detect feature points. In this paper, SURF was used to find key points between two finger vein patterns. This is followed by a right orientation of the key points to ensure a consistent vein pattern alignment between individual images. Matching is then carried out by measuring Euclidean distances between key points found in a new image and those in the database of stored images. The smaller the sum of the Euclidean distances between all pairs of key points in the images under comparison, the better the match.

3   Experimental Results

The experiments used 20 finger vein samples from the SDMULA-HMT finger vein database, where each sample had three images for training and one for testing. The RLT method was used to extract the vein pattern of the fingers and Fig. 4 shows an example of the vein pattern of an image following RLT extraction.

Fig. 4.   Vein pattern obtained following repeated line tracking

Figure 5 shows the result of dimensionality reduction using PCA on finger vein patterns using different numbers of principal components. Figure 5(a) shows the vein pattern with 32 principal components while Fig. 5(b) is the vein pattern obtained with 64 principal components. Although the veins in Fig. 5(a) appear less well defined and the image itself contains less information than in Fig. 5(b), having fewer principal components gives an advantage in terms of reducing the length of the training vector and so shortening training time. Consequently, for a realistic implementation, it will be necessary to determine a suitable compromise between parameter accuracy and computational practicality.

a   b

Fig. 5.   Finger vein patterns constructed with (a) the first 32 principal components, (b) the first 64 principal components

Figure 6 illustrates matches between pairs of images after applying SURF, but without the application of PCA. Figure 6(a) shows an example of finger vein patterns for a pair of images obtained from the same subject. It was found that the matching points obtained were consistent between images from the same subjects and the Euclidean distances were found to be reliably small. Figure 6(b) shows an example of an image pair obtained from different subjects and the Euclidean distances were found to be large in comparison with those obtained from same subject image pairs.

Fig. 6. Comparison of pairs of images without the application of PCA obtained for (a) the same subjects, (b) different subjects

Figure 7 shows the matching obtained between pairs of images after applying SURF, but in this case following the application of PCA. Due to the dimensionality reduction involved in PCA, the images in Fig. 7 are of a lower quality than the corresponding images in Fig. 6.

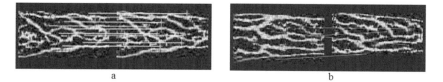

Fig. 7. Comparison of pairs of images including the application of PCA obtained for (a) the same subjects, (b) different subjects

In biometrics, the performance of verification systems can be evaluated by the equal error rate (EER) when the false rejection rate is held at a zero false acceptance rate. The smaller the EER value, the better the performance of the verification system. In this paper, the experiment was carried out for finger vein verification both without PCA and with PCA for different numbers of principal components. Figure 8 shows the receiver operating characteristic (ROC) curves that plot the number of false negative values against the false positive rates. The smaller the area under the curve (AUC) then the better is the accuracy of the identification. 'ROC rand' is a classifier that randomly guesses where the ROC lies along the line connecting the origin and the point (1, 1). In Fig. 8 it can be seen that the performance was improved by using PCA, but remains unsatisfactory for a practical system.

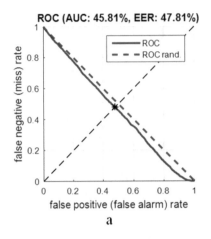

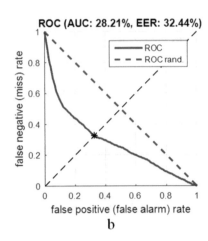

Fig. 8. ROC curve for the finger vein verification system (a) without PCA, (b) with the first 32 principal components

Further experiments were carried out and the performance was significantly improved when using only the first 9 principal components and the first 16 principal components, as shown in Fig. 9. This demonstrates that PCA can reduce the quantity of data required for identification without losing important information and features.

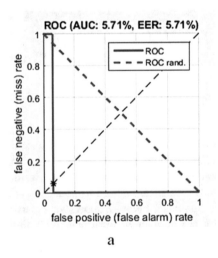

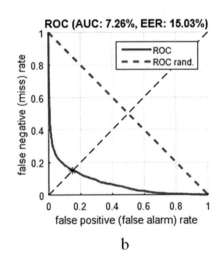

Fig. 9. ROC curve for the finger vein verification system for (a) the first nine principal components, (b) for the first 16 principal components

Table 1 summarizes the performances of the four different methods implemented for the finger vein verification system. It can be seen that the best performance was obtained after the most substantial PCA dimensionality reduction, with the implementation using nine principal components exhibiting both the smallest EER and AUC values.

Table 1. Performance comparison of the finger vein verification approaches

Method	EER (%)	AUC(%)
RLT without PCA	47.81	45.81
RLT with PC = 32	32.44	28.21
RLT with PC = 16	15.03	7.26
RLT with PC = 9	5.71	5.71

4   Conclusion

This paper has proposed an efficient finger vein based verification system using a PCA dimensionality reduction method. Experimental results have indicated that the finger vein verification system when using PCA was able to perform better than a system implemented without PCA. For the PCA verification system, the performance of the system was dependent on the number of principal components used, with the solution using only the first nine principal components performing better than systems using either the first 16 or 32 principal components. Although the vein patterns to which PCA was not applied contained more information than the vein patterns that did undergo PCA dimensionality reduction, PCA was nevertheless able to reduce the data to a set better suited for use in a vein verification system.

Acknowledgements. This work was supported by University Malaysia Pahang and funded by Ministry of Higher Education Malaysia under FRGS Grant RDU160108.

References

1. Gopinath, P.: Human identification based on finger veins-a review. Int. J. Humanit Arts Med. Sci. **2**(1), 29–33 (2014)
2. Lee, E.C., Park, K.R.: Image restoration of skin scattering and optical blurring for finger vein recognition. Opt. Lasers Eng. **49**(7), 816–828 (2011)
3. Park, K.R.: Finger vein recognition by combining global and local features based on SVM. Comput. Inform. **30**(2), 295–309 (2011)
4. Wu, J.-D., Liu, C.-T.: Finger-vein pattern identification using SVM and neural network technique. Expert Syst. Appl. **38**(11), 14284–14289 (2011)
5. Yin, Y., Liu, L., Sun, X.: SDUMLA-HMT: a multimodal biometric database. In: Chinese Conference on Biometric Recognition, pp. 260–268 (2011)
6. Lu, Y., Yoon, S., Park, D. S.: Finger vein recognition based on matching score-level fusion of gabor features. J. Korean. Inst. Commun. Inf. Sci. **38**(2), 178–182 (2013)
7. Miura N., Nagasaka A., Miyatake T.: Extraction of Finger-Vein Patterns Using Maximum Curvature Points in Image Profiles. Mach. Vision, 347–350, 16–18 May 2005
8. Nivas, S., Prakash, P.: Real-time finger-vein recognition system. Eng. Res. Gen. Sci. **2**(5), 580–591 (2014)

Palm Vein Recognition Using Scale Invariant Feature Transform with RANSAC Mismatching Removal

Shi Chuan Soh[1], M.Z. Ibrahim[1(✉)], Marlina Binti Yakno[1], and D.J. Mulvaney[2]

[1] Faculty of Electrical & Electronic Engineering, University Malaysia Pahang, 26600 Pekan, Pahang, Malaysia
sschuan92@gmail.com, {zamri,marlinayakno}@ump.edu.my
[2] School of Electronic, Electrical and Systems Engineering, Loughborough University, Loughborough, LE11 3TU, UK
d.j.mulvaney@Iboro.ac.uk

Abstract. Palm vein recognition has been gaining increasing interest as a biometric method, although there still remains an issue regarding difficulties in obtaining robust signals. In this paper, the effects of random sample consensus point mismatching removal and the use of different wavelengths of illumination on the recognition rate are investigated. The CASIA multi-spectral palm print image database was used to provide input signals and the scale invariant feature transform (SIFT) and random sample consensus (RANSAC) mismatching removal approaches were adopted for vein extraction and point feature matching. The results show that the RANSAC mismatching point removal was able to eliminate outliers while preserving the appropriate SIFT key points and that this led to an improvement in the equal error rate metric, signifying better recognition performance. The palm vein recognition system was found to achieve a better verification rate when infrared illumination in a specific spectral band was used to obtain the palm vein image.

Keywords: Vein recognition · Scale invariant feature transform · Random sample consensus

1 Introduction

Advances in the reliability of biometric authentication, such as using fingerprints, voice identification and iris recognition, has led to the introduction of a number of systems that are able to provide personal authentication. Such personal authentication is generally accepted as providing a more secure means of access rather than relying on what individuals know or carry.

Although hand recognition was one of the first forms of biometrics to be used, relatively little research has been published on the use of this approach for authentication purposes. Of the approaches that use the hand, the recognition of its vein pattern is one of the most promising methods, with the potential to characterize images obtained from fingers, the palm-dorsal or the palm area. The use of the vein pattern has the advantage

© Springer Nature Singapore Pte Ltd. 2018
K.J. Kim et al. (eds.), *IT Convergence and Security 2017*,
Lecture Notes in Electrical Engineering 449,
DOI 10.1007/978-981-10-6451-7_25

that it is harder to forge as the network of blood vessels lies under the skin and so is not immediately visible to the naked eye, unlike many other biometric attributes [1]. Other advantages of using vein patterns for biometrics is that they are believed to be unique to each individual, even identical twins [1], and the pattern does not normally significantly change over an extended period of time. In terms of acquisition, it is possible to design contactless vein pattern acquisition systems; such systems are often more easily accepted by users as any perception of discomfort is minimized and the likelihood of cross contamination is reduced. It is also important to note that the palm is generally the most reliable hand region from which to obtain vein patterns for biometrics, as this region does not normally exhibit hair growth that may affect the quality of the images captured [2].

Based on the current literature relating to palm vein research, vein pattern analysis methods can be classified into four types, namely geometry-based, statistical-based, feature-based and subspace approaches. Geometry-based methods are those that directly use the vein pattern structure information - this approach often has poor discriminatory ability and is highly sensitive to rotation, translation and scales changes [3]. As an example of a geometry-based system, Lee [4] used a two-dimensional Gabor filter with feature matching using Hamming distances for the vein extraction stage. Statistical-based methods calculate characteristics of the vein patterns which are then used in the matching process, but the statistical approach is often sensitive to changes in translation, rotation and scale, making it unsuitable for contactless vein recognition since careful alignment of the capturing equipment would be needed [5]. Feature-based methods include principal components analysis, linear discriminant analysis and independent component analysis, all of which perform dimensionality reduction resulting in a small set of features that can then be used in simplified matching methods requiring a reduced computational overhead. Subspace approaches attempt to reduce the correlation between estimators in an ensemble in a training subset. Subspace methods often use machine learning or artificial intelligence approaches for classification [3].

In this paper, a local invariant feature-based method is proposed that is independent of scale, translational and rotational changes. In the method, key points are extracted as vein pattern features and a Euclidean distance similarity measure is used for authentication. The impacts on the verification rate when performing random sample consensus (RANSAC) mismatching removal and the use of different wavelengths to illuminate the palm are evaluated.

The paper is organized as follows. Section 2 presents the process of the proposed method with subsection image database, preprocessing stage, feature extraction, mismatching point removal, and feature matching. Section 3 gives the experimental results and analysis. Section 4 provides the conclusions.

2   Proposed Method

2.1   Image Database

The CASIA multi-spectral palm vein database was used in this research. The database contains a total of 7200 palm vein images captured from 100 subjects using a multi spectral imaging device. A CCD camera was used to capture images from the hand that was evenly illuminated in turn by light of six different wavelengths, namely 460, 630, 700, 850, and 940 nm as well as white light, at intervals determined by a purpose-built control circuit. Figure 1 shows examples of images obtained at the six different illuminating wavelengths. During capture, the subjects were allowed to vary hand postures to a certain extent in order to simulate typical operational usage, while at the same time allowing the investigation of alternative contactless processing approaches that need to be robust to orientation, scale and translation changes. Illumination wavelengths of 700, 850, and 940 nm were used for the experiments in this paper.

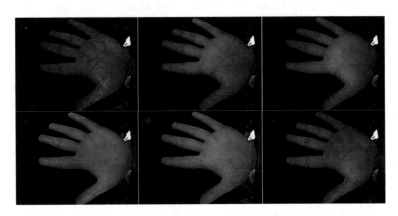

Fig. 1. Images obtained from illumination by the six different wavelengths of light [6]

2.2   Preprocessing

A low-pass Gaussian filter and Otsu threshold [7] were used to provide an approach with a low computational overhead that was able to reduce the presence in the image of high frequency and background noise. The Otsu technique is able to find a suitable threshold value from a bi-modal gray-level histogram, following which segmentation of the image can be performed according to illumination differences. Morphological filtering of the images was also carried out in order to reduce the presence of a small area of white pixels apparent in a number of the images at a point close to the palm region. Region of interest (ROI) extraction was then applied to remove the background, leaving only the palm region, and the palm images were aligned to a specific orientation to ensure the resulting images are invariant to rotation. Three key points were then generated, namely the valley points between the index and middle finger and between the ring and little finger, as well as the position of the palm point. The images are translated so that the center point of the palm region is the same in all images and the locations of the valley points were then simply scaled accordingly [8].

A number of stages of image enhancement were then carried out. Firstly, histogram equalization was used to provide an even distribution of intensities and this was found to make the presence of the veins more apparent to a naked eye observer of the resulting images. In this work, contrast-limited adaptive histogram equalization was applied; this is a popular equalization method that is known to be better than simple histogram equalization at limiting the amplification of noise [9]. The image obtained following ROI extraction was divided into 8 × 8 pixel blocks and stretched to provide a histogram distribution with a limited and uniform contrast [10]. Finally, 'salt and pepper' noise that often appeared near the vein patterns was removed by using a median filter that is able to conserve the edges in the image.

2.3 Feature Extraction

The scale invariant feature transform (SIFT) is a feature detection method that is insensitive to rotation, scale and translation changes and was used in the current work to calculate palm vein features that are invariant to scale changes [9].

The implementation of SIFT involves scale space extreme detection, key point localization, orientation assignment and the generation of key point descriptors [11]. The key points are obtained from the local extrema generated by difference of Gaussian operations applied in scale space to a series of resampled versions of the images [12]. Only those key points that exhibit good contrast performance following a repeated scanning of the image over a range of locations and scale are retained. Orientation histograms are then found in order to generate a gradient magnitude for each key point feature. Finally, a 128-dimensional feature vector is formed to specify the image gradient and orientation for each key point.

2.4 Random Sample Consensus Mismatching Removal

SIFT algorithms use matching between key points in the test and stored images to make authentication decisions. If key points have been incorrectly identified in either of the pairs of an image being compared, then a mismatching is likely to occur and so adversely affect the verification rate. Mismatching is often caused by outliers that result from image processing operations, such as fusion, rotation or resizing. In many cases it is possible to identify when outliers will occur (and so remove them) by estimating the effects that will be caused by applying geometrical transformations to an image.

Random Sample Consensus (RANSAC) is an algorithm to determine the number of inliers that meet the requirements of a pre-defined threshold distance within a certain number of computational iterations [13]. An initial random selection of key point pairs from the SIFT matching results is used to generate a consensus set and its members are assessed to determine whether they fall within the threshold distance. Outliers are removed and the process is repeated iteratively until the consensus set contains the largest number of inliers meeting the threshold criteria is found [14].

2.5 Feature Matching

Distance metrics are a popular approach in determining the similarity between two images using the key point features extracted by SIFT. The minimum Euclidean distance can be calculated by using the straight line distance between corresponding pairs of key point coordinates [15]. A linear support vector machine (SVM) can then be used to determine an overall matching score to assess whether authentication can be confirmed. To assess the performance of the vein recognition system, a receiver operating characteristics (ROC) curve can be calculated, the area under which is a useful performance measure, as is a parameter known as the equal error rate (EER) that can also easily be determined.

3   Experimental Result and Analysis

Experiments were carried out to evaluate the impacts on the recognition rate of using the RANSAC mismatching point removal and the illumination of the palm using different wavelengths of light. In these experiments, 20 images from the CASIA database were used to generate primary results. The effectiveness of RANSAC mismatching point

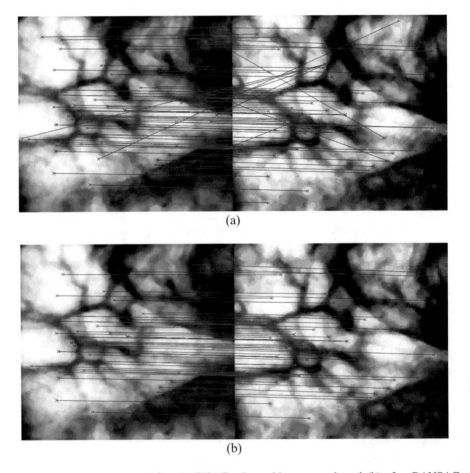

(a)

(b)

Fig. 2. SIFT matching (a) before RANSAC mismatching removal, and (b) after RANSAC mismatching removal

removal and the different illumination wavelengths is accessed using the area under the curve (AUC) of the ROC graph as well as the EER value.

Figure 2(a) and (b) show the matching points found using the SIFT algorithm both before RANSAC mismatching point removal and after point removal, respectively. In Fig. 2(a) it is apparent that a number of mismatching points were present in the image, these are general those lines that cross over a number of the horizontal matches. The outliers that result in these mismatches can easily be identified as lines of significantly longer length than the bulk of the matches found. In order to remove mismatching points, RANSAC removes the outliers and retains only the correctly-matching points [14]. It can be seen in Fig. 2(b) that the mismatching points were removed by RANSAC and only the appropriate matching points were preserved.

Figure 3(a) and (b) show the ROC curve and EER rate obtained before and after the RANSAC mismatching removal was applied, respectively. For Fig. 3(a) it can be seen that the AUC increases following the RANSAC removal operations and in Fig. 3(b) that the EER rate is reduced by RANSAC, both of which indicate that the verification rate will increase after using mismatching removal. From these two figures, it can be concluded that mismatching removal will have a substantial effect on the successful verification rate of a palm vein authentication system.

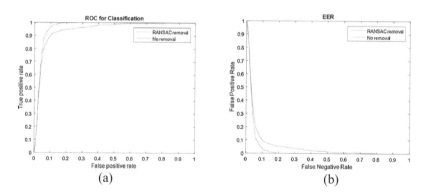

Fig. 3. Results before and after RANSAC (a) ROC curves (b) EER rate

Figure 4 shows the effect on the vein verification rate of using three different wavelengths of light for illuminating the palm. It has been found by other researchers that the most suitable band of wavelengths for extracting the palm vein patterns is normally within the infrared region at a range of 700 nm to 1300 nm and that the absorption peak is normally located at approximately 970 nm [16]. This agrees with the results in Fig. 4(a) and (b), where it can be seen that both the ROC and EER curves show that the verification results are best when using a wavelength of 940 nm compared to the alternative wavelengths of 700 nm and 850 nm. For the sensor and illumination used for extracting the images in the database, it can be concluded that pattern extraction is best performed at the longer infrared wavelength as more information representing the veins can be extracted.

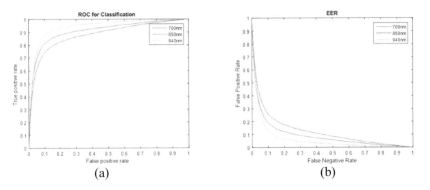

Fig. 4. Recognition results for three illumination wavelengths (**a**) ROC curve (**b**) EER rate

Table 1 shows a summary of the results of the authentication using data obtained both before and after the application of RANSAC and for the three separate illumination wavelengths.

Table 1. Summary of AUC and EER values obtained

Method	AUC %	EER %
SIFT with RANSAC mismatching removal	96.5	7.7
SIFT without mismatching removal	94.3	10.0
SIFT with RANSAC mismatching using 700 nm image	87.7	18.2
SIFT with RANSAC mismatching using 850 nm image	90.8	14.7
SIFT with RANSAC mismatching using 940 nm image	93.8	11.2

4   Conclusions

This paper investigated the effects of palm vein recognition performance of the removal by RANSAC of mismatching points generated by SIFT and of using three different wavelengths of light to illuminate the palm during image acquisition.

Based on the AUC and EER results obtained, the verification rate produced by SIFT was shown to be significantly improved by employing the RANSAC mismatching removal. It is apparent that the removal by RANSAC of the outliers from the consensus set used for matching had a positive impact on the overall verification rate. In addition, it was found that the use of the longer infrared wavelengths for the illumination of the palm allowed the acquisition of images from which it was possible to extract the vein patterns with better quality, thereby giving the potential to simplify the palm vein authentication process.

Acknowledgments.   This work is supported by Faulty of Electrical and Electronic Engineering, University Malaysia Pahang under research grant FRGS Grant RDU160108.

References

1. MacGregor, P., Welford, R.: Veincheck: imaging for security and personnel identification. Adv. Imaging **6**(7), 52–56 (1991)
2. Han, W., Lee, J.: Expert systems with applications palm vein recognition using adaptive Gabor filter. Expert Syst. Appl. **39**(18), 13225–13234 (2012)
3. Kang, W., Wu, Q.: Contactless palm vein recognition using a mutual foreground-based local binary pattern. IEEE Trans. Inf. Forensics Secur. **9**(11), 1974–1985 (2014)
4. Lee, J.C.: A novel biometric system based on palm vein image. Pattern Recogn. Lett. **33**(12), 1520–1528 (2012)
5. Kang, W., Liu, Y., Wu, Q., Yue, X.: Contact-free palm-vein recognition based on local invariant features. PLoS ONE **9**(5), e97548 (2014)
6. Palmprint, C.M., Database, I.: Note on CASIA multi-spectral Palmprint database, pp. 1–4
7. Article, R.: A survey on biometric authentication techniques using palm. J. Glob. Res. Comput. Sci. **5**(8), 2010–2013 (2014). www.jgrcs.info
8. Zhou, Y., Kumar, A., Member, S.: Human identification using palm-vein images. IEEE Trans. Inf. Forensics Secur. **6**(4), 1259–1274 (2011)
9. Pan, M., Kang, W.: Palm vein recognition based on three local invariant feature extraction algorithms. In: CCBR 2011, Beijing, pp. 116–124 (2011)
10. Zhang, H., Tang, C., Li, X., Wai, A., Kong, K.: A study of similarity between genetically identical body vein patterns. In: 2014 IEEE Symposium on Computational Intelligence in Biometrics and Identity Management (CIBIM)
11. Ahmed, M.A., Salem, A.M.: Intelligent techniques for matching palm vein images 2. Palm vein model and related work. IEEE Secur. Priv. **39**(1), 1–14 (2015)
12. Ladoux, P.-O., Rosenberger, C., Dorizzi, B.: Palm vein verification system based on SIFT matching. In: International Conference on Biometrics, pp. 1290–1298 (2009)
13. Wu, Y., Ma, W., Gong, M., Su, L., Jiao, L., Member, S.: A novel point-matching algorithm based on fast sample consensus for image registration. Geosci. Remote Sens. Lett. **12**(1), 43–47 (2015)
14. Vi, C., Vi, W.G.: An integrated RANSAC and graph based mismatch elimination approach for wide-baseline image matching. Int. Arch. Photogramm. Remote Sens. Spat. Inf. Sci. **40**, 23–25 (2015)
15. Michael, G., Connie, T., Teoh, A., Connie, T., Teoh, A.: A Contactless Biometric System Using Palm Print and Palm Vein Features. Image, Rochester (2011)
16. Yin, D., Ding, Z.: Research on finger vein acquisition based on wavelength choice. In: ISCI, pp. 2424–2432 (2015)

Speed Limit Traffic Sign Classification Using Multiple Features Matching

Aryuanto Soetedjo[✉] and I. Komang Somawirata

Department of Electrical Engineering,
National Institute of Technology (ITN) Malang, Malang, Indonesia
aryuanto@gmail.com

Abstract. This paper presents the method to classify the speed limit traffic sign using multiple features, namely histogram of oriented gradient (HOG) and maximally stable extremal regions (MSER) features. The classification process is divided into the outer circular ring matching and the inner part matching. The HOG feature is employed to match the outer circular ring of the sign, while MSER feature is employed to extract the digit number in the inner part of the sign. Both features are extracted from the grayscale image. The algorithm detects the rotation angle of the sign by analyzing the blobs which is extracted using MSER. In the matching process, tested images are matched with the standard reference images by calculating the Euclidean distance. The experimental results show that the proposed method for matching the outer circular ring works properly to recognize the circular sign. Further, the digit number matching achieves the high classification rate of 93.67% for classifying the normal and rotated speed limit signs. The total execution time for classifying six types of speed limit sign is 10.75 ms.

Keywords: Speed limit traffic sign · HOG · MSER · Template matching

1 Introduction

Traffic sign recognition based-on a machine vision is one of the extensive researches in the intelligent transportation system. The system detects and classifies the traffic sign for assisting the driver or employed as an integral part in the autonomous vehicle. The speed limit traffic sign is one of the traffic signs that should be obeyed by the driver to avoid an accident. An automatic system to recognize the speed limit traffic sign is an interesting topic such as addressed by [1–6].

Traffic sign recognition is usually divided into detection stage, where the location of sign is detected from an image, and the classification stage where the detected sign is classified to the reference. In the classification stage, template matching techniques [7–9] and machine learning techniques [1, 3–6, 10, 11] are commonly employed.

In [7], the traffic sign templates consist of the pictograms (black and white images) of traffic signs in the normal position and the rotated one. Instead of the black and white images, the color images were used as the templates [8]. In [7, 8], the correlation technique was employed in the matching process. In [9], the ring partitioned matching was employed to classify the red circular traffic sign. The matching was performed on

© Springer Nature Singapore Pte Ltd. 2018
K.J. Kim et al. (eds.), *IT Convergence and Security 2017*,
Lecture Notes in Electrical Engineering 449,
DOI 10.1007/978-981-10-6451-7_26

each ring with the different weights. To cope with the rotated images, the method employed the image histogram (fuzzy histogram) in the matching process.

The methods commonly used in the machine learning techniques are k-Nearest Neighbors algorithm (kNN), Support Vector Machine (SVM), and Artificial Neural Networks (ANN). In [1], the kNN was employed to classify the rectangle speed limit sign according to the size. Further the optical character recognition (OCR) technique was employed to read the speed number of the sign. The SVM classifier was employed in [3, 10] for classifying the traffic sign. In [3], the property curves of segmented digit number of speed limit sign was used to train the SVM classifier. The histogram of oriented gradient (HOG) was employed as the descriptor of the traffic sign image in the classification process which is performed using the SVM [10, 11]. They concluded that the HOG is an effective descriptor for classifying the traffic sign. In [5], the ANN was employed to recognize the number in the speed limit sign. The method first extracted the digit number of detected speed limit sign to generate the number in the binary image. Then the binary image was used by the ANN in the recognition process. Instead of using the binary image, the grayscale image was used as the input of the ANN [6].

The benefit of template matching technique compared to the machine learning technique is that no training process is required. However, to cope with all possible conditions of traffic signs, more traffic sign references are required as proposed by [7]. Our previous work in [9] overcome the rotation problems by employing the histogram of image which is calculated in each ring area. Since the method is used to classify the circular red sign where there are two colors in the image (red color on the outer part and blue or black color in the inner part), two color thresholding techniques are employed in the preprocessing stage.

In this paper we propose a novel technique to classify the speed limit sign using the multiple features, namely HOG and maximally stable extremal regions (MSER) features extracted from a grayscale image. The method first extracts the HOG feature in the outer border of image to match the circular ring. Then the MSERs are extracted in the inner part of image to obtain the binary image of the digit number of speed limit sign. The template matching is employed to match the number into the predefined one. To cope with the varying rotation problem, an affine transformation is applied in the binary image before the matching process. The main contributions of our work are: (a) it employs the grayscale image, thus there is no complex color conversion; (b) it employs the simple matching technique; (c) the method is rotation invariant.

The rest of paper is organized as follows. Section 2 presents our proposed system. Section 3 presents the experimental results. The conclusion is covered in Sect. 4.

2   Proposed System

2.1   System Overview

Flowchart of proposed system is illustrated in Fig. 1. It starts with the grayscale color conversion to convert RGB image into grayscale image. It is noted that the input image is the detected traffic sign bounded with the rectangle box. Then the HOG feature is extracted in the outer part of the image. Since the outer part of speed limit sign is a

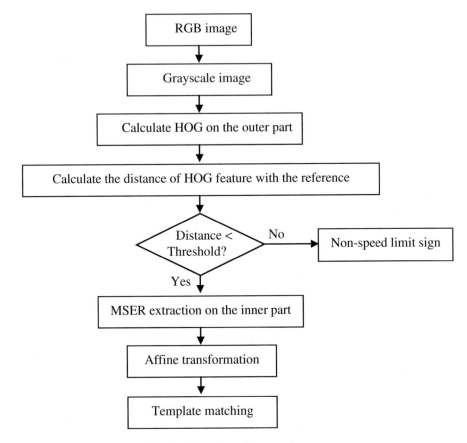

Fig. 1. Flowchart of proposed system

circular ring, the HOG feature is rotation invariant. Thus, the HOG feature could be matched with only one reference circular ring. When it is matched, the next step is to match the digit number, otherwise the sign is classified as the non-speed limit sign.

To match the digit number, it performs the MSER extraction in the inner part of the grayscale image. It yields the connected component of each digit in binary image. Then the contour of each component is found. The ellipse fitting technique is adopted to find the orientation of the component. It is assumed that when the traffic sign rotates, all digits will rotate with the same orientation. Thus we could get the rotation angle from the orientation of the ellipse.

From the detected rotation angle of the sign, an affine transformation is adopted to rotate the number to the normal position (orientation of zero degree). Finally the template matching technique is employed to match the digit number to the predefined reference.

2.2   HOG Feature

The HOG represents the histogram of orientation of gradient in an image which is proposed by [12]. The descriptor is robust to illumination changes. However, it is rotation variant. In this work we adopt it to match the outer part of the speed limit sign (see Fig. 2(a)), where the circular ring exists. Thus, when the speed limit sign is rotated, the HOG feature of the outer part does not change.

To calculate HOG, the image is divided into cells as shown in Fig. 2(a). In this work, the image is resized into 128×128 pixels. The cell's size is 8×8 pixels, thus an image is divided into 16×16 cells. The HOG is calculated on the overlapped blocks, where a block consists of 16×16 pixels. An example of the HOG of the speed limit sign is illustrated in Fig. 2(b), where the orientation of gradient is drawn in the image. To match the HOG feature of the tested image with the reference one, the Euclidean distance method is adopted. Since only the outer part of the image is considered, any types of speed limit signs could be used as the reference. The tested image is classified as the circular sign when the distance is lower than a threshold.

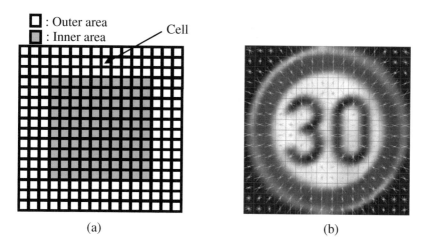

Fig. 2. (a) Outer and inner parts of image; (b) HOG visualization

2.3   MSER Extraction

The MSER is a method proposed by [13] to find the regions which remain the same when the threshold is changed at several values. In the traffic sign recognition system, it is usually employed in the detection stage to find the region of candidate signs as proposed in [11]. They extract the MSER from a grayscale image to detect the traffic sign with the white background. Our method works in the similar approach but in the different way. In the sense that in our method the MSER is used to extract the digit number in the inner part of the speed limit sign for classifying the sign, not detecting the sign. The benefits of our proposed method are twofold. First it extracts the digit number effectively, due to the fact that the MSER is robust against the illumination changes [11]. Second, since the area of image to be extracted is in a small area inside the traffic sign, the computation cost for extracting the MSER could be reduced.

2.4 Rotation Detection and Affine Transformation

To cope with the rotated sign, we propose to detect the rotation angle of the sign by analyzing the blobs (binary image) representing the digit number obtained by the MSER extraction. The examples of blobs are illustrated in Fig. 3, where the image in first column represents the speed limit sign in the normal position, while the images in second and third columns represent the speed limit sign in the rotated position.

To find the rotation angle of the sign, first we find the contour of blobs which represents the boundary of blob. Then the ellipse fitting technique is applied to the contour. The rotation angle of the sign is obtained from the angle of detected ellipse. As illustrated in Fig. 3, the detected ellipse is drawn with the blue color. In the case of speed limit 30 km/h, the ellipses extracted from the number "3" and "0" have the same orientation. The orientation of both ellipses represents the rotation angle of the sign. However, for the speed limit 70 km/h, the ellipses extracted from the number "7" does not represent the rotation angle of the sign. Fortunately, the speed limit signs used in this work have the number "0" in the last digit. Thus we may use the last digit or the rightmost blob to find the rotation angle.

Fig. 3. The blobs and rotation angle

Once the rotation angle is calculated, the next step is to find the affine transformation to transform the image into the normal position (no rotation). The affine transformation for rotation is expressed by

$$\begin{bmatrix} x' \\ y' \\ 1 \end{bmatrix} = \begin{bmatrix} \cos\theta & \sin\theta & cx \times (1 - \cos\theta) - cy \times \sin\theta \\ -\sin\theta & \cos\theta & cx \times \sin\theta + cy \times (1 - \cos\theta) \\ 0 & 0 & 1 \end{bmatrix} \begin{bmatrix} x \\ y \\ 1 \end{bmatrix} \quad (1)$$

where x, y are the pixel coordinates in source image, x', y' are the pixel coordinates after transformation, cx, cy are the center coordinates of rotation, and θ is the rotation angle.

2.5 Template Matching

From the previous stages, the blobs of digit numbers are already in the normal position. Thus in the matching process, we only provide the standard templates of the numbers of the speed limit sign. The template size is the same as the size of inner area shown in

Fig. 4. Templates of digit numbers in the speed limit sign

Fig. 2(a), i.e. 80 × 80 pixels. The templates used in this work are illustrated in Fig. 4. To match the blob with the template, the Euclidean distance is adopted.

3 Experimental Results

The proposed system is tested on a PC, Intel Core i7 3.4 Ghz. The algorithm is implemented using the C++ languange and the OpenCV library [14] for handling the image processing tasks. The speed limit signs used in the experiments are the speed limit of 30, 50, 60, 70, 80, 100 km/h. The tested images are obtained from GTRSB dataset [15]. The tested images consist of six types of speed limit signs as mentioned before, where each type contains 20 images. To examine the rotation problems, we create the rotated versions of the tested images by rotating them by 10°, 15° in both clockwise and counter-clockwise directions. In addition, to verify the algorithm for matching the outer circular ring, we introduce 100 images contain the non circular signs. Therefore there are total 700 tested images used in the experiments.

The results of the circular ring matching using HOG distance are given in Table 1. From the table, the maximum distance of the speed limit sign is 0.1097, while the minimum distance of the non-speed limit sign is 0.1212. The result suggests us to classify the outer circular of the sign properly by setting the threshold value to 0.12. If the distance is below than this value, then it is classified as the speed limit sign. From the experiments, the execution time of the outer circular matching is 2.95 ms.

Table 1. Results of the outer circular matching

Speed limit sign	Minimum matching distance	Maximum matching distance
30 km/h	0.0514	0.1030
50 km/h	0.0589	0.1008
60 km/h	0.0551	0.1062
70 km/h	0.0557	0.1056
80 km/h	0.0570	0.1097
100 km/h	0.0562	0.1084
Non-speed limit sign	0.1212	0.1946

The results of the digit number matching are given in Table 2. The results show that the average classification rate of the sign in normal position (no rotation) is highest. When the signs are rotated by 15° or −15°, the classification rate decreases about 9%. It is worthy to note that even though the classification rate decreases, the proposed algorithm provides a promising method to cope with the rotation problem. The method only requires six templates (and no learning process is required) to classify 600 speed limit signs with the higher classification rate of 93.67%. The performance of the method could be improved in the blob extraction and the template matching techniques, such as by employing the skeleton of digit number in the matching process. It will be addressed as the future work.

We also examine the execution time required during the digit number matching. It requires 46.78 ms to match an image with six reference images, or the matching time is about 7.8 ms for one reference image. Therefore combining with the outer circular matching, the computation time of our proposed classification method is 10.75 ms (2.95 + 7.8).

Table 2. Results of the digit number matching

Speed limit sign	Classification rate				
	Rotation angle of speed limit sign				
	0°	10°	15°	−10°	−15°
30 km/h	100%	100%	90%	100%	85%
50 km/h	100%	100%	95%	90%	100%
60 km/h	100%	90%	90%	90%	90%
70 km/h	95%	100%	95%	80%	80%
80 km/h	100%	90%	80%	95%	85%
100 km/h	100%	95%	95%	100%	100%
Average	99.17%	95.83%	90.83%	92.5%	90.00%
Total average = 93.67%					

4 Conclusion

The multiple features are extracted in the outer and inner parts of the speed limit sign for classifying the sign using the template matching technique. The proposed system first matches the outer part with a circular ring reference. Once it is matched, the digit number in the inner part are extracted and matched with the standard reference signs. The method offers an efficient way in the classification, in the sense that no training process is required. Further the number of sample images is very few. For classifying six types of the speed limit signs, only six reference images are required. The results show that the method achieves the high classification rate and the fast computation time. In future, some improvements of the method will be addressed to increase the performance. Further the method will be extended to cope with the other types of traffic signs.

Acknowledgements. This work is supported by the Research Grant 2017, Competence-based research scheme from Directorate General of Higher Education, Ministry of Research and Technology and Higher Education, Republic of Indonesia, No.: SP DIPA-042.06.1.401516/2017.

References

1. Bilgin, E., Robila, S.: Road sign recognition system on Raspberry Pi. In: 2016 IEEE Long Island Systems, Applications and Technology Conference, New York, USA, pp. 1–5 (2016)
2. Hoang, A.T., Koide, T., Yamamoto, M.: Real-time speed limit traffic sign detection system for robust automotive environments. IEIE Trans. Smart Process. Comput. **4**(4), 237–250 (2015)
3. Biswas, R., Fleyeh, H., Mostakim, M.: Detection and classification of speed limit traffic signs. In: 2014 World Congress on Computer Applications and Information Systems (WCCAIS), Hammamet, Tunisia, pp. 1–6 (2014)
4. Peemen, M., Mesman, B., Corporaal, H.: Speed sign detection and recognition by convolutional neural networks. In: 8th International Automotive Congress, Eindhoven, Netherland, pp. 162–170 (2011)
5. Ali, F.H., Ismail, M.H.: Speed limit road sign detection and recognition system. Int. J. Comput. Appl. **131**(2), 43–50 (2015)
6. Kundu, S.K., Mackens, P.: Speed limit sign recognition using MSER and artificial neural networks. In: 2015 IEEE 18th International Conference on Intelligent Transportation Systems, Las Palmas, Spain, pp. 1849–1854 (2015)
7. Malik, R., Khurshid, J., Ahmad, S.N.: Road sign detection and recognition using colour segmentation, shape analysis and template matching. In: 2007 International Conference on Machine Learning and Cybernetics, Hong Kong, pp. 3556–3560 (2007)
8. Laguna, R., Barrientos, R., Blazquez, L.F., Miguel, L.J.: Traffic sign recognition application based on image processing techniques. In: The 19th World Congress. The International Federation of Automatic Control, Cape Town, South Africa, pp. 104–109 (2014)
9. Soetedjo, A., Yamada, K.: Traffic sign classification using ring-partitioned matching. IEICE Trans. Fundam. **E88**(A9), 2419–2426 (2005)
10. Adam, A., Ioannidis, C.: Automatic road-sign detection and classification based on support vector machines and HOG descriptors. ISPRS Ann. Photogramm. Remote Sens. Spatial Inf. Sci. II **2**(5), 1–7 (2014)
11. Greenhalgh, J., Mirmehdi, M.: Real-time detection and recognition of road traffic signs. IEEE Trans. Intell. Transp. Syst. **13**(4), 1498–1506 (2012)
12. Dalal, N., Triggs, B.: Histograms of oriented gradients for human detection. In: 2005 IEEE Computer Society Conference on Computer Vision and Pattern Recognition (CVPR 2005), San Diego, CA, USA, vol. 1, pp. 886–893 (2005)
13. Matas, J.: Robust wide-baseline stereo from maximally stable extremal regions. Image Vis. Comput. **22**(10), 761–767 (2004)
14. http://opencv.org/
15. Stallkamp, J., Schlipsing, M., Salmen, J., Igel, C.: The German traffic sign recognition benchmark: a multi-class classification competition. In: The IEEE International Joint Conference on Neural Networks, San Jose, CA, USA, pp. 1453–1460 (2011)

Future Network Technology

Big Streaming Data Sampling and Optimization

Abhilash Kancharala[1], Nohjin Park[2], Jongyeop Kim[3],
and Nohpill Park[1(✉)]

[1] Oklahoma State University, Stillwater, OK 74074, USA
npark@cs.okstate.edu
[2] Oklahoma City University, Oklahoma City, OK 73106, USA
[3] Southern Arkansas University, Magnolia, AR 71753, USA

Abstract. This research addresses and resolves the issues with the confidence level of sampled big streaming data that is dynamic with respect to the speed of the streaming data and the dynamically changing sample space. Based on a preliminary work and results from [8], this research focuses more on the confidence level and threshold of dynamic size of the population in order to ensure a better confidence level of the sampled data with respect to a few variables such as speed of the streaming data, population size dynamic over time, sample space (or size), speed of sampling algorithm, size of streaming data, and time duration of data streaming. Theoretical thresholds of the processing of big streaming data with respect to a set of variables as mentioned above are identified in an effort for optimization. Simulation results along with experimental results are provided to validate the efficacy of the proposed theoretical thresholds.

Keywords: Dynamic population · Big streaming data · Sampling

1 Introduction

This research is concerned about the optimization of various big data specific variables while sampling big data streamed through Apache flume.

As the demand for bigger and bigger data is ever increasing and exercised at real time streaming speed, it is in the near-future expected to become a necessity to sample data. A new research need is exigently sought in this context to address and resolve the issue with the sampling process that is unique from conventional sampling process. Conventional sampling process assumes static or stationary population without loss of generality while the sampling process of interest in this work carries a dynamic nature in terms of contents quickly varying over time in particular. Hence, even the sample size, the confidence level of the data can be significantly different depending on the speed of the streaming data and the time duration of the streaming process, to mention a few.

It is important to identify a set of variables in order to establish a sound theoretical foundation to optimize the big streaming data process. The big data specific variables that are taken into consideration in this work are: speed of the streaming data, population size dynamic over time, sample space (or size), speed of sampling algorithm, size of streaming data, and time duration of data streaming.

© Springer Nature Singapore Pte Ltd. 2018
K.J. Kim et al. (eds.), *IT Convergence and Security 2017*,
Lecture Notes in Electrical Engineering 449,
DOI 10.1007/978-981-10-6451-7_27

A preliminary work has been conducted on the similar issue as presented in [8]. In [8], it was the focus of the work how to maximize the confidence level with respect to the sample size, or vice versa, that is, how to minimize the size of the sample samples with respect to a fixed target confidence level. However, the methods and algorithms proposed in this research are applicable to other platforms as well without loss of generality.

Based on the results from [8], the objective of the work is to make efforts to further optimize the processing of the big streaming data by identifying a theoretical threshold dynamic size of the population ensuring a better confidence level with respect to the sampled items and the population size.

The paper is organized as follows: preliminaries and literature review are provided in the following section; then the proposed big streaming data sampling and optimization methods are presented; followed by simulation results; and then concluded.

2   Preliminaries and Literature Review

2.1   Reservoir Based Sampling

The research was presented by Oak Ridge National Laboratory [7], in which the proposed algorithm used the reservoir-based random sampling algorithm with and without replacement. The research starts off with counting the number of data units in the stream, if less than the size of sample space, then the element streamed at that instance is copied to the sample space. If the total number of units in the stream is more than the size of the sample space, then the element streamed at that instance is inserted into the sample space with a probability of m/n, where m represents the sample size and n represents the total population size. And at the same time, an element is evicted from the sample space with a uniform probability. The new element streamed is replaced at the place where the data element is evicted out of the sample space. While in the case of the reservoir based sampling with replacement, a set of Bernoulli trials are performed on the newly observed element and k elements are evicted from the sample space. The k elements that are evicted from the sample space are replaced with k copies of the new data element that was streamed.

The disadvantage of this algorithm is that at any instant of time, the sample space may contain duplicates of the data elements, i.e., the sample space is not unique, which can lead to a biased analysis. In the worst case scenario, sample space might be filled up with duplicates of a single data element. When analysis is done as such, the entire analysis reflects the characteristics of a single unit which likely contradicts the reason for sampling.

The research concludes with the discussion of a faster sampling technique wherein some of the elements in the stream are skipped. The number of consecutive elements skipped is determined beforehand. This algorithm also uses the concept of generation of Bernoulli trails and replacing k copied of the streamed data element with the k evicted elements from the sample space. Duplication of the data elements is also a possibility in this case. When performing analysis wherein there needs to be a unique copy of each of the data element in the sample space, having duplicates in the sample space doesn't produce accurate results as the results would be weighted towards the data elements that have more duplicates.

Note that this algorithm can be employed in this research as a baseline sampling algorithm.

2.2   Sampling Algorithms in a Stream Operator

This research [4] uses the concept similar to that of reservoir-based sampling, wherein the sample space is termed as the reservoir. The algorithm fills up the reservoir with the first n elements, where n is the size of the reservoir. After filling up the reservoir, each iteration generates an independent random variable (r). The next r records of stream input are skipped. The $(r+1)^{th}$ record is placed in the sample space, and the location to be replaced with is determined by the multiple of size of the sample space with a random number between 0 and 1. No duplicates can be present in the sample space, as an element is placed in the sample space only once and the algorithm moves onto the next streamed data unit. The value of 'r' generated by the random variable is between 0 and 'n' (sample space).

2.3   Sampling Content from Online Social Networks

This research [5] is an analysis on comparison of the random sampling vs. expert sampling. Expert sampling is defined as the sampling of tweets that are posted by experts, who are popular users on twitter. The comparisons are done between the random sampled tweets and expert tweets, with conclusions of how the experts affect the entire sampled data. An analysis is given on the comparison of samples with the original population with respect to various tweet attributes like sources, popularity, quality and timelines, and etc.

3   Proposed Big Streaming Data Sampling and Optimization

3.1   Big Streaming Data Variables

The big data specific variables that are taken into consideration are:

1. Speed: Speed with which the stream is being generated. The stream from flume could be filtered with various tags, when no such filter is given the speed would be at its maximum and is most likely limited by the performance of the system. If a filter is provided, speed depends on the speed at which the particular filter related tweets are being generated. Precisely, speed of the stream is not always a constant. It is mostly dependent on whether the streaming is being done with a filter or without a filter. 's' is used to represent the speed of the stream.
2. Dynamic population size: The maximum size of the most recent population that can be considered to increase the confidence level.
3. Sample space: The size of the sample space which is represented by 'S'. The optimal values of the sampling space can be determined by constants and variables that are functionally or theoretically related.
4. Speed of sampling algorithm: The speed of the sampling algorithm is mostly a constant particularly in this work as the entire processing is done on a single

machine in this research. Hence, the speed of processing the sampling is constant. But, if the same algorithm is being performed on a distributed hadoop server instead where machines differ in their configurations then the speed of processing would be a variable.

5. Size: The size represents the total amount of the tweets that are streamed till the current instance, represented by 'N'. The total number of tweets streamed depends on the time the stream has started and also the speed of the stream.
6. Time: Represented by 't', denotes the time elapsed since the start of the stream. Note that this is a monotonically increasing variable.

3.2 Relation of Speed and Time on Population Size

Given, the speed of the stream is 's' and 't' the total time since the start of the stream, the relation between total population size, speed of the stream and the time can be given as follows as shown in [8]:

$$N = s * t \tag{1}$$

3.3 Threshold Value of the Dynamic Population Size

To relate the function of threshold size of the dynamic population and the confidence level, we can solve for Z from the above equation as follows,

$$\sqrt{\frac{1}{N'} * \sum_{i=1}^{N'} (x_i - \mu)^2} = \frac{ME}{Z} * \sqrt{S}$$

Solving for the Z-value, we have

$$Z = \frac{ME}{\sqrt{\frac{1}{N'} * \sum_{i=1}^{N'} (x_i - \mu)^2}} * \sqrt{S}$$

$$Z = \frac{ME * \sqrt{S} * \sqrt{N'}}{\sqrt{\sum_{i=1}^{N'} (x_i - \mu)^2}}$$

The limiting value of Z from the Z-table are from -3.49 to $+3.49$ which corresponds to 0% and 100% confidence levels, respectively. When the sample size is same as the size of the dynamic population, the value of Z would come out to be 3.49 as the entire contents of the sample size will be the same as the total dynamic population size as we can be 100% confident that the mean of the samples represents exactly the mean of the dynamic population. As the size of the dynamic population becomes more than the size of the samples, the confidence level (Z value) keeps decreasing. The worst-case value of the denominator is when half of the values in N' are of the lowest possible number and the other half of the values in N' are of the highest number, thus making the summation of the squared differences the maximum as follows.

$$Z = \frac{ME * \sqrt{S} * \sqrt{N'}}{\sqrt{(x_1 - \mu)^2 + (x_2 - \mu)^2 + \cdots + (x_{N'} - \mu)^2}}$$

For a given value of margin of error and a fixed sample size, the variation of Z value with respect to the dynamic population size can be as,

$$Z \propto \frac{k * \sqrt{N'}}{\sqrt{N'^2}}$$

$$Z \propto \frac{k}{\sqrt{N'}}$$

3.4   Effect of Speed on Threshold Value

As mentioned earlier, the size of the population can be expressed in terms of speed of the stream (s) and time spent on streaming (t).

$$N = s * t \tag{2}$$

The size of the active data denoted by N', can be represented by a similar equation

$$N' = s(t - t') \tag{3}$$

where, t' represents the time at which the first data unit in the active data was streamed. Given a time period, we can determine the maximum allowable speed of the stream. If the speed of the stream increases beyond the maximum value it affects the confidence level of the sampled data. Researches [4–6] focus on the random sampling technique, but this research has focused more on the confidence level and threshold dynamic size of the population ensuring a better confidence level with respect to the sampled items with respect to the population size.

4   Simulation Results

Given a sample size of 300, the confidence level when the dynamic population size is less than or equal to 300 is 100%, as there was never any loss in the elements that are being used to process. Once the value of the dynamic population size is increased more than then sample size, the confidence level keeps monotonically decreasing as there are lesser number of data elements to represent the total population. Using 300 as the sample size, 3% as marginal error and threshold value to 951, we have the confidence level rounding close to 1.96 (95% as per the z-table). A population size of more than 951, would have a confidence level below 1.96 (less than 95%) as shown in Fig. 1.

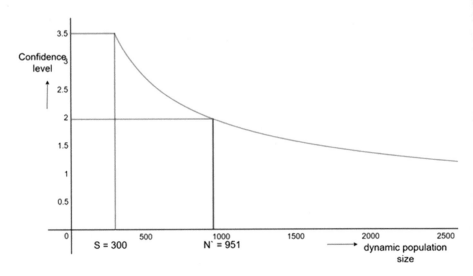

Fig. 1. Threshold dynamic population Size

Figure 1 shows the graph of dynamic population size vs. confidence level but with a change in the areas.

$$Z = \frac{ME * \sqrt{S} * \sqrt{N'}}{\sqrt{(x_1 - \mu)^2 + (x_2 - \mu)^2 + \ldots + (x_{N'} - \mu)^2}}$$

The above equation can be rewritten to solve for N' and express it as a function of Z, ME and S.

$$\frac{\sqrt{N'}}{\sqrt{(x_1 - \mu)^2 + (x_2 - \mu)^2 + \ldots + (x_{N'} - \mu)^2}} = \frac{Z}{ME * \sqrt{S}}$$

$$\cong \frac{\sqrt{N'}}{\sqrt{N'^2}} = \frac{Z}{ME * \sqrt{S}}$$

$$\cong \frac{1}{\sqrt{N'}} = \frac{Z}{ME * \sqrt{S}}$$

$$\cong \sqrt{N'} = \frac{ME * \sqrt{S}}{Z}$$

$$N' = (2/3)$$

$$\cong N' = \left(\frac{ME * \sqrt{S}}{Z}\right)^2$$

Given the margin of error and the sampling size, $ME * \sqrt{S}$ is a constant. We can introduce the proportionality and the above equation can be modified as below:

$$N' \propto \frac{1}{Z^2}$$

Using 300 as the sample size, 3% as the margin of error and z value as 1.96, we have the threshold value of the dynamic population size rounding to 951. Figure 2 depicts the inverse graph shown in Fig. 1.

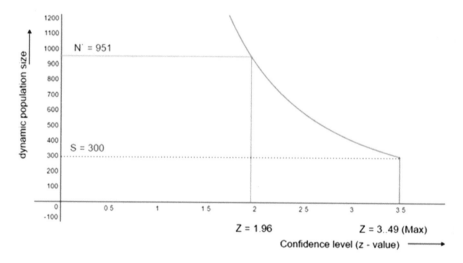

Fig. 2. Inverse function confidence level and dynamic size

The time taken to perform the sampling algorithm for all the scenarios is a factor of $O(S^2)$ as the algorithm needs to scan the entire sample space two times in order to determine the position of the new element in the sample space. The groups/strata that are being constructed based on the characteristics of the input stream should be selected in an appropriate manner such that the sample space value be comparable to the total number of groups/strata produces higher deviations in confidence level. On a similar note, the deviation from the population and the sampled elements decreases as the ratio of the population to the sampled elements increases.

The relation between the Z-value and the dynamic population size is given by

$$Z = \frac{ME * \sqrt{S} * \sqrt{N'}}{\sqrt{(x_1 - \mu)^2 + (x_2 - \mu)^2 + \cdots + (x_{N'} - \mu)^2}}$$

5   Conclusion

The research has proposed various theoretical optimization methods to achieve the best possible results when performing sampling algorithms on dynamic streaming data. In particular, this research has addressed and demonstrated the issues with the confidence level of sampled big data that is dynamic with respect to the speed of the streaming data and the dynamically changing sampling space. Based on a preliminary work and results from [8], this research focuses more on the confidence level and threshold dynamic size of the population ensuring a better confidence level with respect to the sampled items with respect to a few variables such as speed of the streaming data, population size dynamic over time, sample space (or size), speed of sampling algorithm, size of streaming data, and time duration of data streaming. The threshold value of the dynamic population size and the effect of speed on threshold value have been identified and studies based on the theoretical relationship between total population size, speed of the stream and the time as identified in [8]. The theoretical threshold values have been simulated and studied as shown in the previous section.

References

1. Tang, F., Li, L., Barolli, L., Tang, C.: An efficient sampling and classification approach for flow detection in SDN-based big data centers. Journal 2(5), 99–110 (2016)
2. Gadepally, V., Herr, T., Johnson, L., Milechin, L., Milosavljevic, M., Miller, BA.: Sampling operations on big data. In: 2015 49th Asilomar Conference on Signals, Systems and Computers, 8 November 2015
3. Xu, K., Wang, F., Jia, X., Wang, H.: The impact of sampling on big data analysis of social media: a case study on Flu and Ebola. In: 2015 49th Asilomar Conference on Signals, Systems and Computers, 6 December 2015
4. Johnson, T., Muralikrishnan, S., Rozenbaum, I.: Sampling algorithms in a stream operator. In: SIGMOD Conference (2005)
5. Zafar, M.B., Bhattacharya, P., Ganguly, N., Gummadi, K.P., Ghosh, S.: Sampling content from online social networks: comparing random vs. xpert sampling of the twitter stream. ACM Trans. Web 9(3), 12 (2015)
6. Teddlie, C., Yu, F.: Mixed methods sampling: a topology with examples. J. Mixed Methods Res. 1(1), 77–100 (2007)
7. Park, B.H., Ostrouchov, G., Samatova, N.F., Geist, A.: Reservoir based random sampling with replacement from data stream. In: Proceedings of 2004 SIAM International Conference on Data Mining (2015)
8. Kancharla, A., Kim, J., Park, N.-J., Park, N.: Big streaming data buffering optimization. In: International Conference on Computational Science/Intelligence/Applied Informatics (2016)

Artificial Intelligence and Robotics

Fuzzy Model for the Average Delay Time on a Road Ending with a Traffic Light

Zsolt Csaba Johanyák$^{(\boxtimes)}$ (ID) and Rafael Pedro Alvarez Gil (ID)

Department of Information Technology,
GAMF Faculty of Engineering and Computer Science,
Pallasz Athéné University, Izsáki út 10, Kecskemét 6000, Hungary
{johanyak.csaba, alvarez.rafael}@gamf.kefo.hu

Abstract. Urban traffic is continuously increasing and therefore especially in peak-hours an optimized traffic light system can provide significant advantages. As a step towards developing such a system this paper presents a fuzzy model that estimates the average delay times on a road that ends at an intersection with traffic lights. The model was created based on data obtained using a validated microscopic traffic simulator that is based on the Intelligent Driver Model. Simulations were carried out for different traffic flow, traffic signal cycles, and green period values. The newly developed fuzzy model can be used as a module in a traffic light optimization system.

Keywords: Fuzzy model · Average delay time · Traffic light · Microscopic traffic simulator

1 Introduction

The optimization of traffic lights cycles has been intensively investigated in the last decade. The calculation or measurement of delays caused by a road ending in a junction/intersection with traffic lights is one of the key issues in this field. Different models and methods have been used to solve this task so far, and the results may be significantly different from each other depending on whether the intersection is in saturated or near saturated conditions.

Computational intelligence based solutions are widely recognized as suitable tools for control, classification, and other purposes (e.g. [1–6]). Fuzzy controllers and other computational intelligence based methods have been proposed in several works for the optimization of a traffic light at an intersection (e.g. [7–14]). In order to provide a traffic light system with a control solution that determines the optimal traffic signal cycle length and the green period ratio depending on the actual traffic flow values on the intersecting roads, first, one needs to develop a model that describes the behavior of the traffic on a road ending with a traffic light. Hence the goal of the research reported in this paper was to create a fuzzy system that describes the relationship between the average delay time and its parameters, which are traffic signal cycle length, green period ratio, and traffic flow. The data necessary for building the model was obtained using the IDM based microscopic traffic simulator IT MICROSIM [15].

K.J. Kim et al. (eds.), *IT Convergence and Security 2017*,
Lecture Notes in Electrical Engineering 449,
DOI 10.1007/978-981-10-6451-7_28

The rest of this paper is organized as follows. Section 2 describes the parameters and results of the simulation runs, which served as a data source for the model development. Section 3 presents the creation and optimization of our new fuzzy model and the conclusions are drawn in Sect. 4.

2 Simulation Based Investigation of Average Delay Time Values on a Road

The aim of the experiments was to determine the average delay time values depending on the traffic flow, green period ratio, and traffic signal cycle using the microscopic traffic simulator IT MICROSIM [15]. In course of the simulation we considered a 500 m long road ending in an intersection with a traffic light (Fig. 1). The traffic flow was created with randomly arriving vehicles.

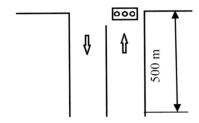

Fig. 1. Road ending in an intersection with a traffic light

The parameters of the simulations were defined as follows. Five values were tried as traffic signal cycle length, i.e. 50, 76, 100, 150, and 200 s, respectively. The traffic signal cycle was composed from two periods with the same durations, the first one was the green period and the second one was the red period. The green period included green light time, 3 s yellow light and 2 s red light, while the red period contained only red light that simulated the time when the cross road gets the green period. Simulations were carried out with a duration of the green period of 25%, 33%, 50%, 67%, and 75% of the total duration of the traffic signal cycle.

In case of the traffic flow eleven distinct values were chosen, i.e. 200, 400, 600, 800, 850, 900, 950, 1000, 1050, 1100, and 1200 vehicle/hour, respectively. Delay times were obtained by subtracting the free-flow travel times from the measured travel times. Free-flow travel times are the travel time values without traffic lights. In the micro-simulator, their value depended on the actual traffic, similar to the real-world situation. Table 1 shows the free-flow travel time values for the traffic flow values used in course of the simulations.

In order to get a detailed picture about the relation between the above-mentioned parameters and the delay times a full factorial experimental design plan was used, i.e. each value of each parameter was tried with each value of the other parameters. It needed $n = 5*5*11 = 275$ simulation runs. For the case of 50% green period ratio the results are shown in Table 2 as well as in Fig. 2.

Table 1. Free-flow travel times

Flow [vehicle/h]	Free-flow travel times [s]	Flow [vehicle/h]	Free-flow travel times [s]
200	34.58	950	37.30
400	35.59	1000	37.41
600	36.19	1050	37.50
800	36.86	1100	37.53
850	37.03	1200	37.58
900	37.17		

Table 2. Average delay times for the 50% green time period ratio [s]

Flow [v/h]	Traffic light cycle length [s]				
	50	76	100	150	200
200	10.63	14.90	17.23	24.42	31.28
400	12.81	17.03	19.60	27.12	35.65
600	16.22	19.62	22.34	31.53	39.41
800	44.45	26.45	27.62	35.58	45.52
850	185.68	31.67	30.27	38.11	46.94
900	218.98	45.83	35.21	42.09	49.38
950	217.97	162.33	56.06	47.76	52.22
1000	221.38	191.40	155.29	62.90	61.67
1050	220.07	191.98	176.02	144.25	95.30
1100	221.18	192.36	177.33	162.73	153.34
1200	222.56	189.96	174.87	162.02	152.21

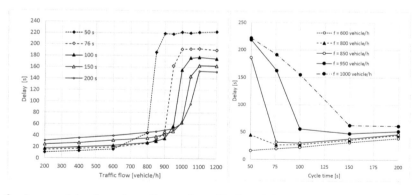

Fig. 2. Average delay times depending on the traffic flow (left) and the traffic light cycle length (right) for the 50% green period ratio

Investigating the results in the left part of Fig. 2 one can recognize that the delay time values can be described by curves that are convex, non-negative, non-decreasing and which depend on the current traffic. For traffic flow values greater than the traffic light cycle capacity, the delay increases until it reaches its maximum value when the road is filled with waiting vehicles. The capacity of a traffic signal cycle is near the value of the abscissa of the inflection point of the graph corresponding to the cycle. One can see that the higher the traffic signal cycle length is, the larger the capacity of the traffic signal cycle becomes.

3 Fuzzy Model for the Average Delay Time

The fuzzy model created is a MISO (Multiple Input Single Output) fuzzy system with three inputs, i.e. green period ratio (GPR), traffic light cycle (TLC), and traffic flow (TF). Its schematic structure is presented in Fig. 3. The range for the three input variables was defined as follows GPR $\in$ [25, 75], TLC $\in$ [50, 200], and TF $\in$ [200, 1200].

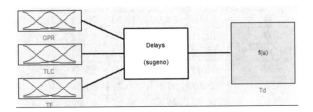

Fig. 3. Fuzzy system

The output of the system is the average delay time (Td). Its range was estimated based on the simulations as $T_d \in$ [0, 850].The system applies Takagi-Sugeno-Kang (TSK) [16] type fuzzy inference and to maintain simplicity triangle shaped membership function type is used for each antecedent fuzzy set. In course of the modeling process several set numbers were tried in each antecedent dimension starting from three up until nine. The best results were obtained with five-five fuzzy sets in the first two input dimensions and eight fuzzy sets in the third one, respectively.

TSK fuzzy inference requires a full coverage of the input space by rule antecedents. Thus, in the actual case the number of necessary rules was $N_R = 5*5*8 = 200$. We opted for a zero order TSK system in order to ensure the fast computation of the output. At the beginning the output of each rule was set to zero. The final parameters of the input fuzzy sets and the consequents of the rules were obtained through a Particle Swarm Optimization (PSO) [17] based tuning process.

At the beginning 275 data tuples formed the whole data set measured during the simulations. From this set 50 data tuples were selected for testing purposes and 225 ones were kept for training the fuzzy system. The tuning aimed the determination of the position of the input sets (support), the relative position of their cores and the output value of the rules. Owing to the fact that the number of parameters to be modified was quite high $N_p = 2*(5 + 5+8) + 200 = 236$, we did not try to find the optimal value for each parameter at the same time, but we modified only one parameter at a time and we repeated the process several times. The algorithm of this iterative tuning is detailed below.

```
For each iteration cycle
  For each input dimension
    For each set of the current dimension
      Find with PSO the optimal position of the support
  For each input dimension
    For each set of the current dimension
      Find with PSO the optimal position of the core
  For each rule
    Find with PSO the optimal consequent value
```

The parameters of PSO were defined as follows. The number of particles in the swarm was 40. The self-confidence coefficient was 2.05, the social coefficient was 2.05 and $c_3 = 0.729$. The number of allowed generations was 10. The number of iteration cycles was 100. The performance of the system was measured by the root mean square of the error expressed in percentage of the output range (RMSEP). At the end of the tuning this value was 0.5706% in case of the training data and 0.5248% in case of the test data, respectively. The resulting input partitions are presented in Figs. 4 and 5.

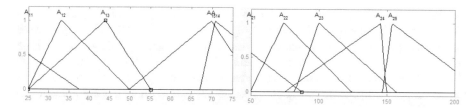

Fig. 4. Input partitions for GPR (left) and TLC (right)

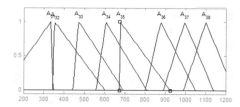

Fig. 5. TF

Figure 6 shows the average delay values measured through simulations (circles) and the T_d values calculated by the fuzzy system in case of the training data set and the test data set, respectively.

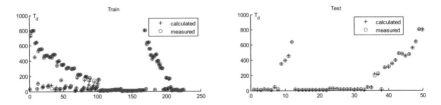

Fig. 6. Average delay time for training (left) and test (right) data sets

4   Conclusions

IT MICROSIM ensures a proper architecture to determine the optimal cycle time values and green period ratios for any traffic flow value by using an arbitrary search technique or design of experiments (DOE) methodology in case of a static traffic flow. However, the application of IT MICROSIM in a real-time system with continuously varying traffic flows is not feasible owing to the high time demand of the simulations. As a first step towards developing a traffic light optimization system that would automatically adjust the green period ratios and traffic light cycle conform to the actual traffic flow in order to achieve an optimal total delay one needs a tool that can calculate the average delay time fast depending on its parameters. This model combined with an efficient computational intelligence based search method (e.g. [18–20]) can be used later to determine the optimal settings for green period ratios and traffic signal cycle.

In this paper, a fuzzy model was presented that applies Takagi-Sugeno-Kang inference and provides a good approximation of the average delay time. The model was created based on data obtained using a validated microscopic traffic simulator that is based on the Intelligent Driver Model. Simulations were carried out for different traffic flow, traffic signal cycle, and green period values. The newly developed fuzzy model can be used as a module in a traffic light optimization system independently from the actual type of the intersection.

Further research will consider the application of a fuzzy rule interpolation based interpolation technique (e.g. [21–23]) in order to reduce the complexity of the rule base, which at its turn determines the time demand of the calculation as well as the time necessary for parameter optimization.

Acknowledgement. This research is supported by EFOP-3.6.1-16-2016-00006 "The development and enhancement of the research potential at Pallasz Athéné University" project. The Project is supported by the Hungarian Government and co-financed by the European Social Fund. The research was also supported by ShiwaForce Ltd., Andrews IT Engineering Ltd., and the Foundation for the Development of Automation in Machinery Industry.

References

1. Devasenapati, S.B., Ramachandran, K.I.: Hybrid fuzzy model based expert sytem for misfire detection in automobile engines. Int. J. Artif. Intell. **7**(A11), 47–62 (2011)
2. Portik, T., Pokorádi, L.: Fuzzy rule based risk assessment with summarized defuzzyfication. In: Proceedings of the XIIIth Conference on Mathematics and its Applications, Timisoara, pp. 277–282 (2013)
3. Škrjanc, I., Blažič, S., Matko, D.: Direct fuzzy model-reference adaptive control. Int. J. Intell. Syst. **17**(10), 943–963 (2002)
4. Vaščák, J.: Approaches in adaptation of fuzzy cognitive maps for navigation purposes. In: Proceedings of the 8th International Symposium on Applied Machine Intelligence and Informatics—SAMI, Herľany, pp. 31–36 (2010)
5. Yorita, A., Botzheim, J., Kubota, N.: Self-efficacy using fuzzy control for long-term communication in robot-assisted language learning. Proceedings of the International Conference on Intelligent Robots and Systems (IROS), Tokyo, pp. 5708–5715 (2013)
6. Pelusi, D., Mascella, R., Tallini, L., Vazquez, L.: Control of Drum Boiler dynamics via an optimized fuzzy controller. Int. J. Simul. Syst. Sci. Technol. **17**(33), 1–7 (2016)
7. Pappis, C., Mamdani, E.: A fuzzy logic controller for a traffic junction. IEEE Trans. Syst. Man. Cybern. **7**(10), 707–717 (1977)
8. Karakuzu, C., Demirci, O.: Fuzzy logic based smart traffic light simulator design and hardware implementation. Appl. Soft Comput. **10**, 66–73 (2010)
9. Postorinoa, M.N., Versacia, M.: Upgrading urban traffic flow by a demand-responsive fuzzy-based traffic lights model. Int. J. Model. Simul. **34**(2), 102–109 (2014)
10. Baydokht, R.N., Noori, S., Azhangzad, A.: Presenting a fuzzy model to control and schedule traffic lights. J. Intell. Fuzzy Syst. **26**(2), 1007–1016 (2014)
11. Murat, Y.Z., Cakici, Z., Yaslan, G.: Use of fuzzy logic traffic signal control approach as dual lane ramp metering model for freeways. In: Online Conference on Soft Computing in Industrial Applications Anywhere on Earth, pp. 10–21, December 2012
12. Castán-Rocha, J.A., Ibarra-Martínez, S., Laria-Menchaca, J., Castan, E.R.: An implementation of case-based reasoning to control traffic light signals. In: Proceedings of the World Congress on Engineering, vol. 1. WCE, London, UK, 2–4 July 2014
13. Teo, K.T.K., Kow, W.Y., Chin, Y.K.: Optimization of traffic flow within an urban traffic light intersection with genetic algorithm. In: Proceedings of 2010 Second International Conference on Computational Intelligence, Modelling and Simulation (CIMSiM 2010), pp. 172–177 (2010). doi:10.1109/CIMSiM.2010.95
14. Khanjary, M., Navidi, H.: Optimizing traffic light of an intersection by using game theory. In: Proceedings of 3rd World Conference on Information Technology (WCIT-2012), AWERProcedia Information Technology & Computer Science, vol. 3, pp. 1163–1168 (2012)
15. Kovács, T., AvarezGil, R.P., Bolla, K., Csizmás, E., Fábián, C., Kovács, L., Medgyes, K., Osztényi, J., Végh, A.: Parameters of the intelligent driver model in signalized intersections. Tech. Gaz. **23**(5), 1469–1474 (2016)
16. Sugeno, M.: Industrial Applications of Fuzzy Control. Elsevier, Japan (1985)
17. Kennedy, J., Eberhart, R.: Particle swarm optimization. In: Proceedings of IEEE International Conference on Neural Networks IV. Perth, pp. 1942–1948 (1995)
18. David, R.-C., Precup, R.-E., Petriu, E.M., Rădac, M.-B., Preitl, S.: Gravitational search algorithm-based design of fuzzy control systems with a reduced parametric sensitivit. Inf. Sci. **247**, 154–173 (2013)

19. Precup, R.-E., David, R.-C., Petriu, E.M., Preitl, S., Rădac, M.-B.: Novel adaptive gravitational search algorithm for fuzzy controlled servo systems. IEEE Trans. Ind. Inform. **8**(4), 791–800 (2012)
20. Pelusi, D., Mascella, R., Tallini, L.: Revised gravitational search algorithms based on evolutionary-fuzzy systems. Algorithms **10**(2), 44 (2017)
21. Johanyák, Z.C.: Performance improvement of the fuzzy rule interpolation method LESFRI. In: Proceedings of the 12th IEEE International Symposium on Computational Intelligence and Informatics, Budapest, pp. 271–276 (2011)
22. Kovács, L., Ratsaby, J.: Analysis of linear interpolation of fuzzy sets with entropy-based distances. Acta. Polytech. Hung. **10**(3), 51–64 (2013)
23. Vincze, D., Kovács, S.: Performance optimization of the fuzzy rule interpolation method FIVE. J. Adv. Comput. Intell. Intell. Inform. **15**(3), 313–320 (2011)

Characteristics of Magnetorheological Fluids Applied to Prosthesis for Lower Limbs with Active Damping

Oscar Arteaga[1(✉)], Diego Camacho[1], Segundo M. Espín[2], Maria I. Erazo[1],
Victor H. Andaluz[1], M. Mounir Bou-Ali[3], Joanes Berasategi[3], Alvaro Velasco[1],
and Erick Mera[1]

[1] Universidad de las Fuerzas Armadas ESPE, Sangolquí, Ecuador
{obarteaga,docamacho,mierazo,vhandaluz1,apvelasco,
epmera1}@espe.edu.ec
[2] Universidad Técnica de Ambato, Ambato, Ecuador
sespin@uta.edu.ec
[3] MGEP Mondragon Goi Eskola Politeknikoa, Mondragon, Spain
{mbouali,jberasategui}@mondragon.edu

Abstract. The presence of people with amputations makes imminent the necessity of devices that replace the limb in both the aesthetic and functional, for which we propose the design and control of a robotic prosthesis with active damping using Magnetoreological Fluids (MRF) that are a new type of intelligent materials that have been characterized obtaining a shear yield stress of 41,65 kPa applying a controlled magnetic field of 0,8 T with a temperature of 20 °C, in order to have the active damping in the prosthesis will be necessary the use of a Magnetoreological (MR) damper which underwent to a dynamic damping behavior test reflecting 250 N as the maximum damping force. Control with Magnetorheological technology allows to have an instantaneous response in function of the signals obtained from the sensors, so that the patients could have a natural gait.

Keywords: Magnetoreological fluids · Robotics · Prosthesis

1 Introduction

The diseases, wars, traffic accidents, industrial injuries, and natural disasters, are the main cause of the increase by tens of thousands the number of amputees every year [1], having the body incomplete is not the only inconvenience, considering a sample of 134 people who had a permanent limb loss, 61 of these have used prostheses during the 2 years and almost the half had psychological damages [2]. Due to the current state of medical treatments, limbs cannot regenerate. Therefore, the only way for lower limb amputees to recover the function and appearance of their limbs is to use prosthesis [1], the same ones that have improved and have had significant technological advances in order to improve the quality of life of the patient [3].

Robotics comprises different disciplines in the design of a multidisciplinary system [4], such as a robotic prosthesis which can better restore missing limb functions by

© Springer Nature Singapore Pte Ltd. 2018
K.J. Kim et al. (eds.), *IT Convergence and Security 2017*,
Lecture Notes in Electrical Engineering 449,
DOI 10.1007/978-981-10-6451-7_29

adjusting impedance parameters at different stages of gait and also has the possibility to regulate the dynamic parameters according to the medium field where the patient is using the prosthesis [5].

According to the literature of robotic systems for assistance and rehabilitation, it focusses on providing missing movements and sensing, with environments that are safe and make regaining movement-related function easier and faster. Robotic prosthetics and exoskeletons will provide natural mobility, dexterity and sense of touch to paralyzed or missing limbs [6–8].

It has been possible to minimize the incidence caused by the absence of the lower extremity proposing a robotic prosthetic for lower limbs using MRF whose mechanical properties allows a complete damping control on the walking, the ones that have liquid-like properties without magnetic fields and forms chain-like structures when external magnetic fields are applied [9].

This paper has 5 Sections including the Introduction. The Sect. 2 describes the MRF's properties. The Sect. 3 explain de design of the robotic prosthesis. The Sect. 4 show the experimental results of the MR damper. Finally, the Conclusions are presented in the Sect. 5.

2   MRF's Properties

Magnetorheological fluids become a category of magnetically controllable fluids, usually formed by iron particles in a liquid carrier. The basic phenomena that are generated in the magnetoreology are related to the possibility of controlling the structure and, consequently, the rheological behavior of the plastic biphasic fluid through the use of relatively moderate magnetic fields [10].

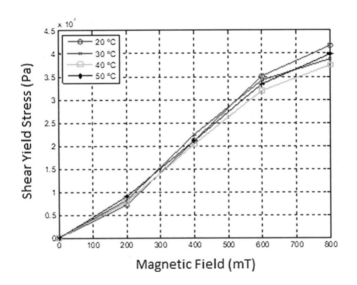

Fig. 1. Shear yield stress in function of the magnetic field and temperature.

The employment use of MRF as key components of several high-tech applications involving vibration control or torque transmission, lead to improved performance with new MRF formulations as is the LORD MRF-140CG fluid. Through the magnetic field is controlled the yield stress, shear thinning, and viscosity, that viscoelastic response, are important factors to consider on the MRF [10].

In Fig. 1. is shown the shear yield stress obtained by the MRF characterization at ranges of ranges of different temperatures on 20 °C to 50 °C and magnetic fields of 0 T to 0, 8 T, was performed using the MRC-501 rotational rheometer (Anton Paar Physica) with the MRD-70/1T cell coupled for the application and control of the magnetic field, and for the control of temperature was used a thermostatic Julabo F-25 bath.

3   Prosthesis Design

For the patients of 25 to 35 years old of the Percentile 90 [11] presents the anthropomorphic characteristics that are showing on Table 1.

Table 1. Parameters of design.

Height [mm]	Weight [kg]
1.700	70

According to Fig. 2. the type of amputation that is taken for the design is a cut of 50[*mm*] above the knee; taking the standard measures according to the design parameters [11], a knee length up to the ground of 485[*mm*] is determined and the 50[*mm*] knee amputees should be taken into account, the appropriate design weight calculated is proportional weights, *i.e.*, 4.15 [Kg].

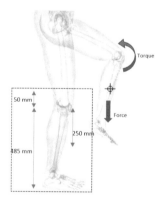

Fig. 2. Anthropometric dimensions.

Considering the design parameters and properties of MRF, the transfemoral prosthesis with active damping is designed as shown in Fig. 3.

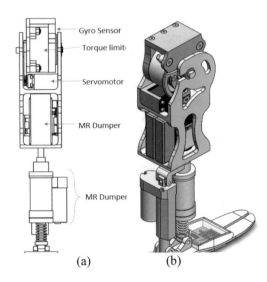

Gyro Sensor

Torque limit

Servomotor

MR Dumper

MR Dumper

(a)   (b)

Fig. 3. Prosthesis design: (a) the three-dimensional structure and (b) the three-dimensional

Figure 3(a) and (b) respectively show the three-dimensional structure and sectional view of the prosthesis, which was designed and developed according to the principle shown in Fig. 3 [12], the prosthesis is mainly composed by a torque limiter that works as a knee providing the necessary angle for walking, the MR damper that gives enough cushioning acting like normal leg and avoiding several future damages in the other leg.

The DC power supplies are required for the current amplifier, servomotor, MR dumper and gyro sensor. The desired knee joint angle (∅) and the necessary force require by the MR damper (F) are predicted from the hip joint angle which is measured by the gyro sensor. [12] The actual knee joint angle is directly measured by the hall sensor. Therefore, the prosthesis is a closed single input and single output closed-loop system including the nonlinear PD controller as shown in Fig. 4. Therefore, the control input is completely different during the swing phase from that during the stance phase. In the swing phase, control signal input is turned off so that the Torque limiter does not resist the rotation of the servomotor. However, in the stance phase the torque limiter supports the body weight and controls the knee joint angle with force from the servomotor. The force sensor provides the necessary information about the force that changes according to the magnetic flux density (B) that MR dumper requires to cushion the prosthesis properly.

The transmission of the torque of the torque limiter depends largely on the yield limit and the viscosity of the fluid. The Bingham plastic model is used to describe the behavior of MR in the disk-shaped actuator, [13] and is given by:

$$\tau(r, \omega, H) = \tau_y(H) + \eta \frac{\omega r}{h} \tag{1}$$

Where τ (r, ω, H) is the shear force, which depends on some factors, $\tau_y(H)$ is the dynamic yield limit that depends on the magnetic flux density, η is the viscosity of the fluid, ω is

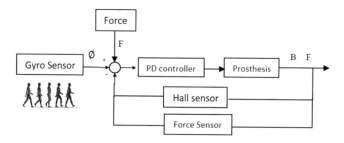

Fig. 4. Block diagram of the prosthesis

the angular velocity of the rotating disk, h is the fluid interval and r is the position on the surface of the disk as shown in Fig. 5. The resistive torque can be found derived from the shear off along the surface of the plate [13].

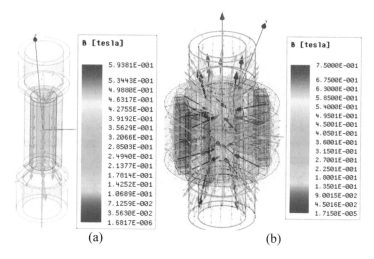

(a)   (b)

Fig. 5. Magnetic Field analysis: (a) venturi analyzed coil made of 1000 turns with parallel magnetic direction, (b) venturi analyzed with an external coil that produce a perpendicular magnetic direction

$$T = 2 \int_{R_i}^{R_0} (\tau_y(H)2\pi r)rdr \tag{2}$$

Where T is the resistive torque, R_i and R_0 are the inner and outer radius, respectively, substituting Eq. 2 in 1 the torque is given by:

$$T = \frac{4\pi}{3}\tau_y(H)\left(R_0^3 - R_i^3\right) + \frac{\eta\omega\pi}{h}\left(R_0^4 - R_i^4\right) \tag{3}$$

As can be seen in Eq. 3, the torque consists of two terms: the one is due to the effects of the MRF and the other is due to the viscous flow. The MR torque limiter is designed to work at very small rotational speeds, so the contribution of viscosity is very small and can be ruled out compared to the effects of the MR so the design is only based on the first term. The MRF used in the actuator is MRF-140CG, supplied by Lord Corporation. The maximum elastic limit in the linear working area is $\tau_y = 42$ kPa and the corresponding magnetic field is H $= 100$ kAm^{-1}. The shaft diameter is chosen to be 12.7 mm or Ri $= 6.35$ mm. As a result, the torque is 4.7 Nm.

On the other hand, for the MR dumper the flow mode consists of a housing, piston, (MRF-140CG), solenoid coil, oil seal, and magnetic core, the magnetic core is made of pure steel for high permeability, and the piston and housing are made of aluminum for low permeability. Thus, the magnetic field, which is generated by the solenoid coil, flows through the magnetic core. According to the Bingham plastic model of plates, the damping force, F_D is divided into the induced yield stress F_τ and viscous components F_η. The damping force is written as follows [12].

$$F_D = F_\tau + F_\eta \tag{4}$$

According to the plate model of Bingham plastic model, the damping force, F_D, can be divided into an induced yield stress F_τ and viscous F_η component.

4 Magnetic Field Analysis

The Fig. 5. proves the theory of magnetism, the item (a) shows the flux density parallel to the fluid, where it is not enough to generates the required force in the item (b) the flux density is perpendicular to the fluid resulting in the magnetism necessary to produce the required force for cushioning.

The damping force for the FEM modeling is calculated using the magnetic flux density as determined at different current levels. The yield shear stress τ_y relationship with the magnetic flux density (B) for MR fluid MRF-140 EG. For the magnetic flux densities computed in the previous section, the corresponding yield shear stress is found by the rheometer analysis [12].

$$F_d = F_\eta + F_\tau = A_p \left(\frac{12\eta l_p Q}{d_i \pi d_o^3} + c \frac{l_m}{d_o} \tau_y \right) sign(v), \tag{5}$$

Where,

$$Q = A_p |v| \tag{6}$$

In the above, Q is the volumetric flow rate of the MRF, A_p is the effective cross-sectional area of the piston, do is the inner diameter of the housing, d_i is the diameter of magnetic core, d is the diameter of the piston rod, v is the piston velocity, τ_y is the yield stress of the MRF, η is the offstate no magnetic field viscosity of the MRF, l_m is the effective axial pole length, l_p is the annular duct length, and sign(v) is used to consider the semi-active force. In order to analyze the magnetic field between the yield stress of the MRF and input current the finite element method is used, it is analyzed in a CAE software that display the intensity and direction of the magnetic field [12].

5   Experiments and Results

In order to bring about an effective MR dumper, the coil is made for producing the necessary magnetic camp to cushion the prosthesis, Fig. 6. the red perpendicular arrows show the principle of keeping the majority of the field in the rod developing a comfortable cushion during the walking, while the torque limiter produce the effective angle to generate a correct movement of the prosthesis.

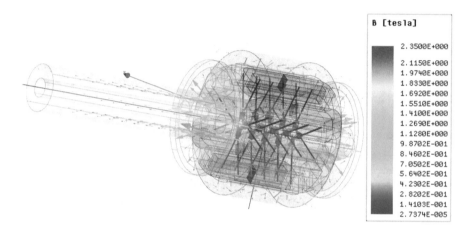

Fig. 6. Piston rod with a perpendicular flux density

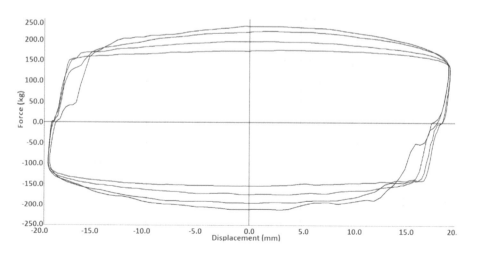

Fig. 7. Force vs Displacement

The test made on the bench Dyno-shock 11, indicates that the damping force is nonlinearly increased by increasing flux density. Figure 7 shows the displacement of the piston versus damping force. The maximum damping force is approximately 250 N. It is remarked here that in this work the MR damper is designed with a high viscosity term depending on the angular velocity of the knee joint angle because the bending angular velocity is fast when the person falls down while walking.

6   Conclusions

The control of the prosthesis was done in such way that the patient has an adequate gait according to the environment in which moves and the activity that performs, providing comfort and autonomy to the patient.

The effectiveness of the robotic prosthesis is demonstrated experimentally, obtaining the active damping, a MR damper was used in which its field-dependent characteristics of the MR damper were measured, such as the damping force where it can be seen that to generates an optimal damping is necessary for the walk the induction of a smaller magnetic field and for running requires a high field.

References

1. Xu, L., Wang, D.H., Fu, O., Yuan, G., Hu, L.Z.: A novel four-bar linkage prosthetic knee based on magnetorheological effect: principle, structure, simulation and control. Smart Mater. Struct. **25**, 12 (2016). doi:10.1088/0964-1726/25/11/115007
2. Price, P.: The diabetic foot: quality of life. Clin. Infect. Dis. **39**, S129–S131 (2004). doi: 10.1086/383274
3. Young, A.J., Simon, A.M., Fey, N.P., Hargrove, L.J.: Intent recognition in a powered lower limb prosthesis using time history information. Ann. Biomed. Eng. **42**(3), 631–641 (2014). doi:10.1007/s10439-013-0909-0
4. Ragusilaa, V., Emami, M.R.: Mechatronics by analogy and application to legged locomotion, Mechatronics. Inf. Process **000**, 1–19 (2016). doi:10.1016/j.mechatronics.2016.02.007
5. Zheng, E., Wang, Q.: Noncontact capacitive sensing based locomotion transition recognition for amputees with robotic transtibial prostheses. IEEE Trans. Neural Syst. Rehabil. Eng. **25**, 161–170 (2016). doi:10.1109/TNSRE.2016.2529581
6. Dellon, B., Matsuoka, Y.: Prosthetics, exoskeletons, and rehabilitation. IEEE Robot. Autom. Mag. **14**, 30–34 (2007). doi:10.1109/MRA.2007.339622
7. Lawson, B.E., Mitchell, J., Truex, D., Shultz, A., Ledoux, E., Goldfarb, M.: A robotic leg prosthesis: design, control, and implementation. IEEE Robot. Autom. Mag. **21**, 70–81 (2014). doi:10.1109/MRA.2014.2360303
8. Lawson, B.E., Ledoux, E.D., Goldfarb, M.: A robotic lower limb prosthesis for efficient bicyclin. IEEE Trans. Rob. **33**, 432–445 (2017). doi:10.1109/TRO.2016.2636844
9. Park, J., Yoon, G.H., Kang, J.W., Choi, S.B.: Design and control of a prosthetic leg for above-knee amputees operated in semiactive and active modes. Smart Mat. Struct. **25**, 13 (2016). doi:10.1088/0964-1726/25/8/085009
10. Susan-Resiga, D., Vékás, L.: Yield stress and flow behavior of concentrated ferrofluid-based magnetorheological fluids: the influence of composition, pp. 645–653. Springer, Heidelberg (2014)

11. Serrano-Sanchez, J.A., Lera-Navarro, A., Espino-Torón, L.: Physical activity and differences of functional fitness and quality of life in older males. Revista Internacional de Medicina y Ciencias de la Actividad Física y del Deporte (2016). doi:10.1093/gerona/56.suppl_2.23
12. Xu, L., Wang, D.H., Fu, Q., Yuan, G., Hu, L.Z.: A novel four-bar linkage prosthetic knee based on magnetorheological effect: principle, structure, simulation and control. Smart Mater. Struct. 25(11), 115007 (2016). doi:10.1088/1361-665X/aa61f1
13. Sarkar, C., Hirani, H.: Theoretical and experimental studies on a magnetorheological brake operating under compression plus shear mode. Smart Mater. Struct. 22(11), 115032 (2013). doi:10.1088/0964-1726/22/11/115032

Multi-Objective Shape Optimization in Generative Design: Art Deco Double Clip Brooch Jewelry Design

Sunisa Sansri[1] and Somlak Wannarumon Kielarova[2(✉)]

[1] Department of Industrial Engineering, Faculty of Engineering, Naresuan University, Phitsanulok, Thailand
sunisasan.nu@gmail.com
[2] iD3 -Industrial Design, Decision and Development Research Unit, Faculty of Engineering, Naresuan University, Phitsanulok, Thailand
somlakw@nu.ac.th

Abstract. This paper proposes multi-objective optimization generative design (MOOGD) system for generating shapes and optimizing two design objectives. The framework of this paper covers parametric modeling of the product shape configuration using Grasshopper plug-in in Rhinoceros software as well as multi-objective optimization developed using Octopus plug-in on Grasshopper. This framework is applied onto the case study of Art Deco double clip brooch jewelry design. The main goals of the study are to design and to optimize shapes of the double clip brooch in two objectives. The first objective is to apply golden ratio to the generating shapes. The second one is to minimize the use of metal referring to weight of the brooch. In the system, MOOGD finally generates a Pareto front to show the optimal solutions, which artists or designers could further use in conceptual product design process. The illustration of the proposed system is provided in this paper.

Keywords: Generative design · Multi-objective genetic algorithm · Multi-objective optimization · Pareto-optimal front · Art deco double clip brooch

1 Introduction

Interactive evolutionary computation (IEC) is a method that based on subjective human evaluation. This method becomes important in product design problems because human designers are able to express emotion, preference or feeling onto the generated designs. This paper aims to develop multi-objective optimization algorithm to be used in generative design system with the case study of Art Deco double clip brooch jewelry design.

The paper is organized in five main sections. The next section, the related works such as IEC, multi-objective optimization, and SPEA-II method are provided. The framework of the proposed system is explained in the Sect. 3. The Sect. 4 illustrated the case study of Art Deco double clip brooch jewelry design. Lastly, the Sect. 5 provides the conclusions and the future directions.

© Springer Nature Singapore Pte Ltd. 2018
K.J. Kim et al. (eds.), *IT Convergence and Security 2017,*
Lecture Notes in Electrical Engineering 449,
DOI 10.1007/978-981-10-6451-7_30

2   Related Works

2.1   Interactive Evolutionary Computation

Interactive evolutionary computation (IEC) [1] is one of the methods of evolutionary computation (EC) for optimization based on subjective human evaluation. Fitness function in IEC is replaced by human evaluation or user evaluation. Therefore, the target of IEC system is to obtain the outputs based on human guides such as emotion, preference or feeling. There are various applications applied using IEC systems [2–4] include artistic applications Kielarova et al. [5] developed an interactive genetic algorithm (IGA) integrated with multi-objective optimization for designing perfume bottle. The qualitative fitness is obtained by user preference, while the quantitative fitness is determined by volume and height of bottle. Brintrup et al. [6] proposed an IGA for plant layout design optimization. In this paper, the sequential IGA and multi-objective IGA were compared. García-Hernández et al. [7] proposed multi-objective interactive genetic algorithm used for facility layout design of the ovine slaughterhouse and the carton recycling plant problem. Brintrup et al. [8] developed an IGA for ergonomic chair design by fusing qualitative and quantitative criteria. They compared three IGA algorithms: sequential IGA, parallel IGA and multi-objective IGA in several criteria. Brintrup et al. [9] developed IEC with fuzzy systems for handling qualitative in design optimization of house floor planning with two objectives: minimize the building cost and maximize the subjective user evaluation.

2.2   Multi-Objective Optimization

Optimization problems [10] are divided into two types according to the objective functions. The first type is single objective optimization, and the second one is multi-objective optimization (also called multi-criteria optimization or vector optimization problem). In this paper, multi-objective optimization is used for solving the problems.

There are various methods of MOP. The popular methods are weighted-sum approach and Pareto-based approach [11]. Weighted-sum approach is a traditional method that sum all objectives to a single one. Then, it uses this single solution optimization to find an optimum solution, which is based on the weights assigned by user. Considering multi-objective optimization problem, it is impossible to find only one best solution, therefore, the Pareto-based approach [10] was invented. It aims to find the trade-offs of all objectives simultaneously. In this paper, based on characteristics of the problems, Pareto-based approach was chosen to develop the MOOGD.

Zhang et al. [12] developed a multi-objective genetic algorithm for shape optimization of free-form buildings based on solar radiation gain and space efficiency in the severe cold zones of China. The Octopus plug-in on Grasshopper was used as the optimization tool. Octopus generated Pareto frontier to show the optimal solutions to user. Boonlong et al. [12] presented the improved compressed objective genetic algorithm (COGA-II). They used rank assignment for screening non-dominated solutions. It is the best one for approximating Pareto front from non-dominated solutions. When COGA-II benchmarked with a non-dominated sorting genetic algorithm II (NSGA-II) and an improved strength Pareto genetic algorithm (SPEA-II), the results proved that COGA-II is better than NSGA-II and SPEA-II. Chutima et al. [13] proposed combinatorial optimization with coincidence algorithm (COIN) in mixed-model u-shaped assembly line balancing problems with three objectives. COIN was tested with a well-known algorithm namely non-dominated sorting genetic algorithm II (NSGA-II) and MNSGA-II (a memetic version of NSGA-II). Their experimental showed that COIN is better than NSGA-II.

Multi-objective evolutionary algorithm (MOEA) was derived from single-objective evolutionary algorithm (EA) optimization and multi-objective problem (MOP) domains. Carlos et al. [10] present the structures of various MOEAs.

The NSGA-II and SPEA-II are the most prominent and the most popular of MOEAs. Zitzler et al. [14] suggested that the NSGA-II and SPEA-II express the best overall performance, but the SPEA-II has advantages over the NSGA-II in higher dimensional objective spaces. For this reason, in this work, the SPEA-II technique is used for MOP, which will be further explained in the next section.

2.3 SPEA-II

The strength Pareto evolutionary algorithm (SPEA) was introduced by Eckart Zitzler et al. [15]. Characteristics of the SPEA are as follows: (1) storing non-dominated solutions externally, (2) evaluating fitness of individual dependent on the number of external non-dominated points that dominate it, (3) using the Pareto dominance relationship for preserving population diversity, and (4) reduce the non-dominated set by incorporating a clustering procedure.

Zitzler et al. [14] improved the SPEA namely SPEA-II. They had attempted to eliminate the weaknesses of the previous SPEA and to design a powerful and up-to-date EMO algorithm. The SPEA-II has three main differences from the previous SPEA [10] are as follows: (1) fitness assignment which takes into account for all individual by incorporates a fine-grained, (2) guides search by uses a nearest neighbor density estimation technique, and (3) has an enhanced archive truncation method.

3 The Framework of MOOGD

This paper proposes the framework of multi-objective optimization used in generative design (MOOGD) for shape optimization to be used in product design problem.

In this paper, a Pareto-optimal front is constructed for optimizing two objectives. It aims to find reasonable compromises or trade-offs between two objectives. The Pareto-optimal front is constructed by plotting all non-dominated solutions, which are not dominated by other solutions in the solution space. In this work, SPEA-II method is chosen to develop MOOGD for shape optimization. SPEA-II begins with initializing random population, and then sorting non-dominated. The algorithm assigns fitness to each individual in the current generation. If the termination criteria are not yet satisfied, the genetic operators (crossover and mutation) will applied in the next generation. Then the system combines the current population and non-dominated solutions in the previous generation and sorts them. If the non-dominated solution is bigger than the limit, it will activate the Nearest Neighborhood Technique to remove some members of non-dominated solutions. The termination criteria used are the average change in the spread of the Pareto front over the stall generations or the maximum of generations.

In the proposed generative design system, the user firstly defines the inputs of product shape modeling according to personal preferences. Those inputs are used to construct parametric modeling with Grasshopper plug-in in Rhinoceros, and perform factor calculations. Then, it is optimized two objectives by SPEA-II method. After the optimization process, the system are finally generated the results in the Pareto-optimal front for the user.

4 A Case Study: Art Deco Double Clip Brooch Jewelry Design

4.1 Problem Definition

The framework of this paper was applied to jewelry design in a case study of Art Deco double clip brooch or dress clips as shown in Fig. 1. The double clip brooch first appeared around the year 1927. Louis Cartier had invented the double clip brooch by assembling two clips into a main body brooch [16].

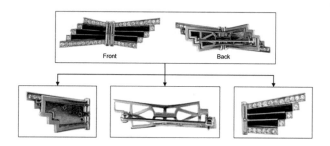

Fig. 1. Art Deco double clip brooch or dress clips [17].

In the case study of double clip brooch design, the metal body of jewelry is considered excluding gemstones. The quantitative fitness is evaluated the ratio and weight of the generated double clip brooches. The goal of the system is to optimize the shapes of the double clip brooches within two objectives: minimization of the proportion of the brooch shape to golden ratio and minimization of the brooch weight. The metal part of the brooch is made from silver. The golden ratio is connected to aesthetic appeal in art and design [18]. Wannarumon et al. [19] had applied the golden ratio to jewelry design.

4.2 Parameters and Constraints

In this work, the shape of the double clip brooch can be modifed by changing shape parameters. There are six parameters for generating shapes and controlling the proportion and the weight of the double clip brooch as shown in Fig. 2, and those parameters are encoded in real-valued chromosomes as illustrated in Table 1.

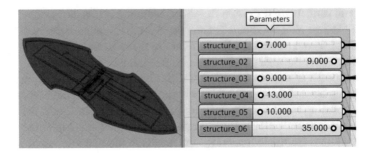

Fig. 2. Shape parameters for generating double clip brooch.

Table 1. Table of chromosome for controlling shape proportion and brooch weight.

Structure_01	Structure_02	Structure_03	Structure_04	Structure_05	Structure_06
7.00–12.00	3.00–9.00	9.00–18.00	13.00–16.00	10.00–12.00	20.50–35.00

Two objective functions are derived from calculating the proportion and the weight of the double clip brooches. One hundred samplings of the double clip brooches are randomly collected according to statistical survey methodology [20]. From the survey results, the acceptable range of the proportion of the brooches is 0.86–5.00, while of the weight is 10.75–76.20 grams. The bound constraints of six decision variables are shown in Table 1.

4.3   Setup and Optimization Results

The MOOGD prototype was developed using Grasshopper in Rhinoceros 5.0 on a computer workstation with Windows 7 system (Intel Xeon 2.8 GHz processor, 8G RAM, 64 bit). The Octopus plug-in in Grasshopper [21] was applied to build MOOGD. The followings are the set of GA parameters: the population size = 50 [22], maximum generations = 100, crossover rate = 0.8, mutation probability = 0.1, mutation rate = 0.5 and elitism = 0.5 [12].

MOOGD starts with user inputs or random initialization of the shape parameters. After obtaining the inputs, MOOGD is performing the optimization between the shape proportion and the brooch weight, until the Pareto-optimal front is generated. The framework of the proposed generative design system based on MOOGD is presented in Fig. 3. The Pareto-optimal front is constructed from non-dominated and it is expressed by a curved surface as shown in Fig. 4. All solutions on the Pareto-optimal front could be chosen as the final solution by personal preferences, because each solution on the Pareto-optimal front has its own strong point.

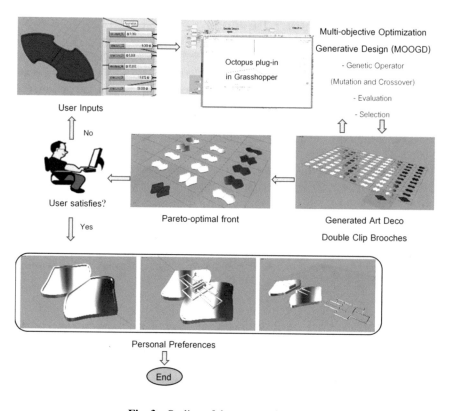

Fig. 3. Outline of the proposed MOOGD.

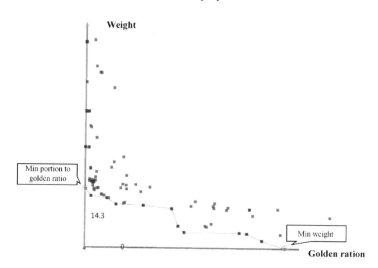

Fig. 4. Pareto-optimal front generated by MOOGD.

5   Conclusions and Future Directions

This paper proposes the framework of multi-objective optimization for shape optimization application named MOOGD, in the case study of Art Deco double clip brooch jewelry design. Two objectives, shape proportion and brooch weight, are optimized by SPEA-II. Shape of the double clip brooch is practically modeled using Grasshopper plug-in in Rhinoceros. MOOGD is developed using Octopus plug-in on Grasshopper. Pareto-optimal front is generated by the optimal solutions, which user can choose them based on his preference. The proposed framework can be applied to other applications.

The future direction of this research will involve the GA parameter study for shape optimization. Moreover, the user interface for interaction between user and system will be further studied.

Acknowledgement. The research has been carried out as part of the research projects funded by National Research Council of Thailand and Naresuan University with Contract No. R2560B005. The author would like to gratefully thank all participants for their collaborations in this research.

References

1. Takagi, H.: Interactive evolutionary computation: Fusion of the capabilities of EC optimization and human evaluation. Proc. IEEE **89**, 1275–1296 (2001)
2. Wei, Y., Wang, M., Qiu, J.: New approach to delay-dependent filtering for discrete-time Markovian jump systems with time-varying delay and incomplete transition descriptions. IET Control Theory Appl. **7**, 684–696 (2013)
3. Wei, Y., Qiu, J., Karimi, H.R., Wang, M.: Model reduction for continuous-time Markovian jump systems with incomplete statistics of mode information. Int. J. Syst. Sci. **45**, 1496–1507 (2014)
4. Wei, Y., Qiu, J., Karimi, H.R., Wang, M.: Filtering design for two-dimensional Markovian jump systems with state-delays and deficient mode information. Inf. Sci. **269**, 316–331 (2014)
5. Kielarova, S.W., Sansri, S.: Shape optimization in product design using interactive genetic algorithm integrated with multi-objective optimization. In: Multi-disciplinary Trends in Artificial Intelligence: 10th International Workshop, MIWAI 2016, Chiang Mai, Thailand, December 7–9, 2016, Proceedings, pp. 76–86. Springer International Publishing (2016)
6. Brintrup, A.M., Ramsden, J., Tiwari, A.: An interactive genetic algorithm-based framework for handling qualitative criteria in design optimization. Comput. Ind. **58**, 279–291 (2007)
7. García-Hernández, L., Arauzo-Azofra, A., Salas-Morera, L., Pierreval, H., Corchado, E.: Facility layout design using a multi-objective interactive genetic algorithm to support the DM. Expert Syst. **32**, 94–107 (2015)
8. Brintrup, A.M., Ramsden, J., Takagi, H., Tiwari, A.: Ergonomic chair design by fusing qualitative and quantitative criteria using interactive genetic algorithms. IEEE Trans. Evol. Comput. **12**, 343–354 (2008)
9. Brintrup, A.M., Ramsden, J., Tiwari, A.: Integrated qualitativeness in design by multi-objective optimization and interactive evolutionary computation. In: The 2005 IEEE Congress on Evolutionary Computation 2005, pp. 2154–2160. IEEE (2005)
10. Coello, C.A.C., Lamont, G.B., Van Veldhuizen, D.A.: Evolutionary Algorithms for Solving Multi-Objective Problems. Springer, New York (2007)

11. Deb, K.: Multi-objective Optimisation Using Evolutionary Algorithms: An Introduction. Springer, London (2011)
12. Zhang, L., Zhang, L., Wang, Y.: Shape optimization of free-form buildings based on solar radiation gain and space efficiency using a multi-objective genetic algorithm in the severe cold zones of China. Sol. Energy **132**, 38–50 (2016)
13. Chutima, P., Olanviwatchai, P.: Mixed-model U-shaped assembly line balancing problems with coincidence memetic algorithm. J. Softw. Eng. Appl. **3**, 347 (2010)
14. Zitzler, E., Laumanns, M., Thiele, L.: SPEA2: Improving the strength Pareto evolutionary algorithm (2001)
15. Zitzler, E., Thiele, L.: Multiobjective evolutionary algorithms: a comparative case study and the strength Pareto approach. IEEE Trans. Evol. Comput. **3**, 257–271 (1999)
16. http://revivaljewels.com/blog/2014/7/8/how-to-get-additional-wardrobe-mileage-from-your-jewellery
17. https://www.1stdibs.com/search/?q=art%20deco%20double%20clip%20brooch
18. Huntley, H.E.: The Divine Proportion: A Study in Mathematical Beauty. Dover Publications, New York (1970)
19. Wannarumon, S., Bohez, E.L., Annanon, K.: Aesthetic evolutionary algorithm for fractal-based user-centered jewelry design. Artif. Intell. Eng. Des. Anal. Manuf. **22**, 19–39 (2008)
20. Yamane, Taro: Statistics; an Introductory Analysis. Harper & Row, New York (1967)
21. http://www.food4rhino.com/app/octopus
22. Marsault, X.: A multiobjective and interactive genetic algorithm to optimize the building form in early design stages. In: 13th Conference of International Building Performance Simulation Association, pp. 809–816 (2013)

Adaptation of the Bioloid Humanoid as an Auxiliary in the Treatment of Autistic Children

Luis Proaño[1](✉), Vicente Morales[2], Danny Pérez[2],
Víctor H. Andaluz[1], Fabián Baño[2], Ricardo Espín[2], Kelvin Pérez[2],
Esteban Puma[2], Jimmy Sangolquiza[2], and Cesar A. Naranjo[1]

[1] Universidad de las Fuerzas Armadas ESPE, Sangolquí, Ecuador
luis.e.proa@gmail.com,
{vhandaluz1,canaranjo}@espe.edu.ec
[2] Universidad Técnica de Ambato, Ambato, Ecuador
jvmorales99@uta.edu.ec, ddsanty1992@gmail.com,
hbano0123@hotmail.com, ricardoespin81@gmail.com,
kperez0478@hotmail.com, estepuma94@gmail.com,
sjimmy093@hotmail.com, hbano0123@gmail.com

Abstract. This paper describes a system that allows children with autism spectrum disorder - ASD - to be more aware of their emotions and increase their social relation, through the development of two practical games focused on human–robot interaction, based on the DSM-5 protocol.

Keywords: Artificial vision · Autism · Humanoid · Control · Intelligent actuator

1 Introduction

Autism, also known as autism spectrum disorder -TEA-, is a disability that affects participation in social activities in early childhood and adulthood, it is considered as a disorder of neuronal development that affects the way of relating to other people and it is the fastest growing disability worldwide [1–5]. In general, people with ASD tend to have diminished intellectual capacity. Below are described in three main areas of difficulty also known as the triad of impediments which are as follows: (*i*) *communication* are problems related to the verbal and nonverbal language possessed by the individual; (*ii*) *social interaction* is the problem of understanding the emotions of other people; and finally, (*iii*) *repetitive and restricted behavior* that is the difficult adaptation to new places [6]. At present, the estimates presented by the World Health Organization (WHO) indicate that 1 among 160 children suffer from autism disorders, being as a variable depending on the country, such as: in the European Union affects 1 in 150 children, the US affects 1 in 68 children, in South Korea it affects 1 in 38 children, and in low- and middle-income countries this disorder becomes over and over frequent [7, 8].

Today, there are many methods and technologies for the treatment of autism spectrum disorder, which are: (*i*) *Evaluation* these methodologies are techniques that help to

© Springer Nature Singapore Pte Ltd. 2018
K.J. Kim et al. (eds.), *IT Convergence and Security 2017*,
Lecture Notes in Electrical Engineering 449,
DOI 10.1007/978-981-10-6451-7_31

give guidelines for finding new ways to diagnose, evaluate and treat autism through direct interaction child-therapist [9–11]. (*ii*) *Devices* are all the electronic devices like radios, devices of voice, cellular, etc., which through music, videos and images help the recovery of the child with autism, [11]. (*iii*) *Game console* these technologies are much more didactic for the entertainment and learning of the autistic child, helping to be able to interact better with other people, but only through gestures [7]. (*iv*) *Virtual environments* this technology has been growing in recent years, which has environments for manual tasks, math and science courses, helping in a very satisfactory way the child with socialization, learning and communication [13]. (*v*) *Robotics* in recent years has played a very important role in helping medicine and other fields. As for autism, the main objective of the study is to promote interaction and communication with children with ASD by robots. Robots are used as mediators between the child and the experimenter, this type of technology is very beneficial because the child can interact directly with the robot improving their social relationship with others helping to reduce the level of autism [14, 15].

Nevertheless, there may be other methods of treatment such as behavioral programs, education programs, medications, i.e., nowadays, a great variety of robots have been designed for studies of a medical–psychological nature, especially within the field of study in therapies against autism, statistical data feasible in the implementation of the present project, likewise many of these treatments are through the assistance of a wide variety of electronic devices such as computers, console game, IPod among others, effectively complementing the treatments mentioned in the previous paragraph, i.e., there are two major groups for application against this psychological disorder, known as: commercial robots and experimental robots. (*i*) *Commercial robots* are much better known and used for the treatment of this anomaly psychology, because they show satisfactory improvement results in most cases, among some examples of commercial robots we mention: (a) IROMEC is an interactive robot and it has a programmable system, defined with game scenarios that provides several stimuli to promote interaction with the child in different ways, in addition, is intended to play and relate to children with ASD, this robot focuses on different areas of development as: sensory, motor, cognitive, social, emotional, communicative and interaction [16]; (b) AGENT ROBOT NAO is a humanoid robot with an integrated camera, which detects the behavior of children during interaction with the robot and a portable device formed by two dry electrodes made of silicon to detect and measure facial expressions, the main advantage of this portable device is that it does not limit the child's movement and does not always need to be exposed to a camera, thus allowing the child to play freely by improving their social interaction [17, 18]. (*ii*) *Experimental robots* are characterized by being new and complementary proposals in special cases of treatment among which we can mention; (a) KILIRO is a parrot type facial recognition robot inspired by the therapeutic biology and rival teaching model that helps children with ASD improve their learning and social interaction skills [17]; (b) the ROBOSKIN is a robot that has an anthropomorphic 3D skeleton and has a mobile device such as smart phones or tablets, these devices help the child to answer questions, participate in games and control the therapies, and can play music, movies and animations, has different costumes to represent different types of animals, people or characters [8]; (c) KASPAR is programmed to smile, laugh, wink and wiggle arms, built by scientists at the

University of Hertfordshire, Kaspar comes in several versions, even an advanced enough to play the Nintendo Wii [8].

This paper focuses on the implementation of a system in which robotics is adapted to the treatment of children with autism using the human–robot interaction approach based on behavioral treatment protocols described in previous paragraphs. The proposal of the work, is based on the development of two games that allow children with ASD to interact with a humanoid robot, improving their social relationship and allowing them to be more aware of their emotions. The first game has two functions: to develop the motor of the child by manipulating the robot, if the patient requires it, and to arouse the child's interest in the robot, for this, tasks are assigned to the child to position the robot and graphs provided by the system interface, the therapist can analyze and evaluate the patient. In the second game, once you have the attention of the child, the system interprets behavioral signals through facial recognition and reading emotions, performing tasks previously established by the therapist. In addition, about the treatment, it is based on the DSM-5 standard, which is a methodology that allows to diagnose, treat and evaluate the different disorders presented in the patients, therefore, the therapist will be able to choose the modality of action of the robot based on the cognitive emotional limitation of the patient.

This work is organized as follows. The structure of the system is described in Sect. 2. Section 3 shows the design of the control. The design of the command interface is presented in Sect. 4. In Sect. 5 the experimental results are shown and the performance of the system is evaluated. Finally, the conclusions are found in Sect. 6.

2 System Structure

The proposed system is constructed using the human–robot interaction architecture, with activities based on well-known behavioral treatment protocols (DSM-5). In architecture, although there is a contribution of the therapist, the system estimates the extent of the reactivation of children through the interpretation of their behavioral signals after requesting that they perform the elements of training of visual contact and reading of emotions to automatically carry out pre-assigned acts. In addition, the interaction adopts automated technologies capable of performing a repetitive feedback in the robot, while maintaining consistency in the process.

Figure 1 shows the structure of the proposed system where the robot is programmed to perform tasks that stimulate the child, arousing their interest and perception, waiting for responses to this stimulus, the system receives this information through non-verbal channels, which allows to have behavioral keys to perform tasks based on treatment protocols that stimulate the child, and so, in a repetitive way in order to develop social skills in the patient, since people with this disorder are not aware of their mood and they have no empathy, so it is difficult for them to relate to others.

Thus, for the first proposed game, the robot's extremities are placed in a certain position so that the child replicates these positions and thus improves his/her motor skills, arouses his/her interest and is motivated to interact with the robot.

The second proposed game, the system captures the child's emotions through facial recognition, and obtains an estimate of the child's behavior, in which case the system

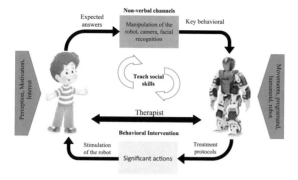

Fig. 1. Assisted behavioral intervention by the robot for the treatment of autism.

interprets it as happy or sad and according to this the robot executes the pre-established tasks of dancing or embracing, ensuring that the child is more aware of their mood and can better perform in their social environment in each therapy (Fig. 2).

3 Controller Design

This section presents the control system developed for children with autistic disorder, which seeks to improve this deficiency by including technologies such as artificial vision and robotics; the proposed system has four modules: (A) vision sensor, (B) artificial vision, (C) control algorithm and (D) robot. These are described below.

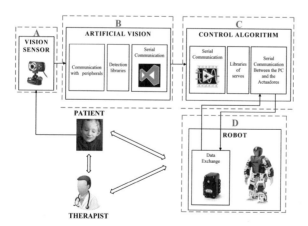

Fig. 2. Intelligent system block diagram

A. Vision sensor: Consists of entering data through a camera that is responsible for capturing the work environment and the profile of the patient, in order to differentiate personal traits. Captured images are considered as inputs to the artificial vision system.

B. Artificial vision: The artificial vision system is able to generate and discriminate data according to the patient's mental state through the child–robot interaction. Therefore, a programming algorithm has been designed based on the particularities, needs and skills affected.

Most children cannot assimilate their moods so that with the help of an integrated camera to the computer or any mobile device can detect their expressions, it should be emphasized that the child exposed to this recognition should be alone because this program is very sensitive and detects other details with facial similarity. The process begins with a face recognition -FR- of the child, so that the program can standardize the image -SI- and thus facilitate the detection of the smile -SD- using the libraries of opencv haarcascade_smile, as shown in Fig. 3.

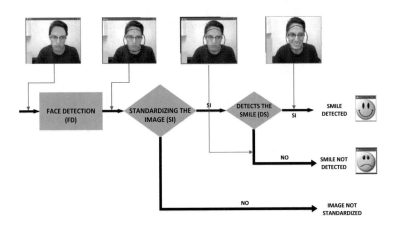

Fig. 3. Facial recognition algorithm block diagram

Additionally, it communicates with the next stage of the general system by sending data, which will vary according to the gesture captured (joy, sadness).

C. Control algorithm: The control algorithm is constituted as the receiver stage and treatment emitted by the artificial vision software, which is responsible for relating the different devices such as: computers, tablets, among others to the robotized assistant. This information serves to carry out the monitoring process in patients, as well as analyze their progress with each therapy. Other functionalities that are provided in this layer are the robots control and the selection of the two games for patients, the movements of the robot will be loaded and stored in vectors of n positions that can be accessed from the user interface.

D. Robot: This last section corresponds to the robot which contains intelligent modular actuators that incorporates a microcontroller which can be handled through commands, most of which set or read parameters that define its behavior, such as the ability to control its speed, temperature, position, voltage and load. In addition to having a specialized control circuit for network.

4   Interface Design

4.1   Block Diagram of the General System

This section describes the interface for patient, operator and humanoid interaction that is designed in LabVIEW for ease of development, control, and monitoring of data. This interface has two types of operation, where each one has its respective functionalities, which are described briefly in the block diagram as shown in Fig. 4.

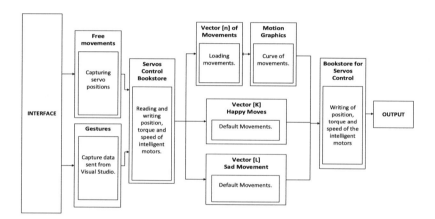

Fig. 4. LabVIEW interface

4.1.1   Free Movement Interface

The free-motion interface is divided into: (A) control panel, (B) motion indicators, (C) program indicators and (D) stop button, as shown in Fig. 5.

A. Control panel: The control panel is designed to allow the operator to view and manipulate basic system settings and controls, such as defining the communication port of the humanoid and its transmission rate, enabling/disabling, loading and unloading of the data towards the humanoid and finally the restart of the movements. It also has a grill dedicated to control the speed of execution of the program.

B. Movement indicators: This module consists of a group of real-time reading indicators emitted by the servomotors that make up the humanoid robot and will serve as support for the operator.

C. Program indicators: As the previous module provides information on the execution of captured movements and a dedicated button for selecting the program which is based on free movements or pre-established movements.

D. Stop button: It stops system continuity.

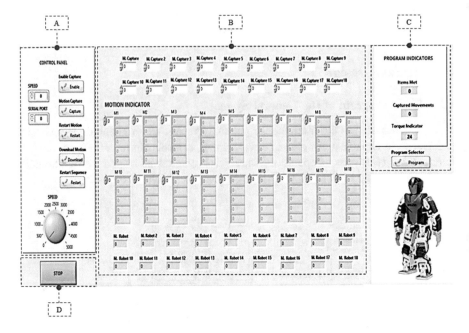

Fig. 5. Graphic interface system

4.1.2 Gesture Interface

Figure 6, operates on the basis of obtaining data emitted by the software Visual Studio, which are transformed into bits (0 and 1) and interpreted by the LabVIEW.

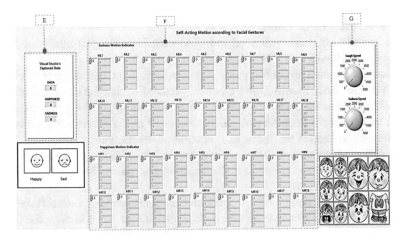

Fig. 6. Facial recognition control

E. Data captured from Visual Studio: This section complies with the work of reporting on the behavior of the captured data and simultaneously with the type of gesture to which the data corresponds.

F. Automatic control according to the facial gestures: This module corresponds to a group of matrices which store the writing data interpreted by the intelligent servomotors, intended to provide the movements of the robot BIOLOID.

G. Execution speed of movements according to gestures: Each knob in this section controls the execution time of each robot animation.

4.1.3   Graph of Completed and Captured Robot Movements

In the last function of the program it is observed the movements that were captured and those that were fulfilled, this will allow to realize the control of compensation depending on the case Fig. 7.

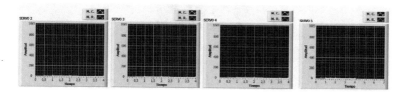

Fig. 7. Graph of the position of the servos in each movement

5   Experimental Results

The procedure aims to develop motor skills and social skills in children with autism through robotics, applied in the Ambato Specialized Educational Unit, whose program is based on the method of Treatment and Education of Autistic Children with Communication Problems (TEACCH) Table 1.

Table 1. Procedure for the treatment and education of autistic children

Stages	Actions
Preparation of therapy	Organize and prepare the work environment
Motivation	Play activities (games, songs, exercises)
Adaptation to the environment	Recognition of objects
Explanation of rules for therapies	Use of pictograms
Introduction of the robot	Use of pictograms
Development of therapy	Child - robot interaction
Closure of therapy	Use of pictograms
Post-therapy	Evaluate the results obtained from therapy

To begin the treatment, the patient is placed in front of the computer in a comfortable position free of distractions, supporting the artificial vision sensor to improve the possibilities of face capture and thus moods (facial gestures).

The artificial vision software discriminates the patient's mood in real time to generate a reference signal interpreted by the control interface or HMI. Figure 8 shows the section dedicated to the communication constituting as the first step for the operation of the total system and the dependence of the following interfaces.

Fig. 8. Control panel buttons

This interface is characterized by having a link with the artificial vision sensor and the humanoid, where it is possible to perform the same based on the type of data interpreted, resulting in the execution of pre-established movements based on the therapist's criteria Fig. 9.

(a)   (b)

Fig. 9. (a) Movements preloaded to joy. (b) Movements preloaded to sadness.

Finally, the third interface fulfills the objective of graphically indicating the captured or generator movements by the patient vs the compliant movements of the humanoid Fig. 10. For the case of servo 2 the therapist recites that the patient must comply with seven movements defined by Table 2.

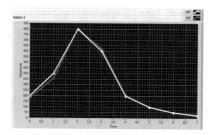

Fig. 10. Capture of movements reached

Table 2. Comparison of two servo positions

Data	Request (MP)	Position (MR)	Error
0	200	181	19
1	400	357	43
2	800	802	2
3	600	623	23
4	200	207	7
5	100	100	0
6	50	47	3
7	24	31	7

6   Conclusions

The results and efficiency of the system with the BIOLOID robot are validated by the therapist, through the use of the applications presented by the project with children who have some degree of autistic disorder allowing to increase their stimulation and attention. For the routine the doctor is in charge of evaluating the state of the patient to establish the treatment and the tasks that the child must fulfill. The system provides two modes of operation, the first allows to capture through a camera the mood of the child and this signal commands movements performed by the BIOLOID robot allowing an analysis of the patient's condition; while the latter mode allows to capture the patient's attention by manipulating the robot and complying with movements that the therapist chooses for therapy thus improving the interpersonal relationship of the patients.

References

1. Stephanidis, C., Antona, M.: Universal Access in Human–Computer Interaction. Springer, Berlin (2007)
2. Della Fina, V., Cera, R.: Protecting the Rights of People with Autism in the Fields of Education and Employment. Springer, Berlin (2015)
3. Yun, S.S., Kim, H., Choi, J., Park, S.K.: A robot-assisted behavioral intervention system for children with autism spectrum disorders. Robot. Auton. Syst. **76**, 58–67 (2016)

4. Park, C.H., Jeon, M., Howard, A.M.: Robotic framework with multi-modal perception for physio-musical interactive therapy for children with autism. In: Joint IEEE International Conference on Development and Learning and Epigenetic Robotics, 2015, pp. 150–151. IEEE (2015)
5. Ennis-Cole, D.L.: Technology for Learners with Autism Spectrum Disorders. Springer (2015). doi:10.1007/978-3-319-05981-5
6. Al Shirian, S., Al Dera, H.: Descriptive characteristics of children with autism at Autism Treatment Center, KSA. Physiol. Behav. **151**, 604–608 (2015)
7. Bernardini, S., Porayska-Pomsta, K., Smith, T.J.: ECHOES: an intelligent serious game for fostering social communication in children with autism. Inf. Sci. **264**, 41–60 (2014)
8. Galán-Mena, J., Ávila, G., Pauta-Pintado, J., Lima-Juma, D., Robles-Bykbaev, V., Quisi-Peralta, D.: An intelligent system based on ontologies and ICT tools to support the diagnosis and intervention of children with autism. In: Biennial Congress of Argentina, pp. 1–5. IEEE (2016)
9. Chen, W., Zhang, J., Ding, J.: Mirror system based therapy for autism spectrum disorders. Front. Med. China **2**(4), 344–347 (2008)
10. Lajoie, S.P., Vivet, M.: Artificial Intelligence in Education. IOS Press, Amsterdam (2002)
11. Carrington, S.J., Kent, R.G., Maljaars, J., Le Couteur, A., Gould, J., Wing, L., Leekam, S.R.: DSM-5 Autism Spectrum Disorder: in search of essential behaviours for diagnosis. Res. Autism Spectr. Disord. **8**(6), 701–715 (2014)
12. Cardon, T.A.: Technology and the Treatment of Children with Autism Spectrum Disorder. Springer (2015). doi:10.1007/978-3-319-20872-5
13. Mourning, R., Tang, Y.: Virtual reality social training for adolescents with high-functioning autism. In: IEEE International Conference on Systems, Man, and Cybernetics (SMC), pp. 004848–004853, October 2016
14. Costa, S.C., Soares, F.O., Pereira, A.P., Moreira, F.: Constraints in the design of activities focusing on emotion recognition for children with ASD using robotic tools. In: 2012 4th IEEE RAS and EMBS International Conference on Biomedical Robotics and Biomechatronics (BioRob), pp. 1884–1889, June 2012
15. Soares, F., Costa, S., Silva, S., Gonçalves, N., Rodrigues, J., Santos, C. Moreira, M.F.: Robótica-Autismo project: technology for autistic children. In: 2013 IEEE 3rd Portuguese Meeting on Bioengineering (ENBENG), pp. 1–4, February 2013
16. Cimatti, A., Pistore, M., Roveri, M., Traverso, P.: Weak, strong, and strong cyclic planning via symbolic model checking. Artif. Intell. (2003). doi:10.1016/S0004-3702(02)00374-0
17. Bharatharaj, J., Huang, L., Al-Jumaily, A.M., Krageloh, C., Elara, M.R.: Experimental evaluation of parrot-inspired robot and adapted model-rival method for teaching children with autism. doi:10.1109/ICARCV.2016.7838636
18. Hirokawa, M., Funahashi, A., Pan, Y., Itoh, Y., Suzuki, K.: Design of a robotic agent that measures smile and facing behavior of children with Autism Spectrum Disorder. In: 2016 25th IEEE International Symposium on Robot and Human Interactive Communication (RO-MAN), pp. 843–848, August 2016

Autonomous Assistance System for People with Amyotrophic Lateral Sclerosis

Alex Santana G.$^{(\boxtimes)}$, Orfait Ortiz C$^{(\boxtimes)}$, Julio F. Acosta$^{(\boxtimes)}$, and Víctor H. Andaluz$^{(\boxtimes)}$

Universidad de las Fuerzas Armadas ESPE, Sangolqui, Ecuador
{amsantana, oportiz, jfacosta, vhandaluz1}@espe.edu.ec

Abstract. This paper puts forward an autonomous assistance system for people with Amyotrophic Lateral Sclerosis, whose main goal is to improve the quality of life of the user, to move in a sure way in a semi-structured environment. For this purpose, it presents a robust scheme of multilayered where the position of look is captured with eye tracking technology, on an interface human-machine that represents an environment indoor; estimates a trajectory free of risks by means of the algorithm Rapidly-exploring random trees or RRT for a subsequent navigation with a wheelchair, resulting in the move of the user to the desired destination.

Keywords: RRT algorithm · Navigation wheelchair · Autonomous assistance

1 Introduction

Through of years, service robotics evolves exponentially. This is due investigations of big scalability in industrial or domestic environments where the autonomy of robots is required [1]. Involve a development of technologically autonomous application in the field of health has become in one of its principal goals. Improves the most important factor in the quality of life in people victims of progressive neurodegenerative affections, which precludes their mobility due to muscle weakness and of various joints, affecting directly to the capacity of moving themselves [1, 2].

For this reason, are developed and proved different secure technological navigation systems with wheelchairs, which allow people with disabilities in the majority of their extremity motor, the capacity to drive or move independently, to avoid to the maximum collisions and conditions that are unsecure at the moment of displacement from one actual position to the desired destination of the user. In this way, these types of systems with wheelchair are made up of subsystems like: sensory, planning and control which interact through interface human-machine [2].

The processes with which information is acquired are too intrusive and tedious to be done manually by the user. Therefore, it is necessary to detect events in human machine interfaces, so as not to invade the user. These systems are called eye tracking and serve to detect a specific field of vision through a computer [3]. The main proposal of this method is the representation of coordinates with information necessary for the user to select the destination within a structured environment [3, 4].

© Springer Nature Singapore Pte Ltd. 2018
K.J. Kim et al. (eds.), *IT Convergence and Security 2017*,
Lecture Notes in Electrical Engineering 449,
DOI 10.1007/978-981-10-6451-7_32

In this way, planning in this type of systems is the axis of autonomy together with an adequate control architecture applied in the wheelchair; this is given by a layer of Path Planning that aims to create algorithms that allow determining roads considering the Restrictions on the movements of mobile robots, and the dynamics of the working environment. The applications of the path planning are oriented to different tasks that interact with the human being in the ambit of (i) health allows the support in robot tasks for the elderly and surgery; (ii) military is oriented to the supervision of remote-controlled, autonomous robots and intelligent weaponry; (iii) industrial monitoring of robots with artificial intelligence - through the use of mobile robots, among other areas [5, 6]. The autonomy of a mobile robot to reach an end point since an initial point is delimited by the problems of collision with fixed and moving objects in both structured and unstructured environments. As mentioned the Path planning can be classified into two main methods (i) path-free collision, plan a path without considering obstacles in the workspace; And (ii) path planning with evasion of obstacles, which consists of evade the obstacles that appear in the way of the robot, while it gets to reach the goal [7]. In order for a robot to reach a desired goal, have been proposed random methods such as RPP (Random Path Planning) [8, 9] and PPP (Probabilistic Path Planning) [10, 11]. These methods allow to introduce in the restrictions of movement of the systems in the space of configurations, $i.e.$, a found path is not always valid or executable in reference to the restrictions of the movements of the robot (not holonomic) and in the way in That the wheelchair can be moved by avoiding potential risks to the user [12].

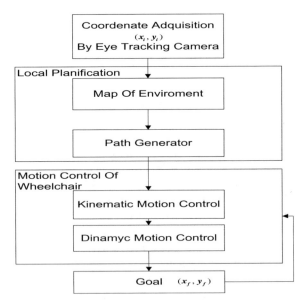

Fig. 1. A multi-layer system that represents the functioning from layers such as sensing, local planning and motion control.

In this paper, present an autonomous assistance system by people with amyotrophic lateral sclerosis, where its main characteristic is to mobilize the user in a safe way, for this a multilayer's system was developed, shown in Fig. 1. And consists of (i) sensing layer by means of eye tracking technology that allows to locate a user's gaze position which represents a coordinate in the image of a room; (ii) planning layer is oriented to the mapping of the structured environment and later stage of path planning applying the algorithm RRT giving result a safe trajectory for the displacement of the wheelchair; (iii) motion control layer here shows the kinematic and dynamic modeling of the wheelchair in addition to a tcp-ip communication to the planning layer which is transparent to the user and serves to measure the robot's current position, The section of desired destination and therefore the control actions pertinent so that the motors move the wheelchair until fulfilling the objective of the system.

2   Kinematic and Dynamic Modeling

2.1   Kinematic Modeling

By obtaining the design of the wheelchair robot controllers, the kinematic model of the instantaneous configuration given by the holonomic robot wheelchair is defined as:

$$\begin{cases} \dot{x} = u \cos \psi \; - \; a\omega \sin \psi \\ \dot{y} = u \sin \psi \; - \; a\omega \cos \psi \\ \dot{\psi} = \omega \end{cases} \tag{1}$$

a compact form that can be defined to the kinematic model:

$$\begin{aligned} \dot{\mathbf{h}} &= \mathbf{J}(\psi)\mathbf{v} \\ \dot{\psi} &= \omega \end{aligned} \tag{2}$$

2.2   Dynamic Modeling

The dynamic model of longitudinal velocity of reference without including disturbances can be defined as:

$$u_{ref} = \varsigma_1 \dot{u} - \varsigma_2 \dot{\omega} + \varsigma_3 u - \varsigma_4 \omega^2 \tag{3}$$

The dynamic model of the reference angular velocity without including perturbations can be defined as:

$$\omega_{ref} = -\varsigma_5 \dot{u} - \varsigma_6 \dot{\omega} + \varsigma_7 \omega - \varsigma_8 u\omega \tag{4}$$

Then (3) and (5) can be represented as follows

$$\mathbf{M}(\varsigma)\dot{\mathbf{v}} + \mathbf{C}(\varsigma, \mathbf{v})\mathbf{v} = \mathbf{v_{ref}} \qquad (5)$$

where, $\mathbf{M}(\varsigma) \in \Re^{nxn}$ with $n = 2$ represents the inertia of the chair user system detailed in the following matrix and $\mathbf{C}(\varsigma, \mathbf{v}) \in \Re^{nxn}$ represents the components of the centripetal forces.

The model presented describes the chair-user system, since it is not necessary to generate torques as inputs of the model both kinematic and dynamic, as is usual to find in commercial robots. The dynamic parameters ς contain the constants of the models considering the constants of the motors and the PD controllers, the constants of models that are not part of the parameters ς are easily obtained [1, 2].

3   Rapidly-exploring Random Trees or RRT

3.1   Mapping Environment

To identify the environment uses 2D images, which represent rooms that are processed by a binary matrix; show information as position and dimension of obstacles identified through of grids thus forming the map with all the elements that compose it as shown in Fig. 2, in addition to apply an algorithm corners detection of Harris that results a stack of corners in the obstacles of the environment.

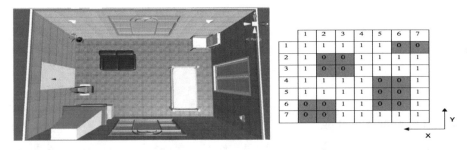

Fig. 2. Representation of an environment through a 2D binary map with grid where the value of 0 represents obstacles and the value of 1 represent the positions that the robot can take in open spaces in the coordinates (x, y) of the map.

For the positioning of the robot, it takes into account a grid composed of four quadrants that simulate the dimensions of the robot despite being taken as a point surface, shown in Fig. 3, this group is used to determine possible directions to which can rotate the robot.

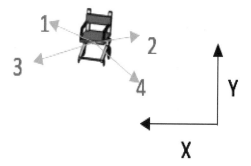

Fig. 3. Representation of the punctual surface that covers the robot in addition to the grids that make up its dimensions on the free spaces in the coordinates (x, y) of the map.

3.2   RRT Algorithm

This algorithm is a variation of RRT path planning inspired in the biological navigation of collective behavior, where the robot is considered as a punctual surface located in the initial conditions of the map, from this a segmented line is created towards the goal whose data of euclidean distance and angle of rotation positioned the robot in front of obstacles presents in the path towards the target. Once the obstacles in the collision range are known, the resultant stack is called of the detection of corners in the previous mapping stage, next to this process display a routine of sub-robots represented by X, which identifies and resizes said corners, the main robot occupies a dimension of four quadrants thus avoiding any future collision, all of this shows in Fig. 4.

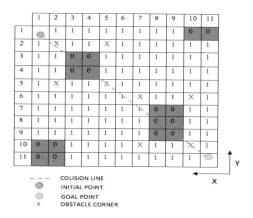

Fig. 4. Representation of the planning algorithm RRT.

The next step is to create a tree structure, which will provide all possible answers to reach the destination. For this, the stack of corners detected is use by the sub-robots routine, which calls the main robot to collect every possible corner shown in Fig. 5, this is store by a repetition structure, which at each step acquires information of two

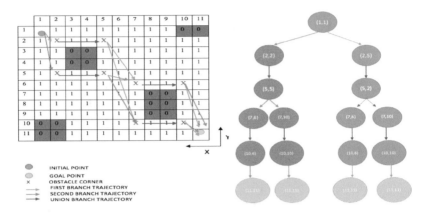

Fig. 5. Creation of the tree structure by collecting data from the main robot and sub-robots.

sub-robots located in the vertex of the figure that forms the obstacle and in the next step collects the information of the sub-robot located in the edge corresponding to the previous vertex thus forming the structure of the tree show in Fig. 5.

The pseudo code of the RRT algorithm is include in Code 1.

Code1.RRT

```
read Initial position (xi, yi), goal position (xf, yf) and the
map
call Procedure obstacle_discriminator (map)
call Procedure corner _detection (obstacle_map)
call Procedure sub_robot_corner _dimension (corner_map)
add to tree initial_position (xi, yi)
  while euclidian_distance (actual_position,
  goal_position)>0 do
if (obstacle=0) then
        add to tree (n) =actual_position (xn, yn)
        add to tree (n+1) =closest_corner_position (xc, yc)
        euclidian_distance (tree, goal)
  else
        euclidian_distance (tree, goal)
find goal_position
call Procedure eliminate_null_branch_tree
add to tree goal_position (xf, yf)
stack_pair_path ←call Procedure pair_tree_position
stack_odd_path ←call Procedure odd_tree_position
optimization_path ←call Procedure shortest_distance_stack
find shortest_stack with optimization_path
```

The optimization process is given only when the tree structure is creating and all the branches that do not end in the desired destination, in this way the stack information is acquire of all the response branches located in even and odd positions, shown in Fig. 6, once the stack is obtained, computes a distance between positions and the result of these summations is compared to find an optimized solution show in Figs. 6 and 7a, b.

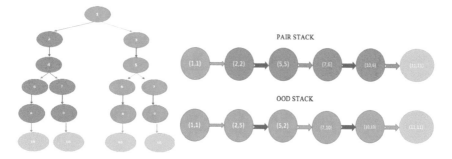

Fig. 6. Optimization of the tree structure collected by the RRT algorithm.

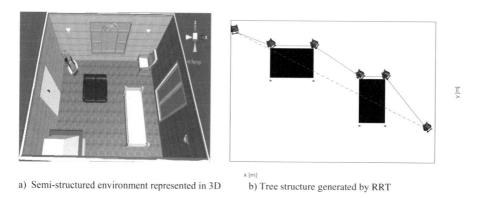

a) Semi-structured environment represented in 3D   b) Tree structure generated by RRT

Fig. 7. Stroboscopic movement of the wheelchair in the path after RRT algorithm

4 Controller Design

In this case, for the wheelchair to comply with the trajectory obtained from RRT, is considered experimentally the problem of path following, in the configuration space $R(X, Y, Z)$, shows in Figs. 8 and 9, where is considered the velocity of the system that depends on the error control, the linear and angular velocity of the wheelchair as shows in Figs. 9, 10 and 11b [5, 6]. For this a kinematic controller is used based on the kinematic model of the wheelchair, $\mathbf{h} = f(\psi)\mathbf{v}$. In this way the following control law, is proposed.

$$\begin{bmatrix} u_c \\ \omega_c \end{bmatrix} = \mathbf{J}^{-1}\left(\begin{bmatrix} |\upsilon_P| \cos(\theta_T) \\ |\upsilon_P| \sin(\theta_T) \end{bmatrix} + \begin{bmatrix} l_x \tanh\left(\frac{k_x}{l_x}\tilde{x}\right) \\ l_y \tanh\left(\frac{k_y}{l_y}\tilde{y}\right) \end{bmatrix} \right) \tag{6}$$

where, υ_P is the reference velocity input, the respective projections $(\dot{x}_P, \dot{y}_P)$ in the $(\mathcal{X}, \mathcal{Y})$ directions, $\mathbf{J}^{-1}$ is the matrix of inverse kinematics for the wheelchair, and X, Y are the position error in their respective directions of the wheelchair for the controller.

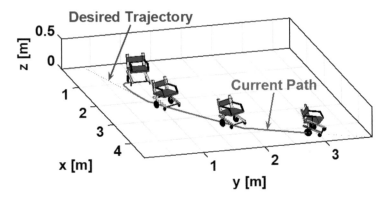

Fig. 8. Stroboscopic movement of the wheelchair mobile platform in the path planning trajectory

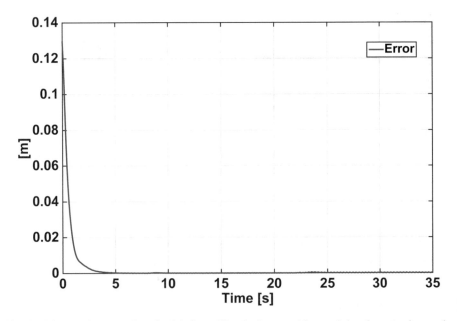

Fig. 9. Distance between the wheelchair mobile platform position and the closest point on the path planning trajectory

The dynamic controller is based in the dynamic model of the wheelchair system also it is use to reduce the velocity error. Hence, the following control law is proposed,

$$\begin{bmatrix} u_{ref} \\ \omega_{ref} \end{bmatrix} = \mathbf{M}\left(\begin{bmatrix} \dot{u}_c \\ \dot{\omega}_c \end{bmatrix} + \begin{bmatrix} l_u \tanh\left(\frac{k_u}{l_u}\tilde{u}\right) \\ l_\omega \tanh\left(\frac{k_\omega}{l_\omega}\tilde{\omega}\right) \end{bmatrix} \right) + \mathbf{C}\begin{bmatrix} u \\ \omega \end{bmatrix} \tag{7}$$

Where $\tilde{u}(t)$ and $\tilde{\omega}(t)$ are the linear and angular velocity; l_u, k_u, l_ω and k_ω are positive gain constants that compensate control error.

In this context is presented an example of stroboscopic movement around the desired trajectory and the current path in Fig. 8. It can be seen that the proposed controller works correctly Fig. 9 the error approach asymptotically to zero; and Fig. 9 shows the velocities of control applied in the human-wheelchair in Fig. 11b [5, 6].

5 Experimental Results

The experimental results correspond to the property of the system to adapt to the user, starting with the detection of gaze position through the training of the user to the interface due to variations in the range of vision as effect of the different user's contexture, shown in the Fig. 10. Also the training in the recognition of events such as focus the eye gaze in a determinate position of the screen to locate the desire point and the eye wink to order the generation of the path planning. They were the principal factors that determinate the skill to adaptability the system with the user.

Fig. 10. Representation of the interaction between the human-machine interface and the user.

Several locations were detected resulting in an accuracy in the planning of trajectories with RRT of 85%, of the tests due to the number of obstacles in formations that not have exit it increasing the time and the amount of memory used in each iteration; the example of planning an indoor environment. Once the optimized RRT, the control architecture linked to the wheelchair is tested by determinate the appropriate linear and angular velocities so that the velocity of the self-contained system so avoids sudden changes when following the trajectory to eliminate some risk in the user, shows in Fig. 11.

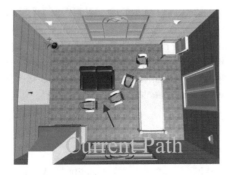
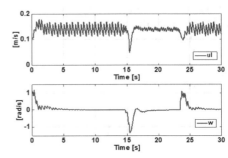

a) Test of the wheelchair controller applied   b) Angular and linear velocity of the wheelchair

Fig. 11. Experimental test of the assistive autonomous system.

6   Conclusions

In this paper, a wheelchair equipped with an eye gaze position detecting camera locates a destination desired by the user. The RRT algorithm employed for planning, it uses the desired position data to generate a trajectory and trough the proposed controller arrives at destination with great accuracy. The concept features whit an innovative multi-layer humanitarian assistance system that can be scaled for use in the hospital system.

References

1. Andaluz, V.H., Ortiz, J.S., Chicaiza, F.A., Varela, J., Espinosa, E.G., Canseco, P.: Adaptive control of the human-wheelchair system through brain signals. In: Kubota, N., Kiguchi, K., Liu, H., Obo, T. (eds.) Intelligent Robotics and Applications. ICIRA 2016. Lecture Notes in Computer Science, vol 9835. Springer (2016)
2. Andaluz, V.H., Canseco, P., Varela, J., Ortiz, J.S., Pérez, M.G., Roberti, F., Carelli, R.: Robust control with dynamic compensation for human-wheelchair system. In: Zhang, X., Liu, H., Chen, Z., Wang, N. (eds.) ICIRA 2014, Part I. LNCS, vol. 8917, pp. 376–389. Springer, Heidelberg (2014). IEEE
3. Hyder, R., Chowdhury, S.S., Fattah, S.A.: Real-time non-intrusive eye-gaze tracking based wheelchair control for the physically challenged. In: 2016 IEEE EMBS Conference on Biomedical Engineering and Sciences (IECBES), pp. 784–787. IEEE, Malaysia (2016)
4. Salehin, M.M., Paul, M.: Human visual field based saliency prediction method using Eye Tracker data for video summarization. In: 2016 IEEE International Conference on Multimedia & Expo Workshops (ICMEW), pp. 1–6. IEEE, Seattle (2016)
5. Andaluz, V.H., Ortiz, J.S., Roberti, F., Carelli, R.: Adaptive cooperative control of multi-mobile manipulators. In: IEEE–IECON Industrial Electronics Society, USA, pp. 2669–2675 (2014)
6. Andaluz, V.H., Ortiz, J.S., Sanchéz, J.S.: Bilateral control of a robotic arm through brain signals. In: De Paolis, L.T., Mongelli, A. (eds.) AVR 2015. LNCS, vol. 9254, pp. 355–368. Springer, Heidelberg (2015)

7. Eraghi, N.O., López-Colino, F., Castro, A., Garrido, J.: Path length comparison in grid maps of planning algorithms: HCTNav, A∗ and Dijkstra. In: Design of Circuits and Integrated Systems, pp. 1–6. IEEE, Madrid (2014)
8. Ganeshmurthy, M., Suresh, G.: Path planning algorithm for autonomous mobile robot in dynamic environment. In: 2015 IEEE International Conference on Communication, vol. 3, pp. 1–6. IEEE, Chennai (2015)
9. Zhao, Y., Felix, J., Kaiguo, Y., Dalta, V.: Path planning for robot-assisted active flexible needle using improved rapidly-exploring random trees. In: 2014 36th Annual International Conference Engineering in Medicine and Biology Society (EMBC), vol. 36, pp. 380–383 (2014)
10. Alessandro, G., Paolo, B., Albano, L., Renato, V.: Path Planning, Path Planning and Trajectory Planning Algorithms: A General Overview, vol. 1, pp. 5–13. Springer, Heidelberg (2015)
11. Palmieri, L., Arras, K.: A novel RRT extend function for efficient and smooth mobile robot motion planning. In: 2014 IEEE/RSJ International Conference on Intelligent Robots and Systems, pp. 205–211. IEEE, Chicago (2014)
12. Biagiotti, L., Melchiorri, C.: A General Overview on Trajectory Planning, Trajectory Planning for Automatic Machines and Robots, vol. 1. Springer, Heidelberg (2008)

Coordinated Control of a Omnidirectional Double Mobile Manipulator

Jessica S. Ortiz[1]($\boxtimes$), María F. Molina[1], Víctor H. Andaluz[1],
José Varela[1], and Vicente Morales[2]

[1] ESPE, Universidad de las Fuerzas Armadas, Sangolquí, Ecuador
{jsortiz4,mfmolina1,vhandaluz1}@espe.edu.ec
[2] Universidad Técnica de Ambato, Ambato, Ecuador
jvmorales99@uta.edu.ec

Abstract. In this paper proposes a coordinated cooperative control algorithm for tracking trajectories applied in two anthropomorphic robotic arms mounted on an omnidirectional platform, which allows the transport of an object in common. For this, the kinematic modelation of the entire coupled system is performed, considering as the position of interest the midpoint of the operative ends of each manipulator and its formation characteristics. The stability of the proposed controller for tracking trajectories using the Lyapunov method is demonstrated analytically, obtaining that the control is asymptotically stable. Finally the results obtained are evaluated in a virtual reality simulation environment.

Keywords: Double mobile manipulator · Kinematic modeling · Lyapunov method · Virtual reality

1 Introduction

With the increasing importance of the robots, the robotics has experienced significant changes, such that the interests of investigation realize from the development of robots for industrial and structured environments and for the development of mobile autonomous robots that they operate in not structured and natural environments, since they execute productive tasks with a high level of complexity [1, 2]. These autonomous robots are applicable in a series of challenging tasks as the cleanliness of dangerous materials, vigilance, rescue and recognition in not structured environments [3].

The different applications of the robotics in activities in those who do not control the human capacities, have attracted the attention of the scientific community in the development of different schemes of control [4]. The execution of the tasks can be carried out of individual form or cooperatively, being of cooperative form of greater efficiency in terms of manipulability, flexibility, accessibility and maneuverability [5]. Considering a group of robots to be capable of carrying out a task autonomously in changing environments, it depends on the planning of the real time path and the dynamics of the environment [6].

The manipulative mobile term refers to the robots that are equipped for one o more robotic arms mounted in a mobile platform with wheels or legs [7]; these robots can

© Springer Nature Singapore Pte Ltd. 2018
K.J. Kim et al. (eds.), *IT Convergence and Security 2017*,
Lecture Notes in Electrical Engineering 449,
DOI 10.1007/978-981-10-6451-7_33

execute several applications, since it is possible to combine the mobility of the mobile platform and the manipulation capacity of the robotic arms [8]. The transport and the manipulation are typical assignments in the industry, where there is interfering the large-scale use of autonomous mobile manipulators [9, 10]. As soon as two or more robots work together, cooperative mobile manipulators become important, which raises new challenges [11]. The study of the cooperative robotics has evolved depending on the technological advances and the different capacities of the robots, allowing to execute this way different missions that require locomotion and manipulation skills [12, 13].

There are different works in the literature with different control techniques, *e.g.*, multiple mobile manipulators, which hold a common object cooperatively to follow a desired path,, the control of the coordinated systems allows to control the object in not limited directions [14, 15]. In [16], they present mobile cooperative robots for the control of loads which movement is restricted in a plane based on a decentralized controller and on a controller who assures the exact follow-up of the twist of load. In [17], they propose a cooperative control decentralized for a set of k mobile manipulators (OMM-k) subject to restrictions holonómicas imposed by the object that is being manipulated. In [18], they present a cooperative control for systems multi-agent of high order with unknown directions of control using adaptive feedback technology.

In this work, presents the kinematic modeling of two anthropomorphic robotic arms of omnidirectional type mounted on a mobile platform, considering as the point of interest the midpoint of the two operative ends. The coordinated cooperative controller is developed using the Lyapunov method, which eliminates the inherent errors of kinematics. The design of the controller for tracking trajectories allows the transport of a common object. The experiments performed in a virtual structure demonstrate that the controller is appropriate for the solution of movement problems.

The article is organized in 5 Sections including the Introduction. Section 2 presents the kinematic modeling of the anthropomorphic arms mounted on an omnidirectional platform. The design and stability analysis of the control algorithm are presented in Sect. 3. The results and discussion are shown in Sect. 4. Finally, the conclusions are presented in Sect. 5.

2 Kinematic Models

To determine the system modeling of a two-arm mobile platform, the virtual point in the X-Y-Z plane between the midpoint of each end effector of the robotic arms is fixed; the virtual point is defined by $\mathbf{P}_F = \begin{bmatrix} h_x & h_y & h_z \end{bmatrix}$ that represents the position of its centroid on the inertial frame $<\mathrm{R}>$,

$$
\begin{cases}
h_x = \dfrac{h_{x1} + h_{x2}}{2} \\[2mm]
h_y = \dfrac{h_{y1} + h_{y2}}{2} \\[2mm]
h_z = \dfrac{h_{z1} + h_{z2}}{2}
\end{cases} \tag{1}
$$

Fig. 1. Mobile manipulator with two robotic arms of 4 DOF for the experiment

where h_{xi}, h_{yi}, h_{zi} with $i = 1, 2$ represents the position of each robotic arm mounted on the omnidirectional mobile platform. The vector structure of the virtual shape is defined by $\mathbf{S}_F = [d_F \quad \theta_F \quad \phi_F]$, where, d represents the distance between the position of the end-effector $\mathbf{h}_1$ and $\mathbf{h}_2$, θ and ϕ represents its orientation with respect to the global Y-axis and Z-axis, respectability on the inertial frame $<\text{R}>$, see Fig. 1.

$$\begin{cases} d_F = \sqrt{(h_{x2} - h_{x1})^2 + (h_{y2} - h_{y1})^2 + (h_{z2} - h_{z1})^2} \\ \theta_F = \arctan\left(\dfrac{h_z}{h_x}\right) \\ \phi_F = \arctan\left(\dfrac{h_y}{h_x}\right) \end{cases} \tag{2}$$

The point of interest of the system is represented in a simplified way as $\mathbf{h} = [\mathbf{P}_F \quad \mathbf{S}_F]$, i.e.,

$$\mathbf{h} = [hx \quad hy \quad hz \quad d \quad \theta \quad \phi]^T. \tag{3}$$

The other hand, the kinematic model of the omniderctional mobile platform is represented as,

$$\begin{bmatrix} \dot{x} \\ \dot{y} \\ \dot{\psi} \end{bmatrix} = \begin{bmatrix} \cos\psi & -\sin\psi & 0 \\ \sin\psi & \cos\psi & 0 \\ 0 & 0 & 1 \end{bmatrix} \begin{bmatrix} u_l \\ u_m \\ \omega \end{bmatrix} \tag{4}$$

where, each linear velocity is directed as one of the axes of the frame $<P>$ attached to the center of gravity of the mobile platform: u_l points to the frontal direction;

u_m points to the left-lateral direction, and the angular velocity ω rotates the referential system $<P>$ counterclockwise, around the axis P_Z (considering the top view).

Now, substituting (1), (2), (4) in (3) and deriving (3), we obtain the kinematic model of the point of interest of the mobile manipulators. Also the kinematic model can be written in compact form as $\dot{\mathbf{h}} = f(\mathbf{q_p}, \mathbf{q_a})\mathbf{v}$, i.e.,

$$\dot{\mathbf{h}}(t) = \mathbf{J}(\mathbf{q_p}, \mathbf{q}_{ai})\mathbf{v}(t) \tag{5}$$

where, $\dot{\mathbf{h}}(t) = \begin{bmatrix} \dot{h}_x & \dot{h}_y & \dot{h}_z & \dot{d}_F & \dot{\theta}_F & \dot{\phi}_F \end{bmatrix}^T$ is the velocity vector of the interest point; $\mathbf{J}(\mathbf{q})$ represents the Jacobian matrix, $v(t) = [u_l\ u_m\ \omega\ \dot{q}_{11}\ldots \dot{q}_{na1}\ \dot{q}_{12}\ \ldots \dot{q}_{na2}^T$ is the control vector of mobility of the interest point.

3   Control Algorithm

Is proposed a control system based on the system kinematics (platform, arms and object). The design of the control algorithm is based on the kinematics of the system obtained in Sect. 2, in which, the vector of the maniobrability $\mathbf{v}$ can be expressed as

$$\mathbf{v}(t) = \mathbf{J}^{\#}(\mathbf{q})\,\dot{\mathbf{h}}(t) \tag{6}$$

Therefore the proposed control law is based on a minimum standard solution, which means that at any time the manipulator will reach its navigation goal with the fewest possible movements

$$\mathbf{v}(t) = \mathbf{J}^{\#}(q)\left[\dot{\mathbf{h}}_\mathbf{d}(t) + \mathbf{K}\tanh\left(\tilde{\mathbf{h}}(t)\right)\right] \tag{7}$$

where, $\mathbf{J}^{\#} = (\mathbf{J^T J})^{-1}\mathbf{J^T}$ is the pseudoinverse Jacobian matrix;

$\dot{\mathbf{h}}_\mathbf{d}(t) = \begin{bmatrix} \dot{h}_{xd} & \dot{h}_{yd} & \dot{h}_{zd} & \dot{d}_d & \dot{\theta}_d & \dot{\phi}_d \end{bmatrix}^T$ is the vector of desired velocities of the point of interest; $\tilde{\mathbf{h}}(t)$ Is the vector of control errors defined as $\tilde{\mathbf{h}}(t) = \mathbf{h}_\mathbf{d}(t) - \mathbf{h}(t)$; $\mathbf{K}$ is a positive definite diagonal matrix that weighs the error vector; and finally $\mathbf{v}(t) = \mathbf{v}(t) = [u_l\ u_m\ \omega\ \dot{q}_{11}\ldots \dot{q}_{na1}\ \dot{q}_{12}\ \dot{q}_{na2}\ldots^T]$ represent the maneuverability vector of the system.

The behavior of the system control error is analyzed by assuming perfect tracking of the speed. The closed loop equation of the system is obtained by substituting (7) in (6)

$$\dot{\tilde{\mathbf{h}}} + \mathbf{K}\tanh\left(\tilde{\mathbf{h}}(t)\right) = \mathbf{0} \tag{8}$$

For stability analysis the following candidate function of Lyapunov it's considered. Its temporary derivative in the trajectories of the system is

$$\dot{V}\left(\tilde{\mathbf{h}}_i\right) = -\tilde{\mathbf{h}}_i^T \mathbf{K}\tanh\left(\mathbf{L}\,\tilde{\mathbf{h}}\right) < 0, \tag{9}$$

This implies that the equilibrium point of the closed loop (9) is asymptotically stable, so that the position error of the i-th final effector verifies $\tilde{\mathbf{h}}_i(t) \to 0$ asymptotically with $t \to \infty$.

4   Results and Discussion

In order to evaluate the performance of the proposed controller, was developed in Matlab a 3D simulator in which the KUKA Youtbot robot is considered, which is made up of an omnidirectional mobile platform with mecanum wheels and two anthropomorphic robotic arms [19].

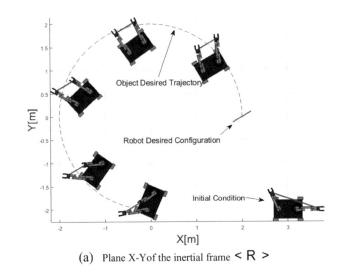

(a)   Plane X-Y of the inertial frame < R >

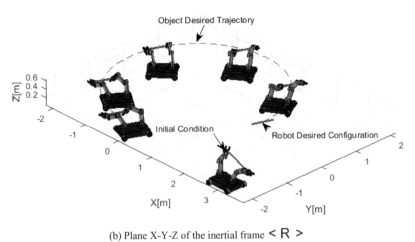

(b) Plane X-Y-Z of the inertial frame < R >

Fig. 2. Stroboscopic movement of the mobile manipulator

The simulation experiments implemented recreate an application of cooperation and coordination between two robotic arms and a mobile platform in order to perform tasks of handling a load. Figure 2 shows the stroboscopic movement in the X-Y-Z space of the reference system $<R>$, which allows to verify that the proposed controller works correctly in cooperation tasks when moving an object in common, *i.e.*, that the task includes in transporting a cooperatively manipulated object between the two robotic arms; in order to evaluate it is considered that the object is in motion and must be reached by the robot with the purpose of following a desired trajectory. The Fig. 3 shows the trajectories described by the operative ends of the robotic arms when performing the task; in the graph you can see the trajectories described by the robot.

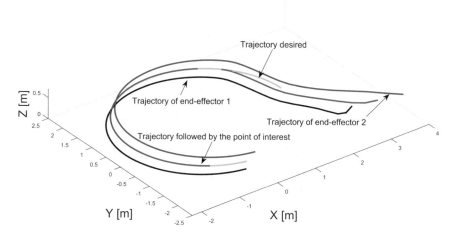

Fig. 3. Path followed by the point of interest.

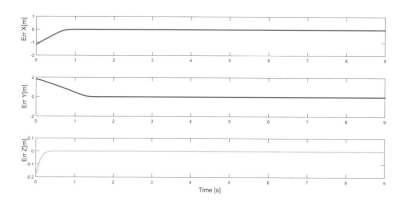

Fig. 4. Errors of position of the point of interest or midpoint of the operative ends.

Figure 4 show that the control errors of the position of the point of interest formed by the ends of the robotic arms; while Fig. 5 illustrates shape and orientation errors, *i.e.,* the distance between the operative ends and the angles forming the object with the arms over the XY and YZ planes with respect to the reference system $<R>$; in the two graphs it can be seen that control errors tend to zero asymptotically when t → ∞.

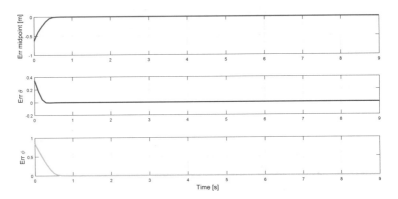

Fig. 5. Errors of shape of the object to be transported

Finally, Figs. 6 and 7 its show the maneuverability commands applied to the omnidirectional platform and the anthropomorphic arms respectively in order to accomplish the task objective.

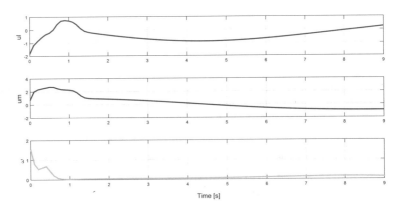

Fig. 6. Omnidirectional platform speeds

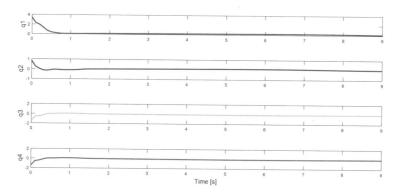

Fig. 7. Robotic arm speeds.

5 Conclusions

In this work the design of a coordinated cooperative controller for the trajectory tracking that allows to transport an object in common was presented. The design of the controller is based on a kinematic control that meets the purpose of movement, reference velocities were defined for the omnidirectional platform and the robotic arms. Stability and robustness are tested by the Lyapunov method. Experiments carried out trough a virtual reality structure confirm the scope of the controller to solve different problems of movement through a strong choice of control references.

References

1. Andaluz, V., Canceso, P., Varela, J., Ortiz, J., Perez, M., Roberti, F., Carelli, R.: Robust Control with Dynamic Compensation for Human-Wheelchair System. Springer, Cham pp. 376–489 (2014)
2. Markus, E., Agee, J., Jimoh, A.: Flat control of industrial robotic manipulators. Robotics Autonomous Syst. **87**, 226–236 (2017)
3. Galicki, M.: An adaptive non-linear constraint control of mobile manipulators. Mechanism Mach. Theor. **88**, 63–85 (2015)
4. Zi, B., Lin, J., Qian, S.: Localization, obstacle avoidance planning and control of a cooperative cable parallel robot for multiple mobile cranes. Robotics Comput.-Integr. Manuf. **34**, 105–123 (2015)
5. Markus, E., Yskander, H., Agee, J., Jimoh, A.: Coordination control of robot manipulators using flat outputs. Robotics Autonomous Syst. **83**, 169–176 (2016)
6. Garcia, C., Cardenas, P., Saltaren, P., Puglisi, L., Aracil, R.: A cooperative multi-agent robotics system: design and modelling. Exp. Syst. Appl. **40**, 4737–4748 (2013)
7. Meng, Z., Liang, X., Andersen, H., Ang, M.: Modelling and Control of a 2-Link Mobile Manipulator with Virtual Prototyping. In: 13th Internacional Conference on Ubiquitous and Ambiente Intelligence (URAI), China, pp. 363–368 (2016)
8. Yang, H., Dongjun, L.: Cooperative Grasping Control of Multiple Mobile Manipulators with Obstacle Avoidance

9. Szabó, R., Gontean, A.: Robotic Arm Autonomous Movement in 3D Space Using Stereo Image Recognition in Linux. In: 11th International Symposium on Electronics and Telecommunications (ISETEC) (2014)

10. Iossifidis, I., Schöner, G.: Autonomous reaching and obstacle avoidance with the anthropomorphic arm of a robotic assistant using the attractor dynamics approach. In: Internacional Conference on Robotics & Automation, pp. 4295–4300 (2004)

11. Schierl, A., Angerer, A., Hoffmann, A., Vistein, M., Reif, W.: A Taxonomy of Distribution for Cooperative Mobile Manipulators. IEEE (2015)

12. Andaluz, V., Roberti, F., Toibero, J., Carelli, R.: Adaptive unified motion of mobile manipulators. Control Eng. Pract. **20**, 1337–1352 (2012)

13. Khatib, O.: Mobile manipulator: the robotic assistant. Robotics Autonomous Syst. **26**, 175–183 (1999)

14. Petitti, A., Franchi, A., Di Paola, D., Rizzo, A.: Decentralized motion control for cooperative manipulation with a team of networked mobile manipulators. In: IEEE International Conference on Robotics and Automation (ICRA), pp. 441–446 (2016)

15. Andaluz, V., Ortiz, J., Perez, M., Roberti, F., Carelli, R.: Adaptive cooperative control of multi-mobile manipulators. In: IECON 2014—40th Annual Conference of the IEEE Industrial Electronics Society (2014)

16. Abbaspour, A., Alipour, K., Jafari, H., Moosavian, A.: Optimal formation and control of cooperative wheeled mobile robots. Comptes Rendus Mécanique **343**, 307–321 (2015)

17. Ponce, A., Castro, J., Guerrero, H., Parra, V., Olguín, E.: Cooperative redundant omnidirectional mobile manipulators: model-free decentralized integral sliding modes and passive velocity fields. In: IEEE IEEE International Conference on Robotics and Automation (ICRA), pp. 2375–2380 (2016)

18. Ma, Q.: Cooperative control of multi-agent systems with unknown control directions. Appl. Math. Comput. **292**, 240–252 (2017)

19. Kuka youBot store. http://www.youbot-store.com/G

Heterogeneous Cooperation for Autonomous Navigation Between Terrestrial and Aerial Robots

Jessica S. Ortiz[✉], Cristhian F. Zapata, Alex D. Vega, Alex Santana G.,
and Víctor H. Andaluz

Universidad de las Fuerzas Armadas ESPE, Sangolquí, Ecuador
{jsortiz4,cfzapata2,advega1,amsantana,vhandaluz1}@espe.edu.ec

Abstract. This work presents a multilayer system that is comprised in four layers, the main layer defines the task, the next performs the image processing and the algorithm of Path Planning in which is considered the evasion of fixed and mobile obstacles in order to be followed By an unmanned land vehicle; The other applies different control algorithms, and finally the last layer is in charge of the interaction, through image processing the speed of mobile objects is estimated within the workspace with the objective that the terrestrial robot does not collide. The stability of the control algorithm is tested through the Lyapunov method. Finally, the experimental results are presented and discussed in which the proposal is validated.

Keywords: Heterogeneous cooperation · Autonomous · UGV · UAV

1 Introduction

Cooperative Control is focused on the use of several robots, this is a field that basically accomplishes tasks that a single vehicle is not able to perform, this generates great advantages, like reduction of costs, greater strength, performance and efficiency [1, 2]. Robots involved in the cooperative control scheme must travel autonomously avoiding obstacles to achieve a certain mission [3]. In recent years, cooperative control has been able to carry out various tasks focused on surveillance, search, environmental monitoring, traffic monitoring, among others, which are applied in various areas of work, such as: (i) industrial field, (ii) military field (iii) Agriculture, (iv) Transit, etc. [4, 5]. One of the most important aspects in cooperative control is software used for information processing and control between robots.

The Robots tasks require a large computational capacity in real time, usually involving subtasks to perform operations such as exploration, avoidance of obstacles, sensing, monitoring and manipulation of objects. The control schemes of multi-robot systems can be classified into: (i) Leader-follower, one robot is designated leader while the others are followers, the leader defines the mass movement group and the other robots are controlled to follow their respective leaders respecting distance and other factors; (ii) behavior-based methods, is defined as a behavioral combination of each member comprising actions and is constructed to achieve a global goal. And (iii) virtual

© Springer Nature Singapore Pte Ltd. 2018
K.J. Kim et al. (eds.), *IT Convergence and Security 2017*,
Lecture Notes in Electrical Engineering 449,
DOI 10.1007/978-981-10-6451-7_34

structures, they involve establishing geometric relations that will remain rigid between the robots and the referential system [1, 2].

The cooperation between robots can be considered heterogeneous due to the variety of sensors, actuators and communication protocols, developed by multiple manufacturers [6]. Heterogeneity combines large and small cores in the computational processors to improve the performance of robots with the help of memory interfaces whose main function is to accelerate applications [7]. There is also heterogeneity when talking about groups of robots that help each other to fulfill a specific function, these can be a set of: unmanned aerial vehicle UAVs, unmanned ground vehicle UGVs, Unmanned underwater vehicles UUVs, or at the same time a combination between them. Heterogeneous systems of robots composed of UAV and a UGV allow several applications through odometry, force and vision so that they can interact with each other and thus fulfill a determined autonomous and cooperative task.

The cooperation between UAV-UGV requires some kind of collective behavior, such as Instance-swarm or flock [8, 9], or special training [10], which allows to perform several tasks in which obstacles can be avoided [11]. In [12], a work is presented that allows a group of fixed-wing UAVs to surround the forming centroid of a UGV. While a UAV with a visual sensor provides additional information (Detection of Negative Obstacles, Surface Irregularities, Land Type, etc.) is presented in [13].

As described, the present work proposes the development of a control scheme for autonomous cooperation between UAV and UGV, where a heuristic algorithm for the Path Planning of unstructured environments is considered through visual feedback. The proposed control scheme consists of 4 layers each functioning as a separate module, which control scheme includes a basic structure defined by the planning layer in which the UAV obtains the coordinates of the starting point of the UGV and defines the task that goes To make; The image processing layer is in charge of capturing the environment and processing it to perform the algorithm of Path Planning, this system will respond to any change of the environment so that the UGV reaches the end point; While in the "Control" layer a road tracking controller is proposed in which it considers a variable speed according to the criterion of road optimization [14]; And finally in the environment layer is considered the interaction of robots with fixed and mobile obstacles.

This article is organized in VI Sections including the Introduction. Section 2 presents the Formulation of the Problem where the multilayer control scheme is described: while in Sect. 3. Image Processing is performed from the capture of the workspace, in Sect. 4 Path Planning, this section describes the generation of Path Planning for a mobile robot to reach an endpoint in the shortest possible time. Section 5 Experimental Results, presents the results of the proposed algorithm and finally Sect. 6 Conclusions.

2   Problem Formulation

Path Planning has been developed in several environments, this has allowed to obtain results in which mobile robots reach an expected place in the shortest time, through algorithms previously defined using visual information as feedback. The control techniques that present feedback are very efficient, but have difficulties when both the initial and desired positions of the robot are very distant [15]. The present work presents the solution to this problem, it considers restrictions in the paths with the help of Path Planning to guarantee the convergence in all the initial configurations.

The Fig. 1 shows the multilayer control scheme to solve the problem of road planning in which the evasion of mobile obstacles in tasks of heterogeneous cooperation between a terrestrial robot and an aerial robot is considered. Each of the layers operates independently, *i.e.*, (i) Layer I is composed of an offline planning that is responsible for obtaining initial coordinates of the robot and a task planning responsible for configuring the final coordinates that the robot must reach; (ii) The layer II is responsible for obtaining the images of the workspace by means of a camera installed in the lower part of the UAV, it emulates the eyes of the terrestrial robot. Objects are identified by image processing; Path Planning is created with a trajectory planning algorithm inspired by the biological navigation of collective behavior; (iii) Layer III is responsible for generating the control signals to the system, so that the UGV fulfills the task defined in the planning layer; And finally (iv) Layer IV represents the environment where all fixed and moving objects are found.

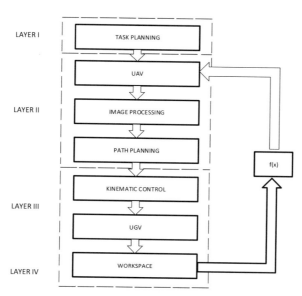

Fig. 1. Multi-layer system

3   Image Processing

This section describes the image processing that is performed for the development of Path Planning and identification of direction and speed of moving objects. For image processing, the acquisition of images is performed by means of a vision camera installed in the lower part of the UAV, parallel to the XY plane of the reference system <R>. The images are processed in order to identify the fixed and mobile objects in the workspace, recognizing the environment and position of the elements so that the UGV can reach from an initial point to an end point without colliding with the objects that are in the environment. The implemented image processing is described in Fig. 2. both for identifying fixed objects and moving objects, each block performs a specific function at the time of image processing.

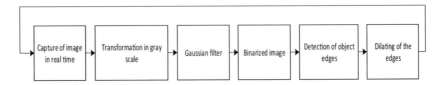

Fig. 2. Block diagram for workspace capture

According to Fig. 2. captured the real-time image of all the elements of the workspace the grayscale transformation is performed, a Gaussian filter is used in order to eliminate the noise in the image. In addition, binarization of the image plane is done in order to differentiate the objects from the UGV workspace, once this is done, the edges of the objects are detected, see Fig. 3(a), then the image is enlarged with the objective That the UGV does not collide with them, Fig. 3(b).

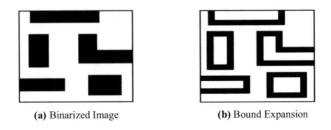

(a) Binarized Image   (b) Bound Expansion

Fig. 3. Expansion process of objects

Figure 4 shows the process that is performed for the identification of moving objects, once the image is captured and binaries, the centroids of all the figures found in the workspace are detected and stored in a vector, the next Sampling is performed the same procedure with the aim of comparing the vectors of the centroids of each figure and analyze if there is a variation of them, if it is greater than 30 pixels means that it is a moving object, in addition you get the speed that will be given In pixels/s since you have

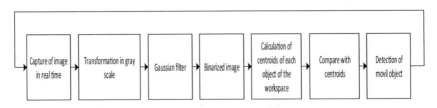

Fig. 4. Block diagram for identification of the mobile object.

time in [s] and position in [pixels], to determine in m/s a conversion of pixels to meters is performed.

4   Path Planning

This section describes the generation of Path Planning whose main objective is to determine the most optimal path that a mobile robot should follow. The ability of autonomous systems to move within environments, structured or not, is achieved by an adequate planning technique; focused on robotics [16, 17].

4.1   Map of the Environment

The environment is identified by images, these are obtained by the UAV and processed to generate a matrix in two dimensions; These show information as about the environment such us: dimension of the obstacles and position that identified through the regions thus forming the workspace with all the elements that compose it as shows in Fig. 5, in addition algorithm expand the obstacle detected with a multi layer in order to protect the surface of the obstacles. For the robot position, it takes into account a region composed of four directions that simulate the potential sense of the robot despite being taken as a point in the region, this group is used to determine possible directions to which can rotate the robot.

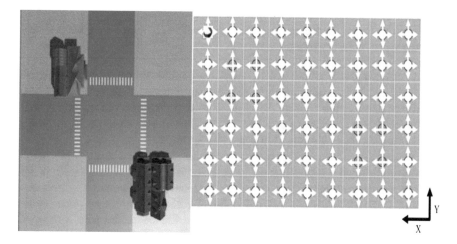

Fig. 5. Representation of an environment through a binary map with regions where the blue regions represents obstacles and the white regions represent the positions that the robot can take in open regions in the reference space of the map and the robot is represented as a point in the surface of the map.

4.2 Path Planning - Algorithm

The algorithm has been inspired in the behavior of navigation in the nature, where the robot is considered as a point located in the initial conditions of the environment, from this a diagonal path is created towards the desired goal whose shows the obstacles in the trajectory together with the data of distance and angular position that located the robot in front of obstacles presents towards the target. The next step is identifying and expand the collision range in the targets, all of this positions are saved in the procedure called expand the target in the previous image processing layer, the robot occupies a dimension of four directions thus avoiding any collision at the time to navigation, all of this shows in Fig. 6.

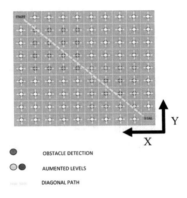

Fig. 6. Representation of the planning algorithm with the respective layers and regions augmented by the image processing stage.

For to create a tree structure, which provide the possible trajectories free of collisions to reach the goal. In this way, it is proposing the multi layers are detected as augmented levels represented in the map as the regions with yellow and red layers around the obstacles that are in a blue layer, all of this used by the robot to avoid collisions in navigation within the environment indicated in the Fig. 6.

The illustration of Fig. 7 represents the process of acquisition, location and planning of trajectories through stages, in which the first step is in charge of loading all the initial conditions to solve the algorithm, the stage of diagonal path calculates the direction and distance of the heading towards the goal in line of view to verify the existence of obstacles. Where it can choose between a path free of collisions and a path with possible collisions. If this is the case, the algorithm is in charge of processing the image that represents the obstacle in such a way that the structure of the tree is loading with the current and new positions, so it is verified again to know if the target was reached in case of the condition is not correct, the process is repeated successively until finding the solution. All this process gives a tree structure that will be optimized in the last stage to reach the desired goal, thus an example of the stroboscopic movement that is shown in Fig. 8.

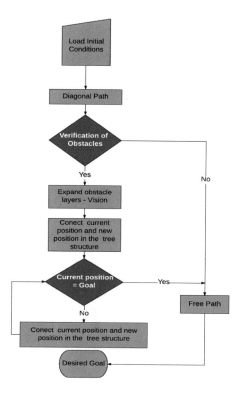

Fig. 7. Flow chart of the creation of the tree structure by collecting data from the path planning.

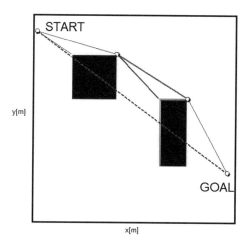

Fig. 8. Stroboscopic movement of the UGV in the path after algorithm

5 Experimental Results

This section presents the simulation results of the proposed algorithm for Path Planning that will be followed by the UGV without colliding with fixed and mobile obstacles. The objective of the simulation is to test the road planning, stability and performance of the proposed controllers. For the communication a TCP/IP protocol is used, the master system is the PC and the slave systems are the UAV with the UGV, the camera installed in the UAV will send the information of the workspace this data will be processed by the master station in order to Generate a path that will be followed by UGV from an initial point to an end point. The first image in Fig. 11 shows the workspace captured by the camera installed on the bottom of the UAV. The second image in Fig. 9. presents the binarization of the workspace. The third image of Fig. 9. shows the dilation result of the objects. In Fig. 10. shows the Path Planning generated from an initial point to an end point. Figure 11. shows the improved Path Planning that will be followed by the UGV.

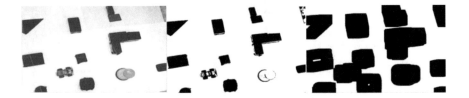

Fig. 9. Workspace image processing

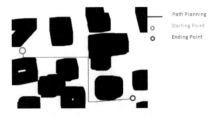

Fig. 10. Path planning

Fig. 11. Improved path planning

The Fig. 12(a) indicates the identification of the mobile object in the vicinity of the generated Path Planning, whereby the UGV will slow down when it approaches the moving object in Fig. 12(b). New images are acquired in the following sampling periods to check that the moving obstacle is not on the desired path and thus increase the speed of the UGV to the end point, see Fig. 12(c). identification end point

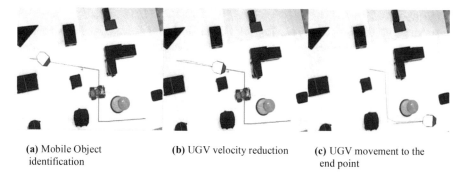

(a) Mobile Object identification   (b) UGV velocity reduction   (c) UGV movement to the end point

Fig. 12. Path Planning Process

Finally, Fig. 13 shows the evolution of the speed of the UGV, this decreases in the presence of large control errors. The simulation parameters used are UGV Maximum Speed of 0.09 m/s, with a gain of 1 and moving object speed 0.5 m/s; this will depend on different environments where the work is tested, the speed of the UGV decreases when approaching the object Mobile with the aim of not colliding

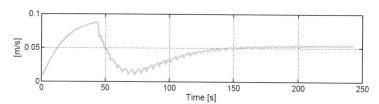

Fig. 13. UGV velocity

6   Conclusions

In this paper a multilayer scheme was presented in order to solve the problem of heterogeneous cooperation between UAV and UGV for displacement tasks in relatively large work spaces. It was proposed a Path Planning algorithm based on visual feedback with the objective of determining the path to be followed by the UGV avoiding the collision of fixed and mobile obstacles, finally proposed a control algorithm for the tracking of roads in which it is considered A speed not constant the same that can depend on the curvature of the road, error or other factors of control.

Acknowledgments. The authors would like to thanks to the Corporación Ecuatoriana para el Desarrollo de la Investigación y la Academia -CEDIA-, for financing the project "Control Coordinado Multi-Operador Aplicado a un Robot Manipulador Aéreo -CEPRA XI-2017-06-".

References

1. Rosales, C., Leica, P., Sarcinelli-Filho, M., Scaglia, G., Carelli, R.: 3D formation control of autonomous vehicles based on null-space. In: Springer, J. Intell. Robot. Syst. **84**(1):453–467 (2016)
2. Andaluz, V., Ortiz, J., Perez, M., Roberti, F., Carelli, R.: Adaptive cooperative control of multi-mobile manipulators. In: IECON 2014 – 40th Annual Conference of the IEEE Industrial Electronics Society (2014)
3. Mohamed, A., Yu, X., Zhang, Y.: Fault-tolerant cooperative control design of multiple wheeled mobile robots. IEEE Trans. Control Syst. Technol., 1–9 (2017)
4. Andaluz, V., López, E., Manobanda, D., Guamushig, F., Chicaiza, F., Sánchez, J., Rivas, D., Pérez, F., Sánchez, C., Morales, V.: Nonlinear controller of quadcopters for agricultural monitoring. Adv. Vis. Comput., 476–487 (2015)
5. Andaluz, V., Chicaiza, F., Meythaler, A., Rivas, D., Chuchico, C.: Construction of a quadcopter for autonomous and teleoperated navigation. In: 2015 Conference on Design of Circuits and Integrated Systems (2015)
6. Naidoo, N., Bright, G., Stopforth, R.: Navigation and control of cooperative mobile robots using a robotic middleware platform. In: 12th IEEE International Conference on Control and Automation (ICCA), pp. 927–932 (2016)
7. Chandramoorthy, N., Tagliavini, G., Irick, K., Pullini, A., Advani, S., Al Habsi, A., Cotter, M., Sampson, J., Narayanan, Y., Benini, L.: Exploring architectural heterogeneity in intelligent vision systems. In: IEEE 21st International Symposium on High Performance Computer Architecture (HPCA), pp. 1–12 (2015)
8. Andaluz, V., Carreli, R., Salinas, L., Roberti, F., Toibero, J.: Visual control with adaptive dynamical compensation for 3D target tracking by mobile manipulators. Mechatronics **22**(4), 491–502 (2012)
9. Herbert, T., Jadbabaie, A., Pappas, G.: Flocking in fixed and switching networks. IEEE Trans. Autom. Control **52**, 863–868 (2007)
10. Fax, J., Murray, R.: Information flow and cooperative control of vehicle formations. IEEE Trans. Autom. Control **49**, 1465–1476 (2004)
11. Gren, P., Leonard, N.: Obstacle avoidance information. In: Proceedings International Conferences Robotics and Automation, pp. 2492–2497 (2003)
12. Tanner, H., Christodoulakis, D.: Cooperation between aerial and ground vehicle groups for reconnaissance missions. In: IEEE Conference on Decision and Control, pp. 5918–5923 (2006)
13. Klodt, L., Khodaverdian, S., Willert, V.: Motion control for UAV-UGV cooperation with visibility constraint. In: IEEE Conference on Control Applications (CCA), pp. 1379–1385 (2015)
14. Andaluz, V., Roberti, F., Toibero, J., Carelli, R.: Adaptive unified motion control of mobile manipulators. Control Eng. Pract. **20**(12), 1337–1352 (2012)
15. Mezouar, Y., Chaumette, F.: Path planning for robust image-based control. IEEE Trans. Robot. Autom. **18**, 534–549 (2002)
16. Eraghi, N.O., López-Colino, F., Castro, A., Garrido, J.: Path length comparison in grid maps of planning algorithms: HCTNav, A∗ and Dijkstra. In: Design of Circuits and Integrated Systems, pp. 1–6. IEEE, Madrid (2014)
17. Jessica, S.O., Cristhian, F.Z., Alex, D.V., Víctor, H.A.: Path planning based on visual feedback between terrestrial and aerial robots cooperation. In: Computational Kinematics - Mechanisms and Machine Science, pp. 96–105. Springer, Netherlands (2017)

Linear Algebra Applied to Kinematic Control of Mobile Manipulators

Víctor H. Andaluz$^{(\boxtimes)}$, Edison R. Sásig, William D. Chicaiza, and Paola M. Velasco

Universidad de las Fuerzas Armadas ESPE, Sangolquí, Ecuador
{vhandaluz1,erssigs,wdchicaiza1,
pmvelasco1}@espe.edu.ec

Abstract. This paper is focused in linear algebra theory applied to control of mobile manipulator robots. In order to design the control algorithm, the kinematic system is approximated using numerical methods. Then, the optimal control actions are obtained through linear algebra approach. The structure of the controller consists in two solutions; a particular solution that allow following the desired trajectory and a homogeneous solution that allow performing secondary objectives as maximum manipulability and avoid static obstacles. In addition, the stability analysis is demonstrated through linear algebra concepts where it is shown that the tracking error tends asymptotically to zero. Finally, experimental results show the effective of proposed control algorithm over the mobile manipulator robot AKASHA.

Keywords: Linear algebra · Numerical methods · Controller design · Model · Mobile manipulator

1 Introduction

Robotics missions have evolved into the service domain where robots are expected to either exploit unknown dynamic environments, interact with human beings or manipulate dangerous products. Mobile manipulators combine these capabilities, thus expanding the working space, often characterized by a high degree of redundancy, combining the manipulation of a fixed base robotic arm with the mobility of a wheeled platform. Such systems allow the most common missions of robotic systems requiring locomotion and manipulation skills [1, 2]. Field or service robotics applications are numerous and all involve robots whose workspace capabilities have to be extended and whose control architecture and strategies must ensure a good overall performance in complex missions [3].

In the last decades, there has been a great deal of interest in mobile robots where, mobile manipulators being an area of great interest to researchers, who are looking for new non-linear control strategies. In [4] solves the trajectory-tracking problem by combining neural networks and robust control. The nonlinear mapping characteristic of neural networks and robust control are integrated in an adaptive control algorithm for mobile manipulator robots with non-linearities, perturbations and non-holonomic constraints all at the simulation level. PD feed-forward non-linear control is developed

© Springer Nature Singapore Pte Ltd. 2018
K.J. Kim et al. (eds.), *IT Convergence and Security 2017*,
Lecture Notes in Electrical Engineering 449,
DOI 10.1007/978-981-10-6451-7_35

in [5], this control makes use of the knowledge of the mathematical model of the system and the measurement of the perturbations of process; is applied in a virtual prototype where the control parameters must be known for the controller to work well; [6] suggests a fuzzy PD controller to adjust parameters online depending on the state of the dynamic system. Other advanced control strategies are implemented, for example in [7] introduces a predictive control algorithm that has restrictions for a holonomic mobile manipulator robot; constraints such as acceleration, velocity, position, and avoiding obstacles are considered [8].

Control based on linear algebra is an innovative technique with main characteristics of the non-necessity of complex calculations to reach the control signal and the simplicity to make mathematical operations. [9–11]. In addition, the algorithm is easy to understand and implement, it allows the direct adaptation to any micro-controller without the need to make use of an external computer [12]. Due the fact that it is not a complex algorithm, this can be run on low processing power drivers, [13] by presenting a high yield on conventional computers, as a result, this algorithm supports savings in processing time and energy at the moment of executing the desired task [14–16]. In [17] presents a control algorithm based on linear algebra for mobile manipulator robots, but this work does not considered the null space configuration and the design is validated using simulation.

This paper proposes an algorithm of control based on linear algebra approach for trajectory tracking tasks of mobile robotic systems, formed by a robotic arm mounted on a mobile platform; the controller is based on kinematics and redundancy of the system. The structure of the control law consists of two solutions: (1) one particular solution that allow to meet the objective of the main task; and (2) a homogeneous solution that allow performing one or more secondary objectives, this work is considered the internal configuration of the mobile manipulator control in order to avoid singular configurations of the system and avoid static obstacles. The designed controller is validated experimentally. In addition, the stability is demonstrated through linear algebra concepts.

This article is organized into 4 Sections. Sect. 2 presents the modeling, design of the control algorithm and the analysis of stability based on approaches of linear algebra for the mobile manipulator. The discussion of experimental results is shown in Sect. 3, and the conclusions of the article in Sect. 4.

2   Modeling and Control Design

In this section, the kinematic model and control law based on linear algebra theory and numerical methods of the mobile manipulator is presented. In addition, the stability is obtained through linear algebra concepts.

2.1   Kinematic Model

The instantaneous kinematic model of a mobile manipulator gives the derivative of its end-effector location as a function of the derivatives of both the robotic arm configuration and the location of the mobile platform.

$$\dot{\mathbf{h}}(t) = \mathbf{J}(\mathbf{q})\mathbf{v}(t) \tag{1}$$

where $\dot{\mathbf{h}} = \begin{bmatrix} \dot{h}_1 & \dot{h}_2 & \cdots & \dot{h}_m \end{bmatrix}^T$ is the vector of end-effector velocity, $\mathbf{v} = \begin{bmatrix} v_1 & v_2 & \cdots & v_{\delta_n} \end{bmatrix}^T = \begin{bmatrix} v_p^T & v_a^T \end{bmatrix}^T$ is the vector of mobile manipulator velocities in which contains the linear and angular velocities of the mobile platform and contains the joint velocities of robotic arm and $\mathbf{J}(\mathbf{q})$ is the Jacobian matrix that defines a linear mapping between the vector of the mobile manipulator velocities $\mathbf{v}(t)$ and the vector of the end-effector velocity [17, 18].

2.2   Kinematic Controller

Considering the first order differential equation

$$\dot{h} = f(h, v, t) \operatorname{con} h(0) = h_0$$

where, h represents the output of the system to be controller, $\dot{h}$ first derivative, v the control action and t the time. The values of $h(t)$ in the discrete time $t = k\,T_0$ are called $h(k)$ where, T_0 represents the sampling time and $k \in \{0, 1, 2, 3, 4, 5...\}$. In addition, as mentioned in [17] the use of numerical methods for the calculus of the system evolution is based mainly on the possibility to approximation the state system at the instant time $k + 1$, if the state and the control action on the time instant k are known, this approximation is called Euler method.

$$h(k+1) = h(k) + T_0 f(h, u, t) \tag{2}$$

In order to design the controller, the model (1) can be approximated as (2)

$$\mathbf{h}(k+1) = \mathbf{h}(k) + T_0 \mathbf{J}(\mathbf{q}(k))\mathbf{v}(k) \tag{3}$$

In addition, for that the tracking error tends to zero the following expression is used [17]

$$\mathbf{h}(k+1) = \mathbf{h_d}(k+1) - \mathbf{W}(\mathbf{h_d}(k) - \mathbf{h}(k)) \tag{4}$$

where, $\mathbf{h_d}$ is the desired trajectory $\mathbf{W}$ is a diagonal matrix and its values $0 < \mathbf{diag}(w_{hx}, w_{hy}, w_{hz}) < 1$ are design parameters of the proposed controller.

Remark 1. If a faster response is required, the values should be closer to 0 and if a slower response is required, the values of should be closer to 1.

Now, to generate the system equations consider (3) and (4) then, the system can be rewritten as $\mathbf{Au} = \mathbf{b}$.

$$\underbrace{\mathbf{J}(\mathbf{q}(k))}_{\mathbf{A}}\underbrace{\mathbf{v}(k)}_{\mathbf{u}} = \underbrace{\frac{\mathbf{h_d}(k+1) - \mathbf{W}(\mathbf{h_d}(k) - \mathbf{h}(k)) - \mathbf{h}(k)}{T_0}}_{\mathbf{b}} \tag{5}$$

Note, that the Jacobian matrix has more unknowns that equation ($m < n$) therefore, an infinite solution exists to (5). A viable solution method is to formulate the problem as a constrained linear optimization problem.

$$\frac{1}{2}(\mathbf{v} - \mathbf{v_0})^T(\mathbf{v} - \mathbf{v_0}) = \min \tag{6}$$

Finally, a law of control for our system given by (5) is obtained minimizing (6):

$$\mathbf{v}_{ref} = \underbrace{\mathbf{J}^T(\mathbf{JJ}^T)^{-1}\mathbf{b}}_{\mathbf{v}_p} + \underbrace{\left(\mathbf{I}_n - \mathbf{J}^T(\mathbf{JJ}^T)^{-1}\mathbf{J}\right)\mathbf{v_0}}_{\mathbf{v}_h} \tag{7}$$

where the first term on the left-hand side is the particular solution ($\mathbf{v}_p$) and second term ($\mathbf{v}_h$) of this equation belong to the null space $\mathbf{J}$. In this work two different secondary objectives are considered: the obstacles avoid by the mobile platform and the singular configuration prevention through the systems manipulability control and is given by $\mathbf{v_0} = \mathbf{H}([\,u_{obs}\quad \omega_{obs}\quad (\mathbf{q}_d^T - \mathbf{q}^T)\,])$, where $\mathbf{H}$ is a diagonal matrix allows to increase, reduce or cancel the effect each objective, $(\mathbf{q}_d^T - \mathbf{q}^T)$ are configuration errors of the mobile robotic arm, in such a way that the manipulator joints will be pulled to the desired values that maximize manipulability and u_{obs}, ω_{obs} are the linear velocity and angular velocity of the mobile platform respectively that avoids the static obstacles by resourcing to the null space configuration [18].

2.3 Stability Analysis

Considering the hypothesis of perfect velocity tracking, i.e., $\mathbf{v}_{ref} = \mathbf{v}$, (7) can be substituted into the kinematic model (3) to obtain the following closed-loop equation:

$$\mathbf{h}(k+1) - \mathbf{h}(k) = T_0\mathbf{J}\left(\mathbf{J}^T(\mathbf{JJ}^T)^{-1}\mathbf{b} + \left(\mathbf{I}_n - \mathbf{J}^T(\mathbf{JJ}^T)^{-1}\mathbf{J}\right)\mathbf{v_0}\right) \tag{8}$$

where $\mathbf{JJ}^T(\mathbf{JJ}^T)^{-1} = \mathbf{I}_m$ the Eq. (6) is reduced

$$\mathbf{h}(k+1) - \mathbf{h}(k) = T_0\mathbf{I}_m\mathbf{b} + (\mathbf{JI}_n - \mathbf{I}_m\mathbf{J})\mathbf{v_0} \tag{9}$$

through the properties of an identity matrix is achieved

$$\mathbf{h}(k+1) - \mathbf{h}(k) = T_0 \left(\frac{\mathbf{h_d}(k+1) - \mathbf{W}(\mathbf{h_d}(k) - \mathbf{h}(k)) - \mathbf{h}(k)}{T_0} \right) \tag{10}$$

reducing terms and grouping is calculated the error in the following state $\mathbf{h_d}(k+1) - \mathbf{h}(k+1)$ depends only on the previous error for a gain $\mathbf{W}(\mathbf{h_d}(k) - \mathbf{h}(k))$ to,

$$\begin{bmatrix} e_{hx}(k+1) \\ e_{hy}(k+1) \\ e_{hz}(k+1) \end{bmatrix} = \begin{bmatrix} w_{hx}(e_{hx}(k)) \\ w_{hy}(e_{hy}(k)) \\ w_{hz}(e_{hz}(k)) \end{bmatrix}$$

The errors on the following states comes by

$$e_i(k+1) = w_i e_i(k)$$
$$e_i(k+2) = w_i e_i(k+1) = w_i^2 e_i(k)$$
$$e_i(k+3) = w_i e_i(k+2) = w_i^3 e_i(k)$$
$$\vdots$$
$$e_i(k+n) = w_i e_i(k+n-1) = w_i^n e_i(k)$$

therefore the error tends asymptotically to zero when $0 < w_i < 1$ and $n \to \infty$

3   Experimental Results

In this section, the performance of the proposed control law is tested throught the experimentation over the AKASHA robot which one consist of a robotic arm within 5DOF mount over a unicycle-type mobile platform. This platform allows linear velocity and angular velocity as a reference signal, (Fig. 1)

Fig. 1. AKASHA: Manipulator Mobile Robot

The experiment corresponds to the control system presents in (7). The desired trajectory for the end-effector of the mobile manipulator is described by $\mathbf{h_d} = \begin{bmatrix} h_{xd} & h_{yd} & h_{zd} \end{bmatrix}^T$, where $h_{xd} = 0.5 + 0.2t$, $h_{yd} = 0.3 \sin (0.4t)$ and $h_{zd} = 0.5 + 0.1 \sin (0.4t)$. In this experiment, the mobile platform starts at $\mathbf{q_p} = \begin{bmatrix} 0 \text{ m} & 0 \text{ m} & 0 \text{ rad} \end{bmatrix}^T$; the robotic arm at $\mathbf{q_a} = \begin{bmatrix} -\frac{29\pi}{36} \text{ rad} & \frac{2\pi}{3} \text{ rad} & \frac{4\pi}{9}\text{rad} \end{bmatrix}^T$ and $\mathbf{v_0} = \mathbf{H}([0 \quad 0 \quad (0 \text{ rad} - q_1) \\ (1.22 \text{ rad} - q_2)(-1.05 \text{ rad} - q_3)])$.

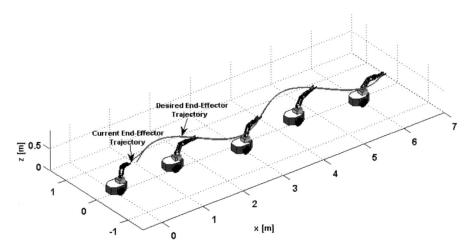

Fig. 2. Stroboscopic movement of the mobile manipulator in the trajectory tracking and internal configuration the of arm.

Figures 2, 3, 4 and 5 represent the experimental results. Figure 2, shows the desired trajectory and the current trajectory of the end-effector. It can be seen that the proposed controller presents a good performance while, considers the internal configuration of arm. Figure 3, shows the evolution of the tracking errors, which remain close to zero, while Figs. 4 and 5 show the optimal control actions.

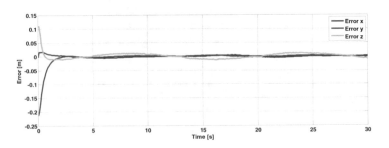

Fig. 3. Control errors of the mobile manipulator

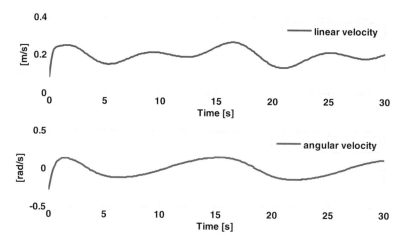

Fig. 4. Velocity commands to the mobile platform

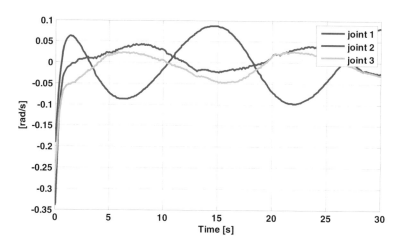

Fig. 5. Joint velocity commands to the robotic arm

For the second experiment, it's consider the avoidance of obstacles and maxima manipulability where, the environment is formed of a straight line and two static obstacles, $\mathbf{v_0} = \mathbf{H}([\,u_{obs} \quad \omega_{obs} \quad (\mathbf{q_d^T} - \mathbf{q^T})\,])$ with $\mathbf{q_d} = [-0.698 \quad 2.27 \quad -2.27]$ to maintain maxima manipulability of the mobile manipulator during task execution, the mobile platform starts at $\mathbf{q_p} = [0 \text{ m} \quad -1 \text{ m} \quad 0 \text{ rad}]^{\mathrm{T}}$ and robotic arm at $\mathbf{q_a} = [0 \text{ rad} \quad 0 \text{ rad} \quad 0 \text{ rad}]^{\mathrm{T}}$.

Figures 6 and 7, shows the desired trajectory and the current trajectory of the end-effector. It can be seen that the proposed controller allow to meet the desired tracking trajectory while, avoiding static obstacles and considers the maxima manipulability. Figure 8, shows the evolution of the tracking errors, which remain close to zero.

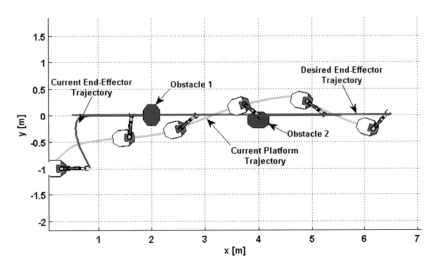

Fig. 6. Stroboscopic movement in two dimensions of the mobile manipulator in the trajectory tracking experiment with avoidance of obstacles and maxima manipulability

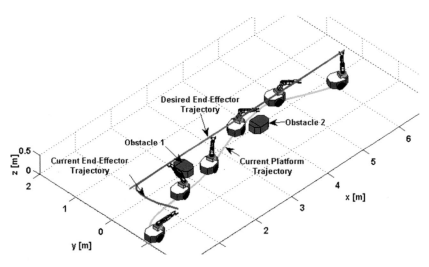

Fig. 7. Stroboscopic movement in three dimension of the mobile manipulator in the trajectory tracking experiment with avoidance of obstacles and maxima manipulability

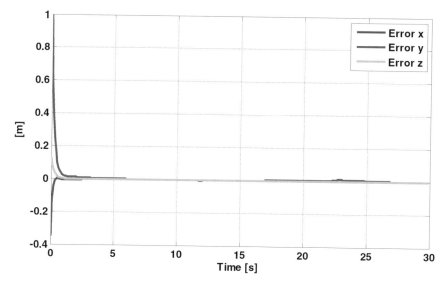

Fig. 8. Control errors of the mobile manipulator with avoid static obstacles

4   Conclusions

In this work a control algorithm based on linear algebra and numerical methods for mobile manipulator robots has been presented An advantage of this controller is its simple implementation in any programming language. The proposed control algorithm stability and performance has been analytically demonstrated trough the linear algebraic concepts. Two types of experiments were performed: one for follow desired trayectory while consider internal configuration of arm and another for follow desired trayectory while consider the avoidance of static obstacles and maxima manipulabilty. Experimental results shown the optimal performance of this control algorithm.

Acknowledgments. The authors would like to thanks to the Consorcio Ecuatoriano para el Desarrollo de Internet Avanzado -CEDIA-, and the Universidad de las Fuerzas Armadas ESPE for financing the project "Tele-Operación Bilateral Cooperativo de Múltiples Manipuladores Móviles – CEPRAIX-2015-05".

References

1. Andaluz, V., Chicaiza, F., Gallardo, C., Quevedo, W., Varela, J., Sánchez, J., Arteaga, O.: Unity3D-MatLab simulator in real time for robotics applications. Lecture Notes in Computer Science, pp. 246–263. Springer, New York (2016)
2. Castellanos, E., García-Sánchez, C., Llanganate, W., Andaluz, V., Quevedo, W.: Robots coordinated control for service tasks in virtual reality environments. Lecture Notes in Computer Science, pp. 164–175. Springer, Cham (2017)

3. Andaluz, V., Quevedo, W., Chicaiza, F., Varela, J., Gallardo, C., Sánchez, J., Arteaga, O.: Transparency of a bilateral tele-operation scheme of a mobile manipulator robot. Lecture Notes in Computer Science, pp. 228–245. Springer, New York (2016)
4. Naijian, C., Hao, Y., Xiangdong, H., Changsheng, A., Chenglong, T., Xiangkui, L.: Adaptive robust control of the mobile manipulator based on neural network. In: 2016 IEEE 11th Conference on Industrial Electronics and Applications (ICIEA) (2016)
5. Meng, Z., Liang, X., Andersen, H., Ang, M.: Modelling and control of a 2-link mobile manipulator with virtual prototyping. In: 2016 13th International Conference on Ubiquitous Robots and Ambient Intelligence (URAI) (2016)
6. Karray, A., Feki, M.: Tracking control of a mobile manipulator with fuzzy PD controller. In: 2015 World Congress on Information Technology and Computer Applications (2015)
7. Avanzini, G., Zanchettin, A., Rocco, P.: Constraint-based model predictive control for holonomic mobile manipulators. In: 2015 IEEE/RSJ International Conference on Intelligent Robots and Systems (IROS) (2015)
8. Fareh, R., Rabie, T.: Tracking trajectory for nonholonomic mobile manipulator using distributed control strategy. In: 2015 10th International Symposium on Mechatronics and its Applications (ISMA) (2015)
9. Scaglia, G., Rosales, A., Quintero, L., Mut, V., Agarwal, R.: A linear-interpolation-based controller design for trajectory tracking of mobile robots. Control Eng. Pract. $18(3)$, 318–329 (2010)
10. Rómoli, S., Serrano, M., Ortiz, O., Vega, J., Eduardo Scaglia, G.: Tracking control of concentration profiles in a fed-batch bioreactor using a linear algebra methodology. ISA Trans. 57, 162–171 (2015)
11. Gandolfo, D., Rosales, C., Patiño, D., Scaglia, G., Jordan, M.: Trajectory tracking control of a PVTOL aircraft based on linear algebra theory. Asian J. Control. $16(6)$, 1849–1858 (2013)
12. Rosales, A., Scaglia, G., Mut, V., di Sciascio, F.: Trajectory tracking of mobile robots in dynamic environments—a linear algebra approach. Robotica $27(07)$, 981–997 (2009)
13. Špinka, O., Hanzálek, Z.: Energy-aware navigation and guidance algorithms for unmanned aerial vehicles. In: 2011 IEEE 17th International Conference on Embedded and Real-Time Computing Systems and Applications, vol. 2, pp. 83–88 (2011)
14. Rosales, C., Gandolfo, D., Scaglia, G., Jordan, M., Carelli, R.: Trajectory tracking of a mini four-rotor helicopter in dynamic environments—a linear algebra approach. Robotica $33(08)$, 1628–1652 (2014)
15. Andaluz, V., Ortiz, J., Perez, M., Roberti, F., Carelli, R.: Adaptive cooperative control of multi-mobile manipulators. In: IECON 2014—40th Annual Conference of the IEEE Industrial Electronics Society. (2014)
16. Romoli, S., Scaglia, G., Serrano, M., Godoy, S., Ortiz, O., Vega, J.: Control of a fed-batch fermenter based on a linear algebra strategy. IEEE Lat. Am. Trans. $12(7)$, 1206–1213 (2014)
17. Andaluz, V., Sásig, E., Chicaiza, W., Velasco, P.: Control Based on Linear Algebra for Mobile Manipulators. Computational Kinematics, pp. 79–86. Springer, Cham (2017)
18. Andaluz, V., Roberti, F., Toibero, J., Carelli, R.: Adaptive unified motion control of mobile manipulators. Control Eng. Pract. $20(12)$, 1337–1352 (2012)

Software Engineering and Knowledge Engineering

Enterprise Requirements Management Knowledge Towards Digital Transformation

Shuichiro Yamamoto[(⊠)]

Nagoya University, Furo-cho, Chikusa-ku, Nagoya, Japan
syamamoto@acm.org

Abstract. As the enterprise services are rapidly digitized by digital transformation, the need to integrated knowledge management of IT requirements and business requirements is increased.

In this paper, the cause of problems on requirements definition is explained to the lack of requirements management knowledge. The types of requirements management knowledge are defined as discovery, recovery, and transformation types from the points of business, application, and technology. A reference model is proposed to integrate different types of requirements management knowledge. A case study on the integration of different requirements guidelines is described to show the effectiveness of the proposed approach.

Keywords: Enterprise requirements management · Digital transformation · Requirements knowledge management · Case study

1 Introduction

There are three types of requirements engineering approaches as shown in Table 1. Fundamental requirements engineering approach is necessary for defining valuable requirements based on user needs without omission (Issue1). For IT modernization to rebuild legacy systems, it is necessary to recover current requirements where experts of legacy systems are absent (Issue2). Moreover, it is necessary to develop future IT requirements for exploring new business values based on new technology (Issue3). An enterprise requirements management is necessary to address these whole Issues. RE-Type I, II, and III try to resolve issues 1, 2, and 3, respectively.

These three types of requirements engineering approaches are corresponding to enterprise architecture (EA) layers [20, 21], i.e. business, application, and technology architecture, respectively. User needs come from business architecture. Legacy systems are used to describe the base line application architecture. New digital technologies affect the target technology architecture. This shows that an enterprise requirements management approach is necessary to treat three types of requirements as a whole.

The rest of the paper is organized as follows. Section 2 describes related work on requirements management approaches. Section 3 proposes a requirements knowledge integration approach. In Sect. 4, an example case study using the proposed approach is presented to integrate the different requirements guideline knowledge structures from a Japanese public institute. Discussions on the effectiveness of the approach are shown in Sect. 5. Our conclusions are presented in Sect. 6.

© Springer Nature Singapore Pte Ltd. 2018
K.J. Kim et al. (eds.), *IT Convergence and Security 2017*,
Lecture Notes in Electrical Engineering 449,
DOI 10.1007/978-981-10-6451-7_36

Table 1. Requirements engineering types

RE-Type	Source	Activity	EA layer	Time scope
I	User needs	Discovery	Business	Now
II	Legacy systems	Recovery	Application	Past
III	Digital technology	Transformation	Technology	Future

2   Related Work

The objectives of requirements management are not only to manage requirements in the process through elicitation, analysis, specification, and validation, but also to manage requirements in the process of implementation and operation and modification. Requirements managements constitutes requirements attribute, requirements change, versioning, and traceability. For example, Davis [8] proposed a requirements management process consists of requirements elicitation, triage, and specification. Requirements triage phase treats priority, implementation cost, negotiation, validation of relationship, and delivery of requirements.

Requirements change control process of traditional requirements engineering approaches [1–10] commonly consists of preparation, establishment of organization, execution, and completion phases. In preparation phase, change control policy and change control plan are defined. The baseline of change requirements is developed. In organization phase, CCB (Change Control Board) is recommended to establish. Activity of CCB is limited to manage requirements change. Plans and policies of CCB are authorized by upper management board. The result of requirements change is reported to the upper management board. In execution phase, requirements change request, analysis of change request, decision of requirements change, requirements change notification, consensus building among stakeholders, requirements change implementation, and validation of requirements change are achieved. In completion phase, decision and report development on requirements change completion are necessary.

Requirements traceability was well recognized from the early stage of requirements engineering. For example, SREM [10] and ARTS [11] have been developed as requirements tracing system in 1970s. So far, definition of tracing [12], effect of tracing [13, 14], tracing metrics [4] have been described. The traceability defined by Ramesh et al. [13] is a system description technique for changing system portions by maintaining the relationship with other system descriptions. Requirements management approaches are explained for specific project in BABOK (Business Analysis Body Of Knowledge) [15], REBOK (Requirements Engineering Body Of Knowledge) [16], and IREB [17, 18]. These BOKs are commonly structured by knowledge domain, process knowledge, product knowledge, and technique knowledge. Each BOK hierarchically decomposes knowledge domains into sub domains and topics.

As mentioned before, existing requirements management knowledge, described current researches, only provides individually for a specific type of software requirements. Therefore, integration of individual requirements engineering knowledge is necessary from the point of whole types of requirements.

3   Requirements Knowledge Integration Approach

By analyzing different requirements engineering knowledge, we found a common reference model as shown in Fig. 1. Each knowledge has objectives, strategy to compose knowledge, and elementary knowledge. Strategy types are issue and process. By using the reference model, different knowledge is able to integrate.

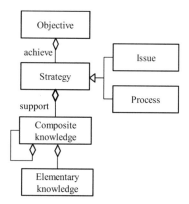

Fig. 1.   Knowledge integration reference model

The knowledge integration method of a set of existing baseline knowledge are as follows. First, all the baseline knowledge is gathered and analyzed by the above knowledge integration reference model. Then, a concept model on baseline knowledge is extracted based on the analysis result. Finally, new integrated knowledge is designed based on the reference model if necessary. This approach enables to evolve the set of baseline knowledge.

4   Case Study

The Software Reliability Enhancement Center of IPA (Information-technology Promotion Agency, Japan) published Requirements definition guide for user [22] and User guide for rebuilding system to the success [23] to address Issue1 and Issue2 in Table 1, respectively. Both requirements guides are analyzed by using the integration reference model. Then develop a conceptual model for integrated guide. Moreover, additional requirements knowledge structure is created for guiding requirements management of digital transformation.

4.1   Requirements Knowledge Integration

Knowledge structure of requirements definition guide for user, and User guide for rebuilding system to the success are analyzed by the reference model. The results are described in Tables 2 and 3, respectively. The conceptual model for the knowledge is

Table 2. Knowledge structure of requirements definition guide for user.

Category	Issues	Practice knowledge
Matters	Validity of system development	IT investment decision criteria, business management view point, technology perspective view point, social view point of nfr, competitive delivery
	Business analysis to prepare system development	Usage situation analysis, elimination of duplication and useless requirements at enterprise level
	Reliable and efficient requirements definition	Consensus building of stakeholders, selection criteria of requirements
	Load of requirements definition	Effort allocation, quality of requirements artifact
	Role collaboration between users, IT divisions, and venders	Contract method, responsibility analysis, budget acquisition
Process	IT development to contribute business	Means-end analysis, goal analysis, requirements systematization, problem-need-issue-requirements relationship analysis
	Control of requirements explosion	Requirements control, excessive requirements evaluation, quantitative metric
	Reduction of business complexity	Adjustment of target business, business improvement, Business package application
	NFR (Non-functional requirements)	NFR grade, NFR elicitation sheet, conflict resolution,
	Consensus building with diverse stakeholders	Stakeholder analysis, rich picture, admissibility, escalation path
	Understanding of baseline business and IT system	Visualization of baseline business/IT system, field work, common understanding
Artifact	Artifact development	Business process diagram, business function relationship table, screen layout, business process definition, ER diagram, entity/data item definition, CRUD table, acceptance test definition, system transition plan, operation requirements definition, NFR definition
	Application of artifact	Vital points to decide requirements artifact, vital points to apply process/data artifact, requirements review, treatment of postponed items, quality improvement of artifact, role divisions to develop artifact

Table 3. Knowledge structure of system rebuilding requirements guide

Category	Issues	Practice knowledge
Method selection	Baseline system analysis	Rebuilding theme, investigation view point
	Target system requirements analysis	Rebuilding theme requirements matrix, target system requirements list, target system requirements analysis diagram
	Rebuilding method selection	Basic selection pattern, rebuilding method selection detail
	Rebuilding method decision	Rebuilding method characteristics, rebuilding risk factor list, rebuilding risk resolution template, rebuilding method decision meeting
Plan design	Requirements validation	Baseline requirements, requirements negotiation meeting, confirmed requirements list
	Baseline specification clarification	Inherited requirements gap analysis, inherited requirements clarification, inheritance policy
	Baseline asset reuse policy definition	Reuse scope decision, document reuse asset, implementation reuse asset, rebuilding method reuse policy matrix
	Insufficient baseline knowledge resolution	Baseline business knowledge, insufficient knowledge complementation plan
	Quality assurance consideration	Business continuity scope, assurance of business continuity, quality assurance policy, quality validation viewpoint, testing characteristics for rebuilding method, testing items, testing completion criteria, after release risk measures
	Decision process definition	Scope of decision process, decision target issue list
	Data transition planning	Data transition validation point
	Rebuilding planning and estimation	Risk remedy based system rebuilding plan, phased rebuilding estimation

also developed based on these tables as shown in Fig. 1. The concept model provides a means to understand the integrated requirements engineering guidelines.

4.2 Requirements Knowledge Extension

A new additional requirements knowledge for digital transformation is designed as shown in Table 4 based on the knowledge integration reference model. Figure 2 shows an integrated requirements management process controlled by Enterprise Requirements Management Board (ERMB) for all three types of requirements. ERMB extends CCB for specific software requirements (Fig. 3).

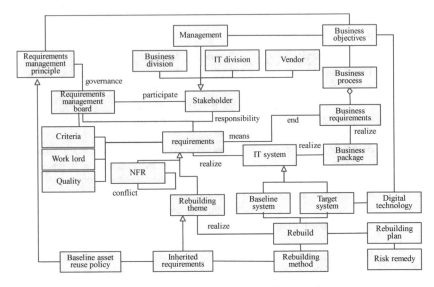

Fig. 2. A concept model to integrate requirements knowledge

Table 4. Knowledge structure of digital transformation requirements guide

Category	Issues	Practice knowledge
Dimension	Scope	Type, transformation target
	Hierarchy	Transformation requirements hierarchy, enterprise transformation requirements
Organization	Enterprise requirements management board	Reference model, transformation requirements management, transformation maturity
	Transformation capability	Transformation capability index, criteria level
Process	Agile	Analysis model, requirements conversation, governance
	EA	Technology architecture requirements, EA model
	Operation	IT management, operation requirements
	Backlog	Requirements arrangement, requirements priority

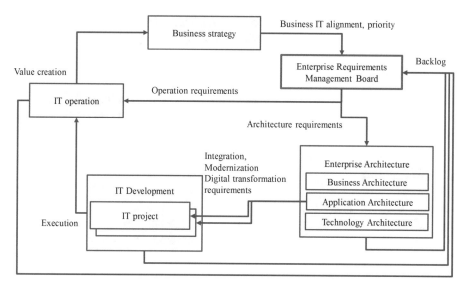

Fig. 3. Enterprise requirements management process

5 Discussion

5.1 Contribution

This paper proposed a reference model to integrate different types of requirements management knowledge. A case study on the integration of different requirements guidelines has been described to show the effectiveness of the proposed approach. A new additional requirements knowledge for digital transformation has also been designed based on the knowledge integration reference model. Especially, ERMB has been proposed by extending existing CCB for specific software requirements. An integrated requirements management process is able to control by ERMB for all three types of requirements.

5.2 Effectiveness

The proposed approach successfully applied to integrate practical requirements knowledge guidelines developed by IPA. The knowledge integration reference model was useful to analyze existing requirements guidelines, because guidelines of many pages can be summarized in concise tables. Moreover, the concept model shown in Fig. 1 integrates primary knowledge entities from different requirements knowledge guidelines in the interrelated fashion.

5.3 Knowledge Evolution

In addition to the requirements knowledge on user needs and rebuilding existing systems, requirements knowledge on digital transformation was integrated by using the proposed knowledge integration reference model. This showed the proposed approach is able to apply for evolving requirements knowledge.

5.4 Limitation

Although the case study showed the usefulness of the proposed method, quantitative evaluation is needed. More number of case studies on knowledge integration are also necessary for sufficient evaluation.

6 Conclusion

Although there are many requirements knowledge guidelines, knowledge integration approach has not been proposed so far. Digital transformation is rapidly spread among IT systems. New requirements management knowledge is also necessary to manage requirements from different sources. This paper proposed an approach to integrate different requirements guidelines based on a common knowledge reference model. Case study on integrating Japanese requirements guidelines showed the effectiveness of the proposed approach. Furthermore, additional requirements knowledge was shown to easily design by using the reference model.

Further study will be necessary to integrate other bodies of requirements knowledge by applying the proposed approach.

References

1. Sommerville, I., Sawyer, P.: Requirements Engineering—A Good Practice Guide. Wiley, Chichester (1997)
2. Leffingwel, D., Widrig, D.: Managing Software Requirements a Unified Approach. Addison-Wesley Professional, Boston (2000)
3. Kotonya, G., Sonnmerville, I.: Requirements Engineering. Wiley, Chichester (2002)
4. Hull, E., Jackson, K., Dick, J.: Requirements Engineering. Springer, Heidelberg (2002)
5. Wiegers, K.: Software Requirements—Practical Techniques for Gathering and Managing Requirements Through the Product Development Cycle. Microsoft Corporation (2003)
6. Aurum, A., Wohlin, C.: Engineering and Managing Software Requirements. Springer, Heidelberg (2005)
7. Davis, A.M.: Software Requirements: Objects, Functions and States. Prentice-Hall, Upper Saddle River (1993)
8. Davis, A.M.: Just Enough Requirements Management—Where Software Development Meets Marketing. Dorset House Publishing, New York (2005)
9. Berenbach, B., Paulish, D., Kazmeier, J., Dudorfeer, A.: Software and Systems Requirements Engineering in Practice. McGraw Hill, New York (2009)
10. Alford, M.: A requirements engineering methodology for real-time processing requirements. IEEE Trans. SE, **SE-3**(1), 66–69 (1977)
11. Dorfman, M., Flynn, R.: ARTS—an automated requirements traceability system. J. Syst. Softw. **4**, 63–74 (1984)
12. IEEE Std. 830-1998: Recommended Practice for Software Requirements Specification. IEEE (1998)

13. Ramesh, B., Jarke, M.: Toward reference models for requirements traceability. IEEE Trans. Softw. Eng. **27**(1), 58–93 (2001)
14. Palmer, J.: Traceability. In: Dorfman M., Thayer, R. (eds.) Software Engineering, pp. 266–276. IEEE (1996)
15. IIBA: A Guide to the Business Analysis Body of Knowledge. Lightning Source Inc. (2013)
16. JISA: Requirements Engineering Body of Knowledge (REBOK), Ver. 1.0, Kindaikagakusha (2011) (In Japanese)
17. Pohl, K., Rupp, C.: Requirements Engineering Fundamentals, A Study Guide for the Certified Professional for Requirements Engineering Exam Fundamental level/IREB compliant. Rockynook (2011)
18. The home of Requirements Engineering. http://www.ireb.org/. Accessed 11 May 2017
19. Jacobson, I., Pan Wei N., McMahon, P., Spence, I., Lidman, S.: The Essence of Software Engineering—Applying the SEMAT Kernel. Addison-Wesley Pearson Education (2013)
20. Josely, A.: TOGAF V.9 A Pocket Guide. The Open Group. Van Haren Publishing (2008)
21. Josely, A.: ArchiMate®2.0, A Pocket Guide. The Open Group. Van Haren Publishing (2013)
22. Software reliability Enhancement Center: Requirements Definition Guide for User. Information-Technology Promotion Agency, Japan (2017). (In Japanese)
23. Software reliability Enhancement Center: User Guide for Rebuilding System to the Success. Information-Technology Promotion Agency, Japan (2017). (In Japanese)

Qualitative Requirements Analysis Process in Organization Goal-Oriented Requirements Engineering (OGORE) for E-Commerce Development

Fransiskus Adikara$^{(\boxtimes)}$ (iD), Sandfreni (iD), Ari Anggarani, and Ernawati

Esa Unggul University, Jl. Terusan Arjuna, Jakarta, Indonesia
fransiskus.adikara@esaunggul.ac.id

Abstract. One of the most important processes in requirements engineering is the requirements analysis process. This paper propose the qualitative requirements analysis process to improving e-commerce system quality using OGORE. The proposed method will completing the previous research with adding qualitative analysis in the process based on AGORA method. By completing the requirements analysis process in OGORE with qualitative analysis, the proposed method can eventually become a new addition to the pre-existing requirements analysis process and can also be used especially in e-commerce system development process. With this qualitative requirements analysis, the Non Functional Requirements (NFRs) of e-commerce system can also be analysis using OGORE.

Keywords: Requirements engineering · Qualitative requirements analysis · E-commerce · Non-functional requirements

1 Introduction

Nowadays in the midst of information and communication technology advantages, internet usage has become an important part in business and trading activities. Almost all aspects of human life impacted by the internet and the business sector is the most perceived by it. E-commerce growth in developing country has significantly increase year by year.

The process of developing information systems including the development of e-commerce system in Indonesia, and other developing countries still face many problems [1]. The most frequent problem occurs when Requirements Engineering (RE) process is not met during information systems development process. Improper and incomplete requirements engineering greatly affects the success rate of information system development. From a previous study it has been concluded that the problems occurring in engineering needs are one of the main causes of an information systems project experiencing budget overruns, delays, and reductions in the scope of work that diminish the ability and effectiveness of the software produced for the company [2, 3].

© Springer Nature Singapore Pte Ltd. 2018
K.J. Kim et al. (eds.), *IT Convergence and Security 2017*,
Lecture Notes in Electrical Engineering 449,
DOI 10.1007/978-981-10-6451-7_37

GORE approach is expected to minimized the increasing requirements that originally come from user interests [4, 5]. In our previous paper [6], we proposes an extension of GORE approach that uses the organization goals (the overall objectives, purpose and general mission of an organization) to elicit system requirements, so the system functions and the resulting requirements can be more qualified and relevant with organizations missions rather than user's interests. In other paper, we also proposes the use of Case-Based Reasoning (CBR) method in requirements engineering process and combine it with AGORA method to refine and analyze requirements [7]. The combination is intended to get the high quality requirements that are based on the reuse of requirements from literatures, best practices, and previous experiences that are recorded in a Case-Based.

Based on the backgrounds described above and continuing previous research on organization goal-oriented requirements engineering (OGORE) [1], this paper propose an requirements analysis process to improving the quality of e-commerce system. This analysis method propose to be applicable in e-commerce systems engineering process. The propose method can eventually become a new addition to the pre-existing requirements analysis methods and become a new method that prioritizes in e-commerce system development process.

This paper is organized as follows. In Sect. 2, the paper starts with the overview about some related works. In the next section, we explain the requirements analysis process in our proposed approach. We discuss the detailed process of our proposed requirements analysis in Sect. 4, and finally the conclusion in Sect. 5.

2   Related Work

The term e-commerce began to emerge in the 1990s through an initiative to convert the paradigm of sale and purchases transactions, and payments from conventional ways into electronic digital forms based on computers and Internet networks [8]. This is very contrary to the conditions before the e-commerce in the world. At that time transactions are made directly through face to face between the provider of goods and services with consumers, even long before the money is created, the transaction is done through a barter process, namely the exchange of goods. The common features of e-commerce functionalities and architecture [9] are consist of two major part: Store Front and Back Office. For Store Front, the e-commerce system should have these function: Registration, Product Catalog, Product Ordering, Payment, and Customer Service. The Back Office of the e-commerce system should handle the ordering and manage to deliver the product to customer, so the system should have these function: Ordering Management, Product Management, Warehouse Management, Payment Management, Customer Relationship Management, and Content Management System.

In previous research [7], based on Case-Based Reasoning (CBR) method theory [10], the proposed requirements engineering process approach is illustrated in Fig. 1. In the previous research about the requirements elicitation process needs [4], we have discussed how to describe the Initial Goal Tree Model, KPIs (*Key Performance Indicators*) and business domains used as initial information for requirements refinement and analysis process.

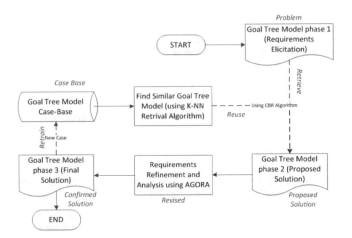

Fig. 1. Flowchart of CBR method approach in organization goal-oriented requirements engineering (OGORE) process [7].

In general, this approach have four steps [7], as the followings:

- *Retrieve*: The objective is to recognize the new cases (problems) using the result of requirements elicitation process. Thus, the case attributes in the propose method consists of: business domain, goal, and KPIs.
- *Reuse*: In this activity, the system uses existing solutions in case-based to address new problems. These existing solutions have some degree of similarities to the new problems. The case-based consists of goal-tree model that is built from best practices, previous solved solutions and literature studies.
- *Revise*: The proposed solutions that were obtained from case-based is re-evaluated to address the new problems. The requirements analysis process is then performed to get the best solution goal tree models for the organization. The analysis process used modified AGORA method to resolve conflicts in the requirements. At this stage, solution goal tree model is produced.
- *Retain*: In this final stage, the refined problem solutions in the form of goal-tree model are kept in the case-based. These solutions are to be used in the future.

To assess the KPIs achievement within a goal, the analysis method modified AGORA [7] as Requirements Analysis Process. In this paper, this process will be add with qualitative requirements analysis to complete the previous method.

Non-functional requirements (NFRs) are commonly characterized from functional requirements by differentiating how the system shall do something in contrast to what the system shall do. A NFR is an attribute of or a constraint on a system, NFRs or named quality attributes, e.g., how to make Web content accessible to people with disabilities [11]. Quality goals (sometimes called "non-functional goals") refer to non-functional concerns such as security, safety, accuracy, usability, performance, cost, or interoperability, in terms of application-specific concepts [12].

3 Research Methodology

In the early stages of the research will develop a requirements analysis method of goal-oriented requirements engineering for e-commerce system. This research is a continuation of previous research that once discussed. With this research, the focus is more on the development of e-commerce system. This process will be more beneficial to the development of e-commerce system.

The first phase of this research is designing requirements analysis process by taking these preparatory steps: literature study and doing analysis of existing requirements analysis techniques. This designing activity is undertaken in conjunction with stakeholders and requirements engineering experts.

The results of first phase will be used in the next phase to modifying the requirements analysis method specially for e-commerce system engineering process. Once developed, the requirements analysis methods can be implemented in the development of e-commerce system.

4 Requirements Analysis Process for E-Commerce System

Continuing previous research using CBR to conduct refinement and analysis process, then in this research focus on requirement analysis process by doing modified activity especially to get non functional requirement. Non-functional requirements (NFRs) are commonly distinguished from functional requirements by differentiating how the system shall do something in contrast to what the system shall do [12]. NFRs are usually documented separately from functional requirements, without quantitative measures, and with relatively vague descriptions. NFRs difficult to analyze and test. In e-Commerce there NFRs are things that related to network infrastructure, the minimum hardware requirements, legality, and regulation for e-Commerce.

The complete requirements analysis process for e-commerce based on the modification of AGORA [13] method that we propose in this paper have two types of analysis: Quantitative Analysis and Qualitative Analysis. Not only quantitative analysis that already proposed in previous research [7], in this paper we propose Qualitative Requirement Analysis on the OGORE method to be used to gauge the level of rationality defined on each task that wants to fulfill its purpose. With this rationality measurements, we can know which one of the requirements that can be met with or without regard to the functionality of the system.

How to conduct an assessment for this qualitative analysis by writing down **the rationale statement** connected to an attribute or being on a node or edge that illustrates the reason why the analyst describes the goal to be a sub-goal, and/or answers the question of why a sub-goal is assigned a particular attribute. No measurement is bound to determine whether a rationalization is acceptable or not. Stakeholders can sit together to discuss and determine which non-functional needs are to be used or not based on beliefs owned by the organization.

In addition of Qualitative Analysis process we also propose the refinement of quantitative analysis process in previous research. Quantitative Requirement Analysis on the proposed method is used to assess the level of preference of high-level stakeholders against defined goals so as to analyze possible conflicts among stakeholders regarding the point of view of the goal. Our quantitative analysis is measure the ability of the task to fulfill the goal. Measuring tool used is as follows:

1. **Preference Matrix (PM):** connected to the node of the KPI and the goal, indicating the level of preference of each stakeholder against the goals and KPIs established so as to indicate the level of satisfaction of the stakeholders against the goal and established the KPI.
2. **Contribution Value:** connected to the edge and represents the contribution rate of the goal/task in achieving the goal and the KPI from the parent goal.

The Preferences Matrix assessment begins in the following way:

1. Each pre-defined goal will be assigned a preference value by each stakeholder. The first stakeholder gives an assessment of the Preferences Matrix of its preference level or satisfaction to the existing goals and KPIs, the value is given using a scale from −10 to 10. The lowest value is given, if the stakeholder is less satisfied and the highest value, if it feels very satisfied with the goal and KPI - his.
2. In addition to assessing itself, each stakeholder must also provide a satisfied value (self estimate) of each other stakeholder, for example to stakeholder O, must give the value of satisfied stakeholders to the second and also the third stakeholder of the goal and KPI is based on subjective assessment of the first stakeholders. And then the second stakeholder and also the third stakeholder to fill the overall matrix of preference in each goal do the same thing.

After that the stakeholders who act as the necessary engineers will give contribution value from each end/edge in achieving goal and KPI from the parent. Scale value from −10 to 10. The lowest value means to contribute negatively or not contribute to the goal whereas the highest score means that the edge has a good contribution to the achievement of the goal. The following Table 1 show the example of Preference Matrix of a goal, O refer as Owner of the Organization, BA as Business Analyst, and SA as System Analyst.

Table 1. Example of preference matrix.

	O	BA	SA
O	5	−6	4
BA	−5	3	1
SA	0	2	10

The analysis of the Preferences Matrix is done by looking at the variance value of the first stakeholder column compared to the variance of the second stakeholder column. If there is a big difference in the value of variance then it shows that the first stakeholder has a different view with the second stakeholder on the determined goal and KPI. Based on this data, the analyst of the needs engineer can find out which parts are still not understood by the first stakeholder or the second stakeholder. This difference in point of view can be a constraint that inhibits the establishment of system requirements. Therefore it must be resolved by discussing the stakeholders with the necessary engineers, so that in the end the Preferences Matrix already does not have a very wide variation in value variance.

For an analysis of the value of contribution made by looking at the value assigned to a field goal. If the value of the contribution is still negative, then the goal becomes a constraint and needs to be considered to remain a system requirement or not. The solution of Goal Tree Model can be set if there is no more negative contribution value in the overall goal that exists. The example of Quantitative and Qualitative Requirements Analysis in Goal-Tree Model is illustrated in Fig. 2.

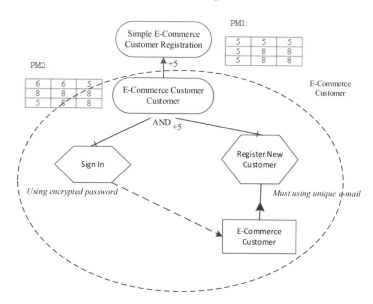

Fig. 2. Example of quantitative and qualitative requirements analysis in goal-tree model.

If there still a negative value on the contribution value and the variance of preference matrix value still very high, then high-level stakeholders should sit in discussions and negotiations on the tasks or goals derived so that the agreement is found in the achievement of the master goal based on the rationale written. When the analysis results show that there is no negative value of Contribution Value that shows the inability of the task or goal to the achievement of the parent goal, the final result of the analysis of Proposed Goal Tree Model is called Solution Goal Tree Model [7].

5   Conclusion and Future Work

To conclude, by the improvement of requirement analysis process in this research, hence requirement analysis processes not only do quantitative analysis, but can also do qualitative analysis. Development of e-commerce system in addition to dealing with the functional needs of the process of sale and purchase transactions and payments, the system also depends on the ability of network infrastructure/internet, hardware capabilities, legality, and security. The qualitative analysis process does not have a standard rule of whether an NFR is acceptable or not, but depends more on the rationality of the process and determined by the stakeholders who determine it. With this proposed analysis process, the CBR process of OGORE to get high quality requirements can be more complete especially in Revised Process, not only doing quantitative requirements analysis, but also can perform qualitative requirements analysis process.

By the process proposed, further research could use the proposal of this method to be implemented in a process of developing an e-commerce system. From the results of the implementation can be assessed the quality of the process and get the results of the needs of e-commerce system.

References

1. Adikara, F., Hendradjaya, B., Sitohang, B.: Integrating KPIs in organization goal-oriented requirements elicitation process to enhance information system. Int. J. Electr. Comput. Eng. **6**(6), 3188–3196 (2016)
2. Adikara, F., Wijaya, P.D., Hendradjaya, B., Sitohang, B.: Information system design based on the result of organization goal-oriented. In: ICISA 2016, vol. 376 (2016)
3. Sandfreni, S., Surendro, K.: Requirements engineering for cloud computing in university using i*(iStar) hierarchy method. In: Information Science and Applications, pp. 885–890 (2015)
4. Adikara, F., Sitohang, B., Hendradjaya, B.: The emergence of user requirements risk in information system development for industry needs. In: 6th International Seminar on Industrial Engineering and Management (2013)
5. Adikara, F., Sitohang, B., Hendradjaya, B.: Penerapan goal oriented requirements engineering (GORE) model (Studi Kasus: Pengembangan Sistem Informasi Penjaminan Mutu Dosen (SIPMD) pada Institusi Pendidikan Tinggi). In: Seminar Nasional Sistem Informasi Indonesia, pp. 230–235 (2013)
6. Adikara, F., Hendradjaya, B., Sitohang, B.: A new proposal for the integration of key performance indicators to requirements elicitation process originating from organization goals. In: International Conference on Data and Software Engineering (2014)
7. Adikara, F., Hendradjaya, B., Sitohang, B.: Requirements refinements and analysis with case-based reasoning techniques to reuse the requirements. In: The 5th International Conference on Electrical Engineering and Informatics 2015, pp. 460–465 (2015)
8. Laudon, K.C., Traver, C.G.: E-commerce: Business, Technology, Society (2014)
9. Laudon, K.C., Guercio Traver, C.: E-commerce: Business, Technology, Society (2007)
10. Aamodt, A.: Case-based reasoning: foundational issues, methodological variations, and system approaches. AI Commun. **7**, 39–59 (1994)
11. Chung, L., Leite, J.D.P.: On non-functional requirements in software engineering. In: Conceptual Modeling: Foundations and Applications, pp. 363–379 (2009)
12. Eckhardt, J., Vogelsang, A., Méndez Fernández, D.: Are 'Non-functional' requirements really non-functional? An investigation of non-functional requirements in practice. In: 38th International Conference on Software Engineering (2016)
13. Kaiya, H., Horai, H., Saeki, M.: AGORA: attributed goal-oriented requirements analysis method, pp. 13–22. IEEE (2002)

An Improvement of Unknown-Item Search for OPAC Using Ontology and Academic Information

Peerasak Intarapaiboon[✉]

Department of Mathematics and Statistics, Faculty of Science and Technology,
Thammasat University, Bangkok 12121, Thailand
peerasak@mathstat.sci.tu.ac.th

Abstract. Many students usually use the unknown-item search strategies, including subject and keyword searches, to retrieve books or other materials provided in library catalogs. However, the success rates for unknown-item searching is relatively low comparing with the known-item search strategies, i.e., title or author searches. In this paper, a framework for improving the unknown-item search is proposed. The main contributions of our framework are concerned with both user's keywords and book indexing: (i) To enhance a user's keyword, the framework will select other relevant terms in a domain-related ontology. (ii) Topics expressed in course description are used as book indexing. A preliminary experiment shows that the traditional OPAC incorporating with the proposed framework gives satisfactory results.

Keywords: Digital library · OPAC · Ontology · Semantic search

1 Introduction

Most of libraries use Online Public Access Cataloging (OPAC) for easy access of books and other materials. Since this article is concerned with books, the explanation of OPAC is relied on this material. There are four basic searching types in OPAC, i.e. Author, Title, Subject, and Keyword searches. We can classify the four search strategies into two groups: known-item search (including author and title searches) and unknown-item search (including keyword and subject searches). Users will use the former group when they have a particular item in mind and they want to determine whether the library holds that item, while those will do the latter when they have an interesting subject in mind. So far, many statistical reports have indicated that the success rates for known-item search is higher than those of unknown-item search [1–3]. One of crucial rationales behind the failure of unknown-item search is that an interesting topic in a user's mind does not match with book's bibliography. More precisely, it might be due to the lack of the user's experience in book indexing on one hand and the old fashion book indexing on the other.

In this work, a novel framework for improving unknown-item search is proposed. The main technical challenges we focus in this work are threefold: (i) How to improve a keyword representing the topic in a user's though; (ii) How to assign more meaningful

© Springer Nature Singapore Pte Ltd. 2018
K.J. Kim et al. (eds.), *IT Convergence and Security 2017*,
Lecture Notes in Electrical Engineering 449,
DOI 10.1007/978-981-10-6451-7_38

indexes to a book; and (iii) How to select compatible books in the library catalog, when a textbook contains the user's interesting topics but it is not in the catalog.

2   Related Works

So far, many statistical reports have indicated that the success rates for known-item search is higher than those of unknown-item search [1–3]. As an evidence, Antell and Huang [1] analyzed the search transaction log of the University of Oklahoma Libraries OPAC. The report revealed that, among all subject search occurrences, 48.8% of them yielded zero results and 10.6% yielded more than five hundred results (In the report, searches that yielded either zero results or more than five hundred results were considered to be unsuccessful).

Many rationale behind the relatively low success rates of unknown-item searches are explored. Some of them are: (i) Since users, particularly undergraduate students, are not familiar with the subject lists used in the libraries, they cannot match terminologies in their mind with the suitable terminologies providing in the subject heading structures [4] (ii) When a concept in user's thought is contains multiple terms, the Boolean operators can be used to make a searching term that is semantically closed the user's thought [5]. However, many users do not well understand the Boolean operators. (iii) The catalog interface does not give users adequate guidance in finding and using LC terms or in revising their searches [6, 7].

Then, several methods have been proposed to raise the rate of success in unknown-item search. Here are some examples: Cousins [8] revealed that the percentages of exact matches between users' queries in historic data and the three wildly-used index systems, i.e., DDC, LCSH, and PRECIS are ranging from 30 to 94% (62.83% on the average). Owning to the low coverage scores, the author claimed that the quality of index systems are inadequate. A new ways to enhance index systems was proposed by indexing bibliography records with selected natural language queries from users in the historical database. The experimental results indicated that natural language enhanced indexing significantly outperform the traditional indexing.

Long [4] introduced 17 guidelines that can be incorporated into OPAC systems to help users perform unknown-item searches more efficiently. Those guidelines, for instance, are (i) The user's search terms should be highlighted in retrieved records; (ii) Users should be told what subject list is used in the library catalog; (iii) The catalog should include search features that incorporate the entire cross-reference structure of subject headings. Based on the guideline, the authors then evaluated the OPAC systems of 31 libraries. The results show that most systems are deficient.

In library science, authority control is a process that organizes bibliographic information. The typical library examples are the set of all books written by an individual author; the set of terms referring to the same object. Utilizing authority files, O'Neill et al. [9] proposed a two-step prototype to improve subject search. In the first step, the authority file is searched to find the appropriate subject heading to the user keyword. Then, in the second step, the bibliographic records are searched to identify the resources with the selected subject heading.

Wood et al. [5] reported that, with a domain specific subject heading, namely Medical Subject Heading (MeSH), medical students can get to a list of articles that are relevant to their searches. For example, the results of searching by "sore throat" do not get the same those by "pharyngitis".

3   Proposed Framework

In this section, we describe our proposed framework for improving unknown-item search of libraries' book searching.

3.1   Framework Overview

Figure 1 shows the proposed framework. When a user submit a keyword, the module GWList will generate a list of terms relating with the user's keyword by using a domain ontology. Then, GCList will search through the curricula in order to retrieve courses relevant to such a word list and create a course list. Based on the course list and a book database, GBList creates a list of books and, finally, each book will be associated with a relevant score. The details of each component are described in the following sections.

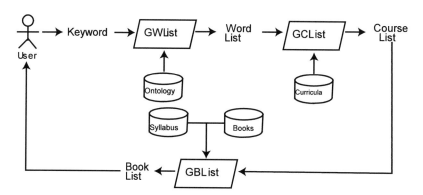

Fig. 1. The proposed framework

3.2   GWList–Word List Generation

The output of this process is a list of words associated with degrees of similarity to the user's keyword. The similarity level is determined relying on the distance in the ontology. Intuitively, the more two concepts are close in terms of their structural properties, the more they are similar. For example, in Fig. 2[1], 'Analytical geometry' is 2 steps away from 'Metric geometry' (i.e., 'Analytical geometry' → 'Geometry' → 'Metric geometry'), while 4 steps away from 'Algebraic topology' (i.e., `Analytical geometry' → 'Geometry' → 'Field of Math' → 'topology' → 'Algebraic topology'). Then, the similarity

---

[1] The full description of this ontology, OntoMathPro, is appeared in [10].

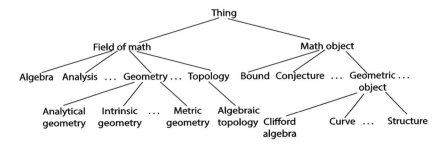

Fig. 2. A part of an ontology

between 'Analytical geometry' and 'Metric geometry' should be higher than that between 'Analytical geometry' and 'Algebraic topology'.

The details of the process is discussed below:

1. Let w_0 be a user's keyword, and W be the set of all words (or concepts) in the ontology in use.
2. Calculate the similarity degree between w_0 and each concept c in the ontology by the following formula:

$$S(w_0, c) = \frac{2D_{LCA}}{D_{w_0} + D_c},\tag{1}$$

where D_{LCA}, D_{w_0} and D_c are the distances from the root node to the least common ancestor of w_0 and c, that to w_0 and that to c, respectively.
3. By a pre-specified threshold α, a list of relevant words with their similarity levels is generated as follows:

$$LW_{w_0}^\alpha = \{(w, S(w_0, w)) | w \in W, S(w_0, w) \geq \alpha\}.\tag{2}$$

Example 1. Consider the ontology in Fig. 2, we have

$$S(\text{Analytical geometry, Metric geometry}) = \frac{2}{3},$$

while

$$S(\text{Analytical geometry, Algebraic topology}) = \frac{1}{3}.$$

The results are satisfactory to our intuition mentioned above.

3.3   GCList–Course List Generation

Given a set of ordered pairs relating with words and their similarity degrees to the user's keyword, w_0,

$$LW_{w_0}^{\alpha} = \{(w, S(w_0, w)) | w \in W, S(w_0, w) \geq \alpha\}.$$

In this process, every course in the course catalogs that its description contains at least one word in $LW_{w_0}^{\alpha}$ is retrieved. Then, the interest scores for the selected courses are determined. The formal steps in this process are details as follows:

1. Denoted by $Des_k = \{d_{k,1}, d_{k,2}, \ldots, d_{k,j_k}\}$ the set of contents in the course description of subject sub_k where j_k is the number of contents in the course description for sub_k.
2. If there exits $d_{k,m} \in Des_k$ such that $d_{k,m} = w_i$ for some $w_i \in LW_{w_0}^{\alpha}$, then the subject sub_k is retrieved.

Denoted by $Score_{w_0}(sub_k)$ the interest score for the retrieved subject sub_k with respect to the keyword w_0, where

$$Score_{w_0}(sub_k) = \frac{\sum_i^{|LW_{w_0}^{\alpha}|} \alpha_i \chi_{Des_k}(w_i)}{\sum_i^{|LW_{w_0}^{\alpha}|} \chi_{Des_k}(w_i)}, \tag{3}$$

$$\chi_{Des_k}(w_i) = \begin{cases} 1, w_i \in Des_k \\ 0, w_i \notin Des_k \end{cases}. \tag{4}$$

3.4   GBList–Book List Generation

For each sub_k selected from the previous process, the textbook's title for that course syllabus is then retrieved. Moreover, we associate such a book with the interest score which is equal to that score of its corresponding course. It means that if $Score_{w_0}(sub_k) = \beta$, then the interest score of the course textbooks is β.

Denoted by

$$BK_{w_0} = \{(b_1, \beta_1), (b_2, \beta_2), \ldots, (b_p, \beta_p)\},$$

the set of books and their interest scores with respect to the user's keyword w_0 when the first and the second entries of each ordered pair are a book title and an interest score, respectively.

Based on the book titles and their call numbers, we will extend BK_{w_0} to discover more interesting books. The book-extension process is detailed below.

1. The title of each book in BK_{w_0} is submitted to the book database. If the book is in the database, then its call number is extracted. If not, the call number of the most

similar book is extracted (Presented in the next section is one title-based method for measuring how close two books are).

2. Other books classified in the same group as obtained call number are retrieved.
3. Finally, every obtained book is associated with an interesting score whose formula will be expressed latter.

3.5   A Title-Based Similarity Measure

In this part, one similarity measure between two books using their own titles is presented. Given $T_1 = \{w_{11}, w_{12}, \ldots, w_{1k}\}$, and $T_2 = \{w_{21}, w_{22}, \ldots, w_{2m}\}$, are the sets of stems (root words)[2], excluding stop words (e.g. 'a', 'the', 'of'), from the titles of books B_1 and B_2, respectively. Based on the Jaccard's similarity measure, the similarity degree of the two books is defined as:

$$Sim(B_1, B_2) = \frac{|T_1 \cap T_2|}{|T_1 \cup T_2|}. \tag{5}$$

Example 2. To measure similarity between the books B_1 and B_2 titled "Data Structures and Algorithms" and "Introduction to Algorithms", by the method explained above, we have

$$T_1 = \{Data, Structure, Algorithm\}, T_2 = \{Introduction, Algorithm\}.$$

Then, $Sim(B_1, B_2) = 0.25$.

3.6   Book Interesting Score Calculation

Recall that $BK_{w_0} = \{(b_1, \beta_1), (b_2, \beta_2), \ldots, (b_p, \beta_p)\}$, is the set of books and their interest scores with respect to the user's keyword w_0 when b_i and β_i are a book title and an interest score, respectively. By the book-extension process described above:

- If b_i is in the book database, then its interesting score is set as β_i.
- If b_i is not in the database and, among the books in the database, d is the closest book to b_i, then the interesting score of d (not b_i) is $sim(b_i, d)$, where the function sim is defined in Eq. (5).

For each of the other books obtained from Step 2 of the process, its score is as equal as the score of its seed book.

---

[2] A stem is the form of a word before any inflectional affixes are added. For example, the words connect, connected, connecting, connections all can be stemmed to the word "connect".

Table 1. Experimental results

Keyword	No. of retrieved books		No. of common books
	OPAC	OPAC+	
Cauchy residue theorem	1	13	0
Differentiable function	1	21	1
Hausdorff space	1	14	1
Laplace transform	28	71	15
Lie algebra	11	27	6
Maclaurin series	7	42	5
Maxwell's equation	13	111	8
Projective plane	2	31	2

4 Preliminary Experimental Results

In this section, more experiments are presented. Eight students having enrolled in mathematical courses were selected. Each of them was asked to present one mathematical keyword. The Eight terms are listed in the first column of Table 1. For our framework, the threshold α in Eq. (2) was set to 0.7 and books of which the interest scores α are greater than 0.8 were selected. After applying the eight keywords to the OPAC system of Thammasat University using the keyword-search option and to the proposed framework, the numbers of retrieved books are summarized in the second column of Table 1. The third column, namely OPAC+, shows the results of our framework. The last column shows the numbers of common books between the two methods. We can see that, for most cases, our framework provides more books than keyword-based searching (After manually looking to the retrieved books, all of them provide contents relevant to keywords). However, from the numbers of common retrieval, it suggests that combing results from the two searching methods would be worthy.

5 Conclusions and Future Works

Two main reasons that yield low success rates of unknown-item search from library catalogs are resolved. Firstly, to help a user who has less experience in subject heading, one module for expanding a user's keyword to other related terms based on domain-specific ontology is introduced. This module is aimed at generating related keywords so as to increase the possibility obtaining obtain the relevant books. Secondly, to associate bibliography records with more information, topics in course description are indirectly embedded into such records.

The explanatory example shows that cooperating between the proposed framework and the traditional keyword search can produce better results than using each of them separately.

As future works, we will extend this framework by using other pieces of academic information such as taxonomy of course catalog. We expect that prerequisite course information facilitate in personal book suggestion. More experiments on users' satisfactory will also be taken into account.

Acknowledgements. This research is supported by Faculty of Science and Technology Fund, Thammasat University.

References

1. Antell, K., Huang, J.: Subject searching success transaction logs, patron perceptions, and implications for library instruction. Inf. Storage Retr. **48**(1), 69–76 (1998)
2. Slone, D.J.: Encounters with the OPAC: on-line searchingin public libraries. J. Am. Soc. Inform. Sci. **51**(8), 757–773 (2000)
3. Tagliacozzo, R., Kochen, M.: Information-seeking behavior of catalog users. Inf. Storage Retr. **6**(5), 363–381 (1970)
4. Long, C.E.: Improving subject searching in web-based OPACs. J. Internet Cat. **2**, 158–186 (2000)
5. Wood, J.A., Smigielski, E.M., Haynes, G.: Case-based approach for improving student MEDLINE searches. Med. Educ. **41**(5), 510–511 (2007)
6. Breeding, M.: Next-generation library catalogs. Libr. Technol. **43**(4), 5–14 (2007)
7. Martell, C.: The absent user: physical use of academic library collections and services continues to decline 1995–2006. J. Acad. Librariansh. **34**(5), 400–407 (2008)
8. Cousins, S.A.: Enhancing subject access to OPACS: controlled vocabulary VS natural language. J. Doc. **48**, 291–309 (1992)
9. O'Neill, E.T., Bennett, R., Kammerer, K.: Using authorities to improve subject searches. Cat. Classif. Q. **52**, 6–19 (2014)
10. Nevzorova, O., Zhiltsov, N., Kirillovich, A., Lipachev, E.: OntoMathPro ontology: a linked data hub for mathematics. Lecture Notes in Computer Science. LNCS (Knowledge Engineering and the Semantic Web), vol. 468, pp. 105–119 (2014)

Activities in Software Project Management Class: An Experience from Flipped Classrooms

Sakgasit Ramingwong[✉] and Lachana Ramingwong

Department of Computer Engineering, Faculty of Engineering,
Chiang Mai University, Chiang Mai, Thailand
sakgasit@eng.cmu.ac.th, lachana@gmail.com

Abstract. Teaching in 21st century mainly focuses on students. It is debatable that whether traditional lecturing or practical workshop is more efficient. This paper describes results of flipped classrooms on software project management which students were assigned to organize one activity at the end of the semester. The result suggests the students' preferences as well as ideas for future innovations.

Keywords: Software engineering · Project management · Simulation · Game-based learning

1 Introduction

The abstract nature of software is one of the major challenges in software engineering. Although software development can be perceived as similar to other engineering products, they are largely differ in many senses. In academia, there have been attempts which professors try to implement games and activities to improve their students' comprehension in certain aspects of the discipline [1, 2]. This method is arguably an efficient alternative to the traditional lecture. Yet, such practices are usually executed from the professor's perspective. Although the feedbacks are generally satisfactory, it may not reflect the read needs from the student's perception.

Department of Computer Engineering, Faculty of Engineering, Chiang Mai University, Thailand encourages their professors to apply innovative means in teaching. The software project management course is one of several example cases which has become largely game-based. During academic year 2016, this subject had been taught three times to students from undergraduate and graduate programs. The total of 76 students participated in the course. At the end of the semester, the students were assigned to form groups and organize or gamify one activity which can promote insights towards software project management. This paper describes and discusses their contribution.

The second section of this paper depicts the overview of the software project management course as well as the implemented workshops. Then, the contributions from the students are described in the second section. The third section discusses the findings. Finally, the fourth section concludes this paper.

© Springer Nature Singapore Pte Ltd. 2018
K.J. Kim et al. (eds.), *IT Convergence and Security 2017*,
Lecture Notes in Electrical Engineering 449,
DOI 10.1007/978-981-10-6451-7_39

2   The Course of Software Project Management

The software project management course at the Department of Computer Engineering, Faculty of Engineering, Chiang Mai University, Thailand is a major elective for undergraduate and postgraduate students in computer engineering programs. On the other hand, it is a core course for undergraduate students in information systems and network engineering program. The subject involves 15 weeks of lecturing. Each period lasts 3 h, resulting in the total of 45 h. Instead of spending all efforts into lecturing, the teaching in this course largely involves in-class workshops. Content-wise, it represents the concept of software project management in a top-down fashion. Firstly, the entire life cycle of software project is exhibited in the conceptual level. Then, the students is given overview on the software business and professional ethics. Finally, each stage of software project life cycle is sequentially instructed. This includes project initiation, project definition, project planning, project implementation and project decommissioning. Most of the contents involve two weeks of lecturing. The contents and their incorporated activities are described in Table 1.

Table 1. Contents and activities in software project management course

Contents	Activities	Duration (minutes)	Type of activity
1. Concept of project management	That PM Game [3]	10	Computerized
2. Software business	Platform Wars [4]	60	Computerized
3. Professional ethics	Case Study	90	Computerized
4. Project initiation	Group SWOT Analysis	60	Physical
5. Project definition	Group Mind Mapping	60	Physical
6. Project planning	ARMI [5]	90	Physical
7. Project implementation	SimSE [6]	90	Computerized
8. Project decommissioning	ECSE [7]	90	Physical

The first part of the course involves general concepts and brief information on phases of software project management. *That PM Game* [3] is chosen to represent a quick overview to human resource management and Gantt chart which is a standard tool in managing of a project. That PM Game is a web-based application with a slight learning curve which is suitable for a classroom. The students need to discuss in a large group before allocating their resources and aim to complete the project on time within the budget. The major drawback of this activity is it cannot be participated by more than one group, therefore lack of competitiveness.

The second part of software project management highlights the importance of software business. Although unfamiliar, the students are instructed on business and economic perspectives and their crucial relationship to software project. *Platform Wars* [4], a single player web-based simulation, is used as the main media for this content.

In this simulation, the students role play as a console company who needs to compete with an AI opponent. This simulation helps the students to understand the relationship of demand and supply as well as several business strategies. The drawback of this game is similar to the previous, i.e. it does not support multiplayer competition.

Ethics of software engineer is the third part in this software project management course. Instead of activities, the students are given video examples of important *Case Studies* in computer business. The rivalry between IBM, Microsoft and Apple is one of several classic example which can efficiently draw attention from the students. After the review of the cases, the students presume the roles of each stakeholders and discuss on their decision making on several sensitive issues.

The actual phases of software project management begin in the fourth part of the course. During this, the students are guided on how to initiate a project from scratches. *SWOT Analysis* [8] is used as a brainstorming tool to draw needs from stakeholders in specific scenarios. Unlike usual individual analysis, the students are instructed to conduct this activity in group, using post-its and large boards instead of tradition pen and paper. At the end of the session, their course project is identified based on the extracted ideas from the analysis.

After the initiation, the students are consequently directed to proceed to the project definition phase. Again, they perform *Mind Mapping* in group using post-its and other basic office equipment. The results of this session include functionalities of the course project as well as tentative technology, estimated duration, budget and other resources. All of which are to be used in the next project phase.

The sixth part of the course involve software project planning. As one of the most important aspect in this phase, *ARMI: A Risk Management Incorporation* [5] is implemented to teach the students on risk management. This physical card-based game simulates undesirable situations in software development. Students need to analyze and manage risks throughout the project life cycle and try to minimize their mitigation budget.

SimSE [6] is a java application which simulate major software development processes. It is used as the main media for teaching concepts of software project implementation. Students can learn the differences, pros, cons and settings of several software processes, i.e. the waterfall model, incremental model, inspection model, rational unified process and extreme programming, from this simulation. Depending on the objectives of the course, other alternatives of activities for project implementation include Problems and Programmers [9], *Kanban Pizza* [10], *Scrum Simulation with LEGO Bricks* [11] and *Plasticine Scrum* [12].

The final part of software project management course highlights the importance of activities which should be performed at the end of the project such as reallocation of staff, reviewing and evaluating, process improvement, and knowledge management. *ECSE: Engineering Construction for Software Engineers* [7] is implemented to wrap-up the entire project essentials. In this workshop, the students use corrugated plastic board to build a toy house based on requirements form a customers. Their task is to find hidden requirements as well as balance their resources and expenditures. Communication is a critical element in this activity. Another substitute for this is *CutIT* [13], a simulation which aims to highlight the importance of process improvement.

It can be seen that at least one game-based activities is integrated to each part of this software project management course and it is arguably balanced between computerized and physical sessions. At the end of the class of 2016, the final assignment given to the students was organizing of an activity which reflects any aspect of software project management from their perspectives. The activity can be either computerized or physically participated. They can adopt the workshops from any media. However, there were encouraged to be original. The time limit for each activity is 60 min.

3 Students' Contributions

As aforementioned, three classes of students participated in software project management course by the Department of Computer Engineering, Faculty of Engineering, Chiang Mai University, Thailand during academic year 2016. This included 25 undergraduate students in computer engineering (elective course), 33 undergraduate students in information systems and network engineering (major course) and 18 postgraduate students in computer engineering (elective course). Each class were taught separately. Major differences between the graduate and postgraduate classes were the level of instructor guidance, assignments and examinations. In this last assignment, which was the only similar tasks amongst the two degrees, the students

Table 2. Activities contributed by students in software project management course

Activity	Student*	Purpose	Duration (minutes)	Type of activity
1. Tower builder	U	Ice breaking	30	Physical
2. Bingo	U	Ice breaking	10	Physical
3. Pair and reflect	U	Ice breaking	10	Physical
4. Auction!	U	Estimation and negotiation	30	Partly computerized
5. Plasticine model	U	Software development	60	Physical
6. Hidden requirements	U	Requirement engineering	30	Physical
7. Snakes and ladders	U	Project life cycle	60	Physical
8. Project life cycle	P	Project life cycle	60	Physical
9. Origami	P	Software development and knowledge management	40	Physical
10. Planning game	P	Human resource management and planning	30	Physical

Note: *U = undergraduate program
P = postgraduate program

were instructed to form groups of 6–10 persons. This resulted in a total of 10 activities. Interestingly, three of them were nothing new while the other seven were rather creative. Table 2 depicts the contributions from the students.

The first three games proposed by the undergraduate groups involved ice breaking activities. *Tower Builder* was a game where students were given papers, paper clips, scissors and tapes to build the tallest tower as possible. *Bingo* was just a traditional bingo game without any twist. *Pair and reflect* was a card game which the player needed to pair up all card in their hands and steal a pen to be the winner. All of these proposed activities are not new but can certainly be used for team building.

The fourth proposed activity was *Auction!* In this game, the players formed groups and attempt to auction for certain products which was shown on the screen. There was a brief of the scenario at the beginning. The real value of the auction items were revealed at the end of the activity.

In *Plasticine Model*, pictures of tentative toy models were shown to the students. The students were assigned to use plasticine to build identical toy models. However, they were given a limited resource therefore needed to manage them properly. There was a handicap system in this game where players can spend money to buy their own fingers in order to build the product. Otherwise, they were allowed to build products without using their hands.

Hidden Requirement was a requirement management game which players need to allocate human resource tokens to specific aspects of software development. At the beginning of the game, the moderator briefly explained the requirements of the product. Then, the players allocated their resource accordingly. The activity ended whenever the players allocated their resource correctly. Elsewise the cycle repeated and the moderator revealed more information on the requirements to the players.

Snakes and Ladders was the most constructive game proposed by the undergraduate students. A giant board of snakes and ladders was built. However, instead of normal numbered spaces, the board was divided into 5 project phases. Each phase included a dozen of project related activities. After the players threw a dice, they would encounter certain project situation in which they might have to solve. The first group of players who reached the end of the board won the game.

Project Life Cycle was essentially another snakes and ladders game proposed by the postgraduate students. Although similar to the previous, this activity came with several more details and challenging factors such as attribute of players, budgeting, progression of product, and risk management. One moderator was needed for each group of players due to its sophisticate calculation system.

Origami highlighted the importance of knowledge management. In this game, the players were tasked to build complex origami (folded paper) products. In order to be able to build the product, the student teams needed to send a representative to train with an expert. Different level of training involved different costs. After that, the representatives conducted a training to their counterparts. A number of defected products were found during the acceptance test due to the loss of communication and failed knowledge management.

The final activity, *Planning Game*, focused on human resource management where the players scouted for their employees and appointed them on the most appropriate tasks. Each employees had three attributes and was more suitable for one activity than another.

In this game, the players needed to negotiate and appropriately allocate their resources in order to be the fastest team to finish the product.

4 Discussion

It can be seen from the previous section that the students can adequately contribute their idea in these flipped classrooms. Despite of the current availability of simulations and activities, this study suggests that are still plenty of gaps and perspectives which could be gamified. It is reasonable that most of the student contributions may not be brilliant at this stage. However, further polish may lead these ideas to perfection. The followings list key observations from this research:

4.1 Hidden Requirements

One common idea amongst the proposed activities is the hidden requirements. This comprehensively reflects the nature of software development which hidden requirements or changes are inevitable. As a result, software engineers need to prepare themselves professionally. Hidden requirements also make the game more challenging. However, in order to appropriately implement this feature, associative development environment must be logical. The moderator can provide essential clues such as background information of the scenario and characteristics of the stakeholders. These clues can subtly depict the hidden requirements and keep the game rational. Gradual release of clues or requirements can be another effective strategy for future software engineering workshops.

4.2 Exploration of Random Events

The snakes and ladders workshops indicate event exploration can be an exciting approach to software engineering activities. Role playing as a team in an ongoing project can provide essential insights for the participants. Despite of its limitation, step-by-step progression of the activity can still be considered as one of the most applicable approaches to such workshop. In addition, randomness helps the activities to be more stimulating. However, in order to keep the workshop fair to every players, uncertainty needs to be carefully moderated. It is preferable for players to encounter the same set of challenges, or at least the same probability of challenge, throughout the activity.

4.3 Human Attributes

Human attributes is another feature which increase depth of the game. Examples of attributes suggested by the students include communication ability, technical ability, and wages. Indeed, activities can be more competitive when the players are allowed to make choice on human or other similar resources with different attributes. Although human attributes included in all contribution in this paper are explicit, having hidden attribute can make the game even more sophisticate and challenging.

4.4   Currency

Currency was used in many of the student contributions. It is not only used for exchanging of in-game elements but also make the activities competitive. However, including currency might slow down certain processes if complex calculation is involved.

4.5   Physical Approach

Despite the balanced ratio of example workshop represented in software project management class, it is surprising that almost all of the student contributions are physical-based activities. This suggests that physical approach can be the more preferable method to draw attention from the students.

5   Conclusion

Gamification has become a popular approach in academia. It can not only effectively draw attention from the students but also subtly provide essential insights which may be difficult to deliver through traditional lecture. This paper describes and discusses a result from an experiment in software project management course which was taught in Department of Computer Engineering, Faculty of Engineering, Chiang Mai University, Thailand. Three classes of students, two undergraduates and one postgraduate, were given an assignment to organize activities which can reflect any aspects in software project management.

A total of ten activities were contributed. Three of them were fully adopted from various sources while the other seven were more original. Despite of the fact that all of these workshops were not flawless, many of them were genuinely promising. Common characteristics found from the student contributions include hidden requirements, exploration of random events, currency, and human attributes. Most of the activities were organized in a traditional physical fashion. These suggest that physical workshops are still largely preferable for the students in the current generation, arguably even more than computerized workshops. All of these findings can be used as an essential guideline to achieve a higher level of gamification of software engineering and software project management in the future.

References

1. Connolly, T.M., Stansfield, M., Hainey, T.: An application of games-based learning within software engineering. Br. J. Educ. Technol. **38**(3), 416–428 (2006)
2. Taran, G.: Using games in software engineering education to teach risk management. In: 20th Conference on Software Engineering Education & Training (CSEET). pp. 211–220 (2007)
3. Charney, R.: The project management game (2017). http://thatpmgame.com

4. MIT Management Sloan School.: Platform wars: simulating the battle for video game supremacy (2017). https://mitsloan.mit.edu/LearningEdge/simulations/platform-wars/Pages/default.aspx
5. Ramingwong, S., Ramingwong, L.: ARMI: a risk management incorporation. In: 2014 11th International Conference on Electrical Engineering/Electronics, Computer, Telecommunications and Information Technology (ECTI-CON). pp. 1–6 (2014)
6. Navarro, E.O., Van Der Hoek, A.: SimSE: an educational simulation game for teaching the Software engineering process. ACM SIGCSE Bull. **36**(3), 233 (2004)
7. Ramingwong, S., Ramingwong, L.: ECSE: a pseudo-SDLC game for software engineering class. Overcoming challenges in software engineering education: delivering non-technical knowledge and skills, pp. 296–309 (2014)
8. Humphrey, A.S.: SWOT analysis for management consulting. SRI Alumni Association Newsletter, pp. 7–8 (2005)
9. Baker A., Navarro E.O., Van Der Hoek, A.: Problems and programmers: an educational software engineering card game. In: 25th international Conference on Software Engineering (ICSE), pp. 614–619 (2003)
10. The Agile42 Team.: Kanban pizza game (2017). http://www.agile42.com/en/training/kanban-pizza-game/
11. Krivitsky, A.: Scrum simulation with LEGO bricks (2009). https://www.scrumalliance.org/system/resource_files/0000/3689/Scrum-Simulation-with-LEGO-Bricks-v2.0.pdf
12. Ramingwong, S., Ramingwong, L.: Plasticine scrum: an alternative solution for simulating scrum software development. Lecture Notes in Electrical Engineering, vol. 339, pp. 851–858 (2015)
13. Ramingwong, S.: CutIT: a game for teaching process improvement in software engineering. In: Third International Conference on Information, Communication and Education Application (ICEA) (2012)

Solo Scrum in Bureaucratic Organization: A Case Study from Thailand

Lachana Ramingwong[✉], Sakgasit Ramingwong, and Pensiri Kusalaporn

Chiang Mai University, Chiang Mai 50200, Thailand
{lachana,sakgasit,pensiri}@eng.cmu.ac.th

Abstract. Scrum becomes the most popular Agile methodology due to its simple principles and ability to adapt to various environment. It offers flexibility to software development and allows teams to have control over the course of development. Although there were several successful attempts of Scrum adoption throughout different parts of the world, there are still challenges in implementing Scrum in bureaucratic organizations. This paper presents findings from a pilot adoption of Solo Scrum, which a single developer employed and adapted Scrum practices in a software project in a Thai bureaucratic organization where several issues turn to be obstacles to the adoption including bureaucratic structure, layers of approvals, culture, etc. Organizational culture is found to be a major obstacle in this case study. Conflicts and solutions are discussed. The challenges and issues related to Mum effects are also addressed. Adoption of Agile methodologies in Thailand is still new topics, which needs to be studied further.

Keywords: Software engineering · Software development · Software process · Scrum · Solo scrum · Agile methodology · Bureaucracy · Mum effects

1 Introduction

Scrum is claimed to be the most popular Agile methodology adopted [1–4]. Scrum continues to gain acceptance due to its simple set of principles [5, 6]. It has received much attention since the majority of software projects are micro and small [7, 8]. Scrum was widely adopted in different parts of the world [9–13].

Scrum was successfully adopted in large and small organizations [14–16]. It was implemented in various kind of organizations including software development company, government entity, Military and many more [17]. Despite its agility and light documentation nature, Scrum has been productively implemented in combination with worldwide quality standards such as ISO 9001 [18] and CMMI [19]. The results show that productivity of Scrum teams was twice of the traditional teams.

Nevertheless, the number of reports on implementation of Scrum in Governmental organizations is rather small [20–22]. Although, several successful stories for Scrum adoption in Government organization were reported [22, 23], there are difficulties and challenges in adopting Scrum in such bureaucratic organizations [14, 21].

K.J. Kim et al. (eds.), *IT Convergence and Security 2017*,
Lecture Notes in Electrical Engineering 449,
DOI 10.1007/978-981-10-6451-7_40

Typical government projects involve substantial documentation, excessive planning and unnecessarily big design up front [24]. Such projects are usually run under complex environment. In addition, their distinctive characteristics, such as strictly sequential chain of command, stakeholders, seniority can be seen as obstacles to the adoption, and, thus, add overheads to the development [25]. Besides those factors, culture also plays an important role in Scrum adoption in government organizations.

In fact, a mainstream of software projects is micro; hence, 1–2 developers can complete the projects. Accordingly, implementing Scrum with a single developer becomes more common [26, 27].

This paper reports how Scrum was adopted in a small software project in a highly bureaucratic environment in Thailand where the waterfall model is normally used in developing software. The project involves a single developer under a chain of commands. Subsequent sections presents findings and discussion from adoption of Scrum in a Thai Government Organization. Section 2 briefly explains Solo Scrum and related work. Section 3 describes the research methodology. Section 4 presents findings and analysis. Section 5 concludes the paper.

2   Solo Scrum and Related Work

Scrum can be described as a project management methodology [10], that can also be applied to other domains, not just software. Shown on Fig. 1 is Scrum's set of flexible and lightweight practices. Scrum promotes transparent action and inspection which can be adapted to fast changing product requirements. Combinations of Scrum and quality assurance standards demonstrate the benefits of these Scrum values [18, 19].

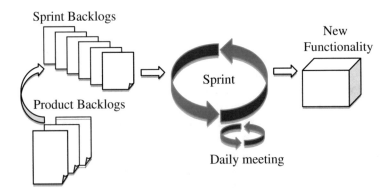

Fig. 1. Scrum process

2.1   Solo Scrum

Much of software development activities can be carried out by single developers, who work as independents or part-time entrepreneurs [28]. This type of Scrum is referred as Solo Scrum [28]. Generally, Scrum promotes small team practices of no more than 10 team members [29]. However, the advancement of software development tools and frameworks simplifies the tasks of building software, and also help individual developers to focus on the simple core principles of Scrum in delivering software on time.

In Solo Scrum, the single developer take on the roles of developer, product owner and scrum master while following adaptation of Scrum essences. In traditional Scrum, product owner assumes customer representative role who communicates customer needs to the team. Scrum master is responsible for making sure that the team follows Scrum values and practices. As there are no extra persons in Solo Scrum to take on product owner and scrum master roles, the single developer communicates with stakeholders and assures that the team is on the right track.

2.2 Scrum Adoption in Bureaucratic Organizations in Thailand

So far, there is no report of Solo Scrum adoption in Thailand published. Morien and Tetiwat (2007) [30] reported that Scrum topics were not included in universities' curriculum at the time, and there were little or no knowledge about Agile development among Thai academics. In 2013, the results from the survey questionnaire showed that Scrum was the most used software development method [31]. The survey revealed that merely 6% of respondents were from government sections, while 90% were from industry. This suggests the popularity of Scrum in industry, but not in government sections. Nanthaamornphong and Wetprasit (2016) [12] informed that most agile-based development projects in Thai government agencies failed as a consequence of the deficiency in collaboration, especially from the members of the agency to the development team, or among the members of the agency itself.

This problem can be linked to Mum Effects, a critical risk that surfaces when a staff member is unwilling to reveal negative information, or hide critical information [32]. This situation in software projects could lead to a serious damage. The risks could be significant higher for offshore projects where communication is one of the most important issues.

Despite encouragement in regular communication through Scrum practices, software projects implemented in Thai bureaucratic organizations could still suffer from Mum effects caused by Thai hierarchical society where a lower rank person tend not to express his negative feelings to a person in higher rank, even though they are work related issues [33].

3 Solo Scrum Implementation in a Thai Government Organization

In this study, one developer was assigned to develop a small web application for internal use in a bureaucratic organization. The developer worked in information technology section of an autonomous university under the supervision of the Royal Thai Government. The adapted Scrum process was referred as Go-Scrum (Government-Scrum). The process comprises the following activities; (1) Management Buy-in, (2) Kick-off meeting & Story discovery, (3) Project planning, (4) Release and Sprint planning and

(5) Sprint with inspection. After the final activity is completed, the new functionality is delivered.

Figure 2 illustrates Go-Scrum. The differences to original Scrum are shown in grey areas of the figure. Prior to Project planning, two meetings are organized to prepare the stakeholders for Go-Scrum. The stakeholders in this project include management team, end-users and solo developer who is responsible for leading and performing development activities.

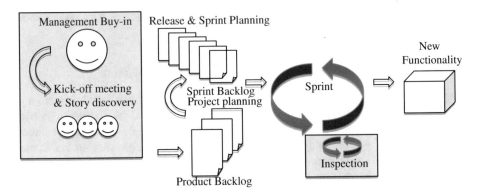

Fig. 2. Go-Scrum process

3.1 Organization Background

The study was conducted at an autonomous academic organization under supervision of the university. The organization, consisted of 7 academic departments and 9 support departments, has 318 employees including academic and support staff. Despite having a dedicated IT department, each department has its own IT personnels. The IT department is responsible for providing software products, solutions and consulting services. Only one member of this department is a programmer who usually works as a sole member of software projects.

3.2 Management Buy-in

A meeting was held to inform everyone involved, especially the management team of the development process, artifacts and deliverables. This is to prepare the management for acceptance of software and to get them to participate in the development effort.

3.3 Kick-off Meeting and Story Discovery

In this activity, all stakeholders were invited to the kick-off meeting where they were brief about the development from the head of the IT section. In the meeting, the stakeholders were encouraged to participate. There were asked questions about the system to be developed. Small user story cards were also given at the meeting. So stakeholders could describe their requirements.

3.4 Project Planning

The solo developer then gathered requirements received from the previous meeting and created product backlogs, which were used to plan the development cycles. Necessary details were recorded.

3.5 Release and Sprint Planning

The backlogs were prioritized and estimated. Sprint backlogs were created. Classic scrum estimation techniques were used to aid in estimation.

3.6 Sprint

There were two inspections every week to track for progress and to plan for the next days instead of implementing a daily meeting everyday.

Each sprint started with a short design session. At the end of the sprint, the sprint review was held. The developer met with the stakeholders to review the work done, discover defects, if any. Burn down charts were also used to show the progress.

After the sprint review, a short retrospective meeting was held to summarize the sprint, the work done, the problems found and suggest for solution.

4 Data Collection, Results and Findings

4.1 Data Collection

The data collected include time spent, meeting records, lines of code, comments from stakeholders in all levels including management. At the end of the project, all stakeholders filled the questionnaire for evaluation of satisfaction in participating in the development using Go-Scrum.

4.2 Results, Findings and Discussion

Early in the process, the stakeholders were unable to clearly identify their needs when filling the user story cards, though this method usually worked well with Agile methodologies observed from the literature review. The stakeholders appear to be able to clearly identify their needs when allowed to speak out. This could be because of differences in types of organization, culture, age and education background. Also it was their first time writing their needs on the card.

Therefore the developer suggested the stakeholders on how to fill the story card by making some examples, and talked to them after getting the cards back to ensure their requirements were understood correctly. Also at the sprint review, the developer verified the system requirements with the stakeholders before proceeding to the next sprint.

Organizing the meetings was somewhat slow in a hierarchical structured organization, as there is a rule that an official letter is needed to appoint the management.

The developer also found that daily meeting was not required as there was no need for communication with other developers. This activity was changed to 2 or 3 inspections per week. For more rapid development where schedule is tight, frequent inspection could be beneficial.

Due to the nature of hierarchical structured organization, follow up meeting and sprint review meeting took longer than expected. Correspondingly, in case that management was absent from a meeting for some reasons, the subsequent meetings would be affected. The project could be delayed because of pending approvals. After the management was notified, there was no problem with upcoming meetings because meeting with other stakeholders can be done informally.

The project was finally completed, though a little later than estimated. The delay was considered acceptable compared to the previous projects implemented with the waterfall model. Besides, stakeholders are more satisfied with the product and the development process. The management explicitly expressed their satisfaction on the process implemented and on the acceptance of software. Using Go-Scrum to develop software in a bureaucratic organization certainly benefits the final software product and project management as a whole. Stakeholders were happy that late requirements could be incorporated to the product.

5   Conclusion

Go-Scrum is an adaptation of original Scrum. Its implementation in a Thai government organization was found to be a success story. Several lessons were learned. It encourages communication among developers and stakeholders. It promotes systematic development process and project tracking. The stakeholders were pleased that the system meets their real needs. Overall, their experience with Go-Scrum was positive.

However, the hierarchical structure in Thai government organization imposed some unnecessary process on top of the original scrum, though it was considered necessary to the management. Therefore, Scrum may not be appropriate to an organization that is rigid and highly structured.

Therefore, adoption of Scrum in a bureaucratic organization must be done with care and adaption should be required. In addition, the overhead should be prepared to introduce the process to management and stakeholders and to gain their acceptance. The team should include a person who understands the structure of such organization to help with coordination and communication.

Mum effects could worsen communication problem already existed in the development process. Solo developers should be made aware of the issue. The issues and challenges discussed in this paper should be considered in designing and adopting Scrum process in bureaucratic organizations.

References

1. Rubin, K.S.: Essential Scrum: a practical guide to the most popular Agile process. Addison-Wesley (2012)
2. Tripp, J.F., Armstrong, D.J.: Exploring the relationship between organizational adoption motives and the tailoring of agile methods. In: IEEE 47th International Conference on System Sciences (HICSS), Hawaii, pp. 4799–4806 (2014)
3. Solinski, A., Petersen, K.: Prioritizing agile benefits and limitations in relation to practice usage. Softw. Qual. J. **24**(2), 447–482 (2016)
4. Matharu, G.S., Mishra, A., Singh, H., Upadhyay, P.: Empirical study of agile software development methodologies: a comparative analysis. ACM SIGSOFT Softw. Eng. Notes **40**(1), 1–6 (2015)
5. West, D.: Water-scrum-fall is the reality of agile for most organizations today. Forrester Research, 26 (2011)
6. Rising, L., Janoff, N.S.: The Scrum software development process for small teams. IEEE Softw. **17**(4), 26–32 (2000)
7. Aguilar, J., Sanchez, M., Fernandez-y-Fernandez, C., Rocha, E., Martinez, D., Figueroa, J.: The size of software projects developed by mexican companies. arXiv preprint arXiv: 1408.1068 (2014)
8. Pagotto, T., Fabri, J.A., Lerario, A., Gonçalves, J.A.: Scrum solo: software process for individual development. In: IEEE 11th Iberian Conference on Information Systems and Technologies (CISTI), pp. 1–6 (2016)
9. Salo, O., Abrahamsson, P.: Agile methods in European embedded software development organisations: a survey on the actual use and usefulness of Extreme Programming and Scrum. IET Softw. **2**(1), 58–64 (2008)
10. Garzás, J., Paulk, M.C.: A case study of software process improvement with CMMI-DEV and Scrum in Spanish companies. J. Softw. Evol. Process **25**(12), 1325–1333 (2013)
11. Mnkandla, E., Dwolatzky, B., Mlotshwa, S.: Tailoring agile methodologies to the Southern African environment. In: International Conference on Extreme Programming and Agile Processes in Software Engineering, pp. 259–262. Springer, Heidelberg (2005)
12. Nanthaamornphong, A., Wetprasit, R.: A case study: adoption of agile in Thailand. In: IEEE International Conference on Advanced Computer Science and Information Systems (ICACSIS), pp. 585–590 (2016)
13. Al-Kautsar, E., Salleh, N., Hoda, R., Asnawi, A.L.: Preliminary analysis on the adoption of agile software development methodology in Indonesia. Development **29**, 25 (2013)
14. Hajjdiab, H., Taleb, A.S., Ali, J.: An industrial case study for scrum adoption. J. Softw. (Oulu) **7**(1) (2012)
15. CSC Homepage, Success Story—Zurich Insurance Group. https://assets1.csc.com/infrastructure_services/downloads/081914_Zurich_Success_Story_PPM_Services.pdf. Accessed 26 May 2017
16. Wan, J., Zhu, Y., Zeng, M.: Case study on critical success factors of running Scrum. J. Softw. Eng. Appl. **6**(2), 59 (2013)
17. Scrum Alliance Homepage: The 2015 State of Scrum Report. http://www.scrumalliance.org/why-scrum/state-of-scrumreport/2015-state-of-scrum. Accessed 26 May 2017
18. McMichael, B., Lombardi, M.: ISO 9001 and Agile development. In: IEEE Agile Conference (AGILE), pp. 262–265 (2007)
19. Sutherland, J., Jakobsen, C.R., Johnson, K.: Scrum and CMMI level 5: the magic potion for code warriors. In: Proceedings of the 41st IEEE Annual Hawaii International Conference on System Sciences, p. 466 (2008)

20. Upender, B.: Staying agile in government software projects. In: Proceedings of IEEE Agile Conference, pp. 153–159 (2005)
21. de Sousa, T.L., Venson, E., da Costa Figueiredo, R.M., Kosloski, R.A., Junior, L.C.M.R.: Using scrum in outsourced government projects: an action research. In: IEEE 49th Hawaii International Conference on System Sciences (HICSS), pp. 5447–5456 (2016)
22. Royce, A.: Agile in government: successful on-time delivery of software. In: 20th Australian Software Engineering Conference: ASWEC 2009, Barton, ACT (2009)
23. Wernham, B.: Agile Project Management for Government Case study Case study: The Success of the FBI The Success of the FBI Sentinel Project Sentinel Project Sentinel Project (2012)
24. Cho, J., Huff, R., Olsen, D.: Management guidelines for scrum agile software development process. Issues Inf. Syst. 12(1), 213–223 (2011)
25. Hajjdiab, H., Taleb, A.S.: Adopting agile software development: issues and challenges. Int. J. Manag. Value Supply Chains (IJMVSC) 2(3), 1–10 (2011)
26. Schloßer, A., Schnitzler, J., Sentis, T., Richenhagen, J.: Agile processes in automotive industry —efficiency and quality in software development. In: 16 Internationales Stuttgarter Symposium, pp. 489–503. Springer Fachmedien Wiesbaden (2016)
27. Pagotto, T., Fabri, J.A., Lerario, A., Gonçalves, J. A.: Scrum solo: software process for individual development. In: IEEE 11th Iberian Conference on Information Systems and Technologies (CISTI), pp. 1–6 (2016)
28. Dent, A., Aitken, A., Rosbotham, S.: From scrum to solo: how small is too small a team to still call it software engineering? In: 19th Australian Software Engineering Conference: ASWEC 2008; Experience Report Proceedings, Engineers Australia, p. 152 (2008)
29. Rising, L., Janoff, N.S.: The Scrum software development process for small teams. IEEE Softw. 17(4), 26–32 (2000)
30. Morien, R., Tetiwat, O.: Agile software development methods adoption in Thailand—a survey of Thai Universities. In: The Proceedings of the Information Systems Education Conference (ISECON), p. 24 (2007)
31. Porrawatpreyakorn, N.: A Survey of early adopters of Agile methods in Thailand. Int. J. Appl. Comput. Technol. Inf. Syst. 3(1) (2016)
32. Ramingwong, S., Snansieng, S.: A survey on mum effect and its influencing factors. Procedia Technol. 9, 618–626 (2013)
33. Thanasankit, T., Corbitt, B.: Understanding Thai culture and its impact on requirements engineering process management during information systems development. Asian Acad. Manag. J. 7(1), 103–126 (2002)

Author Index

© Springer Nature Singapore Pte Ltd. 2018
K.J. Kim et al. (eds.), *IT Convergence and Security 2017*,
Lecture Notes in Electrical Engineering 449,
DOI 10.1007/978-981-10-6451-7

Printed in the United States
By Bookmasters